W9-AZT-591

Managing Information Technology Projects

GRAHAM MCLEOD
DEREK SMITH

UNIVERSITY OF CAPE TOWN

BOYD & FRASER PUBLISHING COMPANY
I(T)P AN INTERNATIONAL THOMSON PUBLISHING COMPANY

DANVERS • ALBANY • BONN • BOSTON • CINCINNATI • DETROIT • LONDON • MADRID • MELBOURNE
MEXICO CITY • NEW YORK • PARIS • SAN FRANCISCO • SINGAPORE • TOKYO • TORONTO • WASHINGTON

To my family
Graham McLeod

To my late brother, David, an inspiration and mentor
Derek Smith

Executive Editor: James H. Edwards
Project Manager: Lisa S. Strite
Production Editor: Barbara Worth
Manufacturing Coordinator: Lisa Flanagan
Marketing Director: William Lisowski
Illustrator: Suzanne Biron
Cover Design: Hannus Design

Printed in the United States of America

This book is printed on recycled, acid-free paper that meets Environmental Protection Agency standards.

For more information, contact boyd & fraser publishing company:

boyd & fraser publishing company
One Corporate Place • Ferncroft Village
Danvers, Massachusetts 01923, USA

International Thomson Publishing Europe
Berkshire House
168-173 High Holborn
London WC1V 7AA United Kingdom

Thomas Nelson Australia
102 Dodds Street
South Melbourne
Victoria 3205 Australia

Nelson Canada
1120 Birchmount Road
Scarborough, Ontario
Canada M1K 5G4

International Thomson Editores
Campos Eliseos 385, Piso 7
Colonia Polanco
11560 México D.F. México

International Thomson Publishing GmbH
Konigswinterer Strasse 418
53227 Bonn, Germany

International Thomson Publishing Asia
Block 211, Henderson Road #08-03
Henderson Industrial Park
Singapore 0315

International Thomson Publishing Japan
Hirakawa-cho Kyowa Building, 3F
2-2-1 Hirakawa-cho, Chiyoda-ku
Tokyo 102 Japan

1 2 3 4 5 6 7 8 9 10 D 9 8 7 6 5
ISBN: 0-7895-0176-7

Brief Contents

Contents

2 • Project Initiation

3 • The Work

4 • The Product

5 • Resources

6 • The Project Lifecycle

7 • Estimating

8 • Project Design

9 • Planning Techniques

10 • Project Management Tools

11 • Project Execution

12 • Measurement

16 • Project Documentation

17 • Communication

18 • Managing People

19 • Implementation

20 • Multiple Project Coordination

21 • Subcontractors

Table of Figures

Tables

Preface

Intended Audience

This book is intended for use by Information Systems (I.S.) and Information Technology (I.T.) project managers in industry and as a text for final year and honors students in I.S. at universities, colleges and technikons. It is assumed that the reader has a fairly thorough knowledge of the system development process and the general I.S./I.T. environment.

Aims of the Text

There are many texts on project management. There are even more on system development, but very few on I.S. project management. There are even fewer on managing the non-development Information Technology (I.T.) project - maintenance, hardware installation, package installation, etc. This text is intended to provide a body of knowledge to support I.T. project managers in the challenge of managing all aspects of a variety of project types.

The aim of the book is to equip the reader with the necessary knowledge (and with practice, the skills) to successfully initiate, plan, manage, control and report on I.T. projects. Ancillary aims are to convey the importance of proper planning, documentation, scope and change control, and quality and risk management.

The book also covers the people skills required in the areas of team selection, structure, motivation, interviewing, presentations, conflict resolution and leadership. The use of automated software tools in support of project management will be introduced, although no particular package is required or assumed. The text does not assume any particular type of project. The techniques presented are applicable to a wide variety of projects, including:

- Systems development
- Package implementation
- End User Computing (EUC) development
- Hardware and software installation
- Major maintenance and enhancements

- Strategic planning
- System conversion
- Business Process Reengineering (BPR)

Most examples quoted will be for a systems development project.

Approach Taken

The text is primarily intended to provide a practical approach which will be easily applied in practice. Care has been taken to ensure that this is based upon a sound academic, empirical and theoretical base. The authors have drawn on their combined 45 years of experience in the industry to identify the issues which are really important, and to balance the amount of coverage given to each area.

Language and Standards

We have striven wherever possible to be nonsexist. When we quote examples, we have used "he" or "she" depending upon the gender of the people actually involved. Elsewhere in the text, we have used the word "he". This is not to imply that all or most project managers are male, but rather to have text that reads easily. Please treat all of these as "genderless" uses of the word, and mentally substitute "she" or whatever you are most comfortable with.

Where examples and explanations use currency, we have chosen to use the generic "dollars" or "$". The actual units involved may have been in another currency, e.g. Rands or Pounds. This has been done in the interests of international portability, since virtually everyone can relate to amounts in dollars. You can assume that amounts quoted have been converted to U.S. dollars.

I.T. constantly introduces new terminology and "buzz words". These help to convey information quickly to the informed, but can be extremely confusing to others. I.T. is particularly rich in acronyms. In all cases, we introduce the term fully the first time it is used in the text, followed by the abbreviation used thereafter. For example, "This is called a Work Breakdown Structure (WBS)". Abbreviations can also be looked up in the Glossary at the end of the book.

Case Studies

We have included four rich case studies as an aid to learning and illustrating the real world application of the principles and techniques in the book. Questions relevant to the material in each chapter follow the individual chapters. The cases are introduced in the next section.

Additional Materials

An instructor package is available to teaching institutions from the publisher. This includes an instructor guide, transparency masters and a data disk in support of case questions and software exercises.

Project Management Software

It is not essential to have access to a package, but it can certainly enhance the learning process, particularly in those chapters dealing with project management techniques, tools and reporting.

Acknowledgments

We would like to acknowledge the contribution of many sources and participants:

Authors: Tom de Marco, Francois Lustman, Tom Gilb, Philip Metzger, Comcon (Piet Opperman), Norden & Putnam, The Butler Cox (now CSC Index) P>E>P Programme, and various Institute of Electrical and Electronics Engineers (IEEE) authors.

Students: Of the University of Cape Town, Faculty Training Institute and *Inspired*, who endured and enriched earlier versions and components of the text.

Colleagues: who encouraged, critiqued and cajoled to produce a better result.

Our reviewers appointed by our publisher, boyd & fraser, who provided many valuable comments and exhorted us to undertake the great labor of cases and an instructor guide.

Our Families: Liz, Simon, Rachel and Daniel Smith and Hilary, Matthew and Zoe McLeod, who put up with us and the late nights.

Jim Edwards III, Executive Editor at boyd & fraser, who provided overall guidance. Lisa Strite and Barbara Worth of boyd & fraser, who managed the production process.

We have sought to obtain permissions wherever possible. In the event that any materials have been overlooked, we apologize and ask that originators contact the publisher so that proper acknowledgment can be made in future editions.

Production Notes

The text for this edition (and complete prior editions) was entirely produced in electronic form by the authors using the Desk Top Publishing (DTP) package Pagestream 2.2 on a Commodore Amiga 4000 computer. Output was to a Postscript Hewlett Packard laser printer. Line art was partly produced by the authors, and partly by boyd & fraser artists. The latter was produced in Aldus Draw using Apple Macintosh equipment, and output to a 1270 dpi image setter.

Contact Address

For comments and suggestions, the authors can be contacted at: Department of Information Systems, University of Cape Town, Private Bag, Rondebosch, 7700, Republic of South Africa. Telephone +27 21 650 2261, Facsimile +27 21 650 4085. You are also welcome to send e-mail to graham@infosys.uct.ac.za or derek@infosys.uct.ac.za.

1 *The Nature of Projects*

Projects are not Routine

Why have a project? To get something done! If we have a situation that we are happy with, and which does not need to change, then we don't need a project. Projects are mounted to achieve some change. This could be the installation of a new hardware configuration, the development of a custom-written software system, or the country-wide installation of an application package, with the associated procedural changes in the business operation.

Definition of a Project

A project is a coordinated effort, using a combination of human, technical, administrative and financial resources, in order to achieve a specific goal within a fixed time period.

Attributes of a Project

Attributes of a project include:

- It has a *goal*
- It has a *start* and a *finish*
- It requires *resources,* including
    - People
    - Money
    - Tools and equipment
    - Administration
- It requires *coordination*
- It is a *temporary structure*
- It is mounted to *achieve change*

Brief History of Project Management

There have been projects since man (as a species, including women) attempted to do more than a single person could accomplish alone. Examples that come to mind include: the pyramids, the Great Wall of China, the cathedrals of Europe, the Roman roads. All of these required a major concerted and coordinated effort involving many people over an extended period of time.

Project management as a discipline, however, is relatively recent. This came about primarily as a result of the Second World War, where the Allied powers (England, U.S.A., etc.) had to respond to the Axis (Germany, Japan, etc.) challenge in the minimum time possible. The stakes were very high - essentially Western civilization as we know it. As a result, the best intellects available were applied to the problems with a very high level of motivation. This resulted in a revolution in the discipline and the creation of many of the techniques which we now use, including Critical Path Method (CPM) and Project Evaluation and Review Technique (PERT).

St Peter's at Rome is unequalled in magnitude and splendor by any other Christian church in the world. It was begun in 1506, and was consecrated in 1626.

The Duomo, Florence, was begun in 1296 and was finished in 1462. The cathedral at Cologne was begun in the middle of the 13th century and only partly finished in 1509, after which work was not resumed on it until 1830. In 1863 the interior was thrown open to the public. In 1880 it was finished.

- *Universal World Reference Encyclopedia*

Cathedrals of Europe *Figure 1.1*

Some Famous Projects

The **Manhattan Project** was the project to develop the atomic bomb. It was a collaborative effort among American, British and Canadian physicists, mathematicians, and technicians and ran over some four years.

The **OS/360 Project** produced the operating system for the first fully compatible range of commercial computers, the IBM 360 series. This was the largest software project ever undertaken at that time, and at the peak employed some 2000 software developers. Fred Brooks, author of "The Mythical Man Month" was a manager on the project. The project ran into major problems, but eventually produced a working operating system, which evolved into OS (Operating System) and finally MVS (Multiple Virtual System).

The **Apollo Project** was the project at the National Aeronautics and Space Administration (NASA) in the United States which placed a man on the moon. This was a national commitment, involving a goal set by John F. Kennedy, and a budget in the billions of dollars. It was also one of the early multi-disciplinary projects involving materials science, rocketry, physics, medicine, electronics, computer science, manufacturing, weather science, and radio telemetry. It was largely due to naval military requirements and the space program that the miniaturization of electronics began, resulting in integrated circuits and the cheap computing power that we have today.

Ultra was a major project undertaken at Bletchley Park in the United Kingdom to break the German machine-encoded ciphers used for secret transmissions during the Second World War. It involved the genius of Alan Turing, whom many remember as the inventor of the Turing Machine, a theoretical sequential computer operating off an instruction tape. The project involved developing hardware and computing algorithms to break the ciphers in response to every improvement the Germans made in their technology.

All of the above projects contributed valuable lessons and techniques in the areas of project management. The discipline has now become more formalized, and college-level courses are widely available. Project management is widely used in engineering and civil engineering. Buildings that previously would have taken many years to complete are now routinely finished in a matter of twelve to eighteen months.

In I.T. projects, there are still a number of unique problems, related mainly to the unclear nature of our objectives. It is very uncommon at the start of a systems project to know exactly what the required system should look like. In this respect, systems projects are much more like research projects. Fortunately, techniques are now evolving to help control these "fuzzy" projects. We will be exploring some of these in the text.

Project versus Ordinary Management

Management is the process of planning, organizing, leading and controlling the efforts of organizational members and the use of other organizational resources in order to achieve stated organizational goals [Bergen, 1986]. It is characterized by a cycle of "Direct, Measure and Control". Directing involves conveying goals, objectives, performance standards and responsibilities to staff. Measurement involves monitoring progress, work results and quality. Control involves applying the necessary changes to priorities, standards, work assignments and allocation of corporate resources to ensure that the goals of the organization are achieved.

In addition to the above, project management also includes the aspects of Initiation and Termination (figure 1.2). Initiation involves defining the Goals and Objectives, setting the scope of the task, determining financial and technical feasibility, designing the process to achieve the objectives, and selecting and building a team. Termination involves delivering the work results to the organization (frequently involving complex implementation, training and adaptation), planning the transition of resources to new assignments, and capturing learning which has taken place in the project for use on subsequent projects.

Like line management, project managers share a responsibility for staff motivation and development during their assignment to the project. In many ways the job of a project manager resembles that of an entrepreneur who has to develop an organization from scratch.

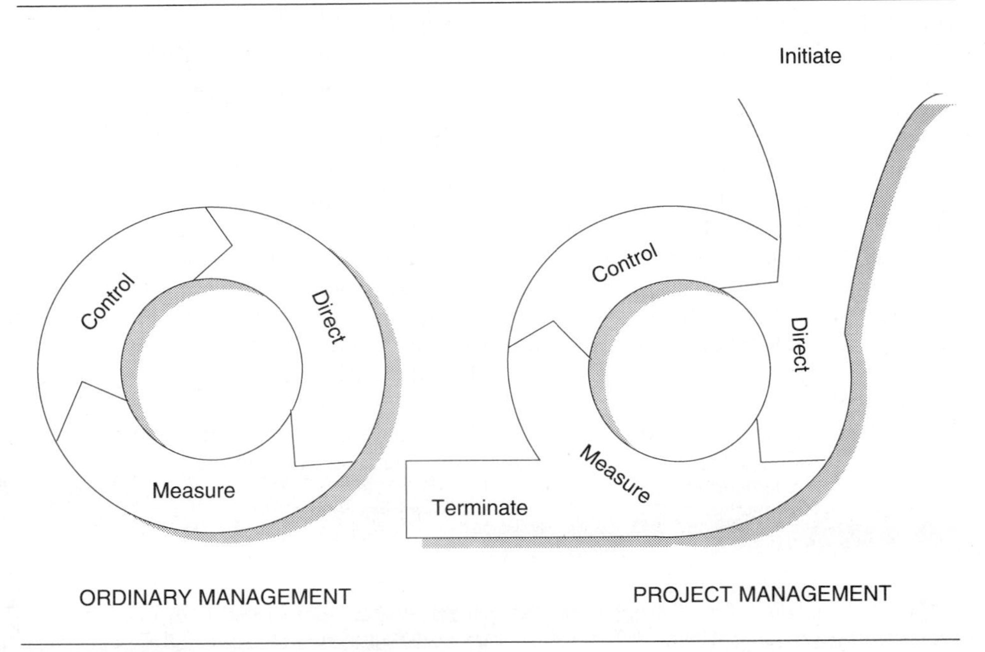

Project vs. Ordinary Management *Figure 1.2*

Project Risk

Unfortunately, I.S. project management is a risky business. Recent figures indicate that very few I.S. projects meet all their objectives. Lodge [1987] in a study among organizations participating in the CSC Index Productivity Enhancement Programme (P>E>P) indicated that, even in these superior organizations, as many as 10 percent of development projects exceeded time and budget estimates by more than one 100 percent. Kerzner [1989] discusses the failure of projects to meet quality requirements. Willbern [1989] and Bentley *et al.* [1991] have identified the phenomenon of "runaway" projects. They say that 30-35% of I.S. projects launched in some 600 companies were "runaways" - so badly out of control that there was no chance of meeting organizational objectives. Ewusi-Mensah and co-authors [1991] describe how many I.S. projects are abandoned, usually at late stages when most of the expenditure has already been incurred. More typically, many I.S. projects deliver the desired result, but very late and way over budget. We will be looking at these problems, their extent, and techniques to minimize them in Chapters 8 and 14.

Project Management Fundamentals

The Project Management Institute (PMI), a U.S.-based organization administering formal Project Management qualifications, identified the following project management fundamentals:

> There is nothing more difficult to plan, more doubtful of success, nor more dangerous to manage than the creation of a new system. For the initiator has the enmity of all who would profit by the preservation of the old system, and merely lukewarm defenders in those who would gain by the new one.
>
> **Count Machiavelli - 1513 AD**

- *Scope* - what is and what is not included. Where are the boundaries?
- *Time and deadlines.* Time is our one irreplaceable resource. Deadlines are dates by which a particular task or product must be complete
- *Human resources.* These are the people who will participate in the project. In I.S. projects, they are often our scarcest and most expensive resource
- *Quality.* Quality of the work done and the products produced is fundamental to achieving the project objectives. It is also vital to achieving productivity
- *Communications.* The project (and the project manager as its chief spokesperson), resides at the center of a web of communications. It is also essential that communication within the project team is effective
- *Risk.* Risk of failure in terms of requirements, budget and deadlines is ever present

These topics will all be dealt with in the text.

Variables in Tension

The project manager will always be balancing three critical variables: Quality, Cost and Time (figure 1.3). These three are connected as if by the sides of a triangle. It is impossible to move one without affecting at least one of the others. If I want higher quality, it will either take longer, or cost more, or both. If I want to reduce the cost, without sacrificing quality, I will need to allow more time. It is very important to remember these relationships when pressure is brought to bear during project execution. The variables and their respective priorities should also be discussed with management at the outset.

Types of I.S. Projects

An I.T. project manager can be called upon to manage a wide variety of project types, including:

- *System development.* A custom-written system is developed from scratch
- *Package implementation.* A pre-written application package is implemented, possibly with modifications

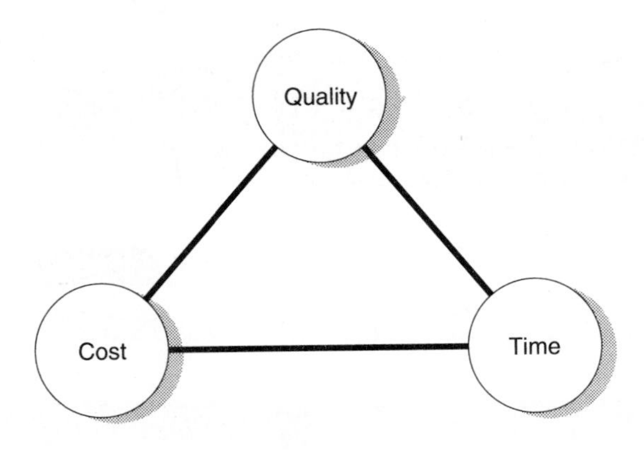

Project Variables in Tension *Figure 1.3*

- *End User Computing* where the target users of the system participate significantly in its development. This is common for Decision Support (DSS) or modeling systems
- *Prototyping* which can be used where unknown technology is to be used, or where requirements are unclear
- *Rapid Application Development* (RAD). Techniques are used to compress the lifecycle. These may include Joint Application Development (JAD), use of Computer Aided Software Engineering (CASE) and timebox methodologies. In the latter, the deadline is fixed, and the scope of the work tackled is scaled to allow the deadline to be achieved
- *Systems Architecture* projects which are used to define the strategic systems plan for an organization. A plan is derived from the business strategy and includes the set of systems which will be implemented by the organization over the planning time frame, normally five years
- *Selection Projects* where application packages or technology are selected to meet the business requirements
- *Projects involving an iterative lifecycle,* where tasks are performed repetitively to approach a goal more exactly. Common for RAD, prototyping and Object Oriented projects
- *Business Re-engineering (BPR) Projects,* which seek new ways of handling business processes to enhance effectiveness
- *Technology Implementation Projects* such as the installation of a network or e-mail system
- *Component Assembly Projects, typically* advocated in Object Oriented (OO) environments. Systems are built from predefined components bought in Class Libraries and "snapped together" with a minimal amount of custom coding and modification to form completed applications

Are I.S. Projects Different?

It is often argued that conventional project management techniques cannot be used in I.S., because our projects are different. Different in what way? In a construction project, it is easy to visualize the finished product in terms of the architect's design - there are well-tried formulae and techniques for determining the materials, sizes and strengths required, and there are established ways of performing the estimating, using stable norms for productivity of various resources. For example, we know how many bricks a bricklayer can be expected to lay in a day. I.S. practitioners claim that our objectives are fuzzy, that they change during construction, that we do not have the norms for calculating requirements, and that we cannot estimate accurately, because the range of productivity levels is vastly different across our resources.

There is some validity in this view, but we contend that the required information will never be available unless we apply some rigor to the process, collect some statistics, and start to build the database, models and norms that we require. This has already started to happen. The typical practitioner may be unaware of the techniques and tools available, because they are recent and derived from many sources. We hope that this text will persuade you that project management principles, some from conventional approaches and some from research methodology, can be applied, and that software projects will one day achieve the same level of professionalism as other construction and engineering endeavors. Bear in mind that the engineers and construction industry have a history of several thousand years, while software projects date back only to the late 1940s.

The management of I.S. staff does have some unique challenges and opportunities. I.S. staff do exhibit personality profiles that are relatively different from the population at large. We will be looking at this later under the topic of People Management.

Is Anything the Same?

Yes, many things are the same. There are a variety of techniques we can borrow from other types of project management, particularly high-tech engineering and research projects. Like all projects, success is heavily dependent upon the skills of the project manager.

Project Manager

What do you need to be a good project manager? Among other attributes you should be (or develop yourself to be):

- A *communicator* to handle the heavy load of person-to-person communications within the team and with external parties, including management, the project sponsor, contractors, support groups and others
- A *manager* to concentrate on the business goals, carry out the activities of directing, measuring, and controlling while motivating staff and developing them while keeping an eye on quality and risk
- An *innovator*, since you will face unique problems requiring innovative solutions
- *Technically competent, respected, and aware.* In managing technical people, it is essential that staff respect your judgment, or you will lose their support. If you ask

them to do things which they know are not possible, you will quickly lose their respect. It is not necessary to be a star technician, but you must at least have a thorough appreciation of the issues at hand, and the feasibility of your requests. If you lack technical skills, draft these into the team, or ask team members to brief you thoroughly before taking decisions

- An *Administrator.* As we have seen, there will be a significant component of administration to keep track of project schedules, resources, deadlines, deliverables, and budgets. If you are not a good administrator, find one and delegate this activity. Using a personal computer (PC) and the right software packages can help a lot
- A *Leader.* You will need to persuade the team, management, the sponsor and support groups to share your objectives, and participate willingly to help you reach them
- *Able to work well under pressure.* Projects are tough. You need to remain calm, keep focused, and not lose sight of objectives, while remaining sensitive to the people, organizational and political factors
- *Goal-oriented,* but not to the exclusion of the human issues
- *Knowledgeable about the company* so that you can keep the project aligned with corporate direction and objectives
- *Senior.* Project managers carry significant responsibilities with respect to corporate resources and finances, as well as the impact that projects have on the organization. Consequently, you need the appropriate authority to get the job done, and seniority to ensure that concerns are heard at the right levels

Responsibilities

The project manager typically has the following responsibilities:

- *Reporting to Senior Management and the Steering Committee.* The project manager represents the needs of the project, requests resources, conveys progress and problems to management
- *Communication with Users* to ensure that the project is achieving their requirements. This involves senior users, such as the sponsor (who is funding the project) as well as operational users, such as terminal operators, who will have to use the completed application. Other "users" include facilities (who will operate a mainframe system), and the group who will maintain the system in production. Each group may have different requirements that must be satisfied
- *Planning and Scheduling.* Deciding upon the best approach to the project. Setting up the goal and objectives, determining feasibility, estimating time and cost, determining resources required and scheduling the activities
- *Obtaining and allocating resources.* A certain amount of marketing skill is needed to communicate the importance of the project in order to obtain the resources from the organization. It is vitally important to allocate the correct resources to each task, and to ensure equitable distribution of work loads across the team
- *Controlling Risk.* The organization will expend a significant amount of money on the project, and divert corporate resources to achieving the project goals. If the goals are not achieved, then it is a waste, and the resources could have been better applied

elsewhere. The project manager must assess risk, eliminate it where possible, and control it where it cannot be eliminated

- *Delivering Results*. This is the ultimate responsibility - the organization chooses an individual whom they believe can deliver the required results
- *People Management.* Project managers have to balance the high task orientation required to deliver results, with sensitivity to the people in the team. If we ignore the human aspects, we run the risk of losing key resources, and ultimately not achieving the results!
- *Coordination* is vital to ensure that all the necessary activities take place in the right sequence and at the right time. This can also involve resources which are not under the project manager's direct control, for example, the Database Administration area. Outside parties, such as equipment vendors or software houses may also be involved. Good relations are thus important
- *Quality Assurance* is vital. There is no point delivering on time if the result is unusable, unreliable or unmaintainable. Quality can never be checked or controlled into the product, it can only be *built in,* at every step of the project. Quality assurance is thus an ongoing task from beginning to end
- *Budget Control* is necessary to ensure that the organization is not exposed to higher commitments than it originally agreed to. The history of I.S. projects in this area is not encouraging, so it is one to which we need to pay particular attention

The Big Picture

Projects do not occur in isolation. Consider figure 1.4.

- There is usually a *sponsor* - the person or group responsible for requesting the project and funding it
- We need to liaise with the *Steering Committee (or Body)* which is the corporate entity responsible for ensuring that Information Systems and Technology are used to good effect in the organization. It normally oversees all projects in progress and will usually have a fixed meeting schedule, typically monthly or quarterly
- There are *Users* at various levels to liaise with. These include: Senior management, line management, operational management, and the people who will directly interact with the system, or use its outputs. Other categories of users include those who will operate the system, or maintain it in production
- The project is normally staffed from the *I.T. function* within the organization, although a number of user staff may be seconded. In this sense, there is a reporting channel to I.T. management
- The project team will interact with *other projects* which we may depend upon, or need to interface with. For example, if we are developing an application system using a new database management system, we will have to work closely with the project to install the new DBMS in the production environment
- *Facilities* is the area which is responsible for the running of operational mainframe systems. We will need to consider their requirements and constraints in the design of our systems, and in the documentation which we produce

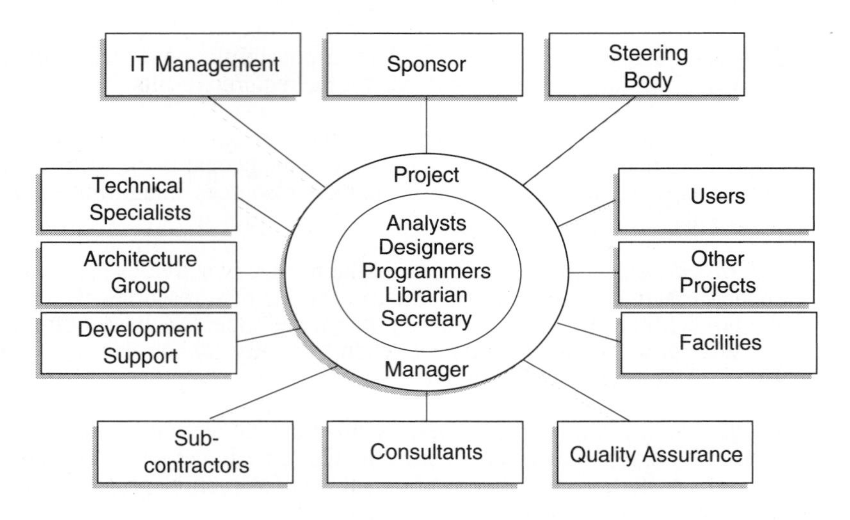

A Web of Communication

The Big Picture *Figure 1.4*

- There is often a corporate *Quality Assurance* function, responsible for setting standards, auditing results, and establishing procedures. We need to ensure that we work closely with them to achieve the required quality levels

- We may not always have the required skills, or the required number of resources. In these instances we may make use of external *consultants or contractors*. These require careful management, since they can be expensive, and can also upset the team who may feel that they are being pushed aside or having their capability criticized

- We may choose to *subcontract* portions of the work to external organizations. This requires that we give them very explicit requirements, and control the contract and relationship carefully

- Some organizations have a *Development Support Group* which can provide valuable assistance with system development methods, CASE tools, development tools and use of new techniques and technologies

- There may be an *Architecture Group* whose job it is to ensure that the overall portfolio of systems produced by all project teams integrates properly. They would be concerned with issues such as shared data, system interfaces, and consistency of user interfaces

- We may need to make use of various *technical specialists* from outside the project, e.g., in the area of networking, performance optimization or facilitating Joint Application Development (JAD) sessions

Within the project team, we can expect to have a variety of individuals in roles such as:

- *Analysts* who will help with feasibility analysis, carry out surveys, interview users, develop conceptual designs, and specify user requirements. They are normally also heavily involved in prototyping, system testing, user training, documentation and system implementation
- *Designers* who are responsible for converting the user requirements into a workable technical design capable of being built using the available resources and technology. Analysts sometimes perform this role. Designers produce detailed system specifications, data designs, and program specifications
- *Programmers* who are responsible for converting detailed designs into working programs which can be executed by the computer. They will perform program design, code the programs, and test their individual programs
- *Librarian and/or Secretary.* The former is responsible for collecting, indexing, storing and issuing project results, including specifications, designs, programs, test data, user documentation and technical documentation. The latter performs the normal secretarial duties, thus relieving project members of telephone answering, setting up meetings, keeping records, typing, writing letters, filing and data capture.

Coordinating and managing this changing web of communication is not easy!

Project Phases

Projects typically comprise several *phases.* Each phase has a set of tasks, expected results and quality checks. There are also some activities which are performed over the whole project lifecycle, viz., People Management, Risk Management and Quality Management. These are illustrated in figure 1.5. In addition, it may be possible to overlap phases to reduce the overall time taken for the project. This has attendant risks, however, since we may find that we need to change work performed based upon a specification which subsequently changes.

Strategic Fit

Projects must be consistent with the overall objectives and direction of the organization if they are to receive the support and resources that they need. They should be derived from a considered corporate planning process, which sets priorities and identifies preliminary goals and boundaries for the projects to be tackled. This relationship is illustrated in figure 1.6. The project manager should understand the business context of the project he is managing.

Distribution of Effort

Most of the effort on a system development project goes into the so-called *build* phase (once specifications are complete). The build phase includes detailed design, coding, testing and documentation. We can save ourselves a lot of pain and money by ensuring that we have a solid base upon which to work before entering this phase. Later in the text we will be examining empirical evidence that illustrates this.

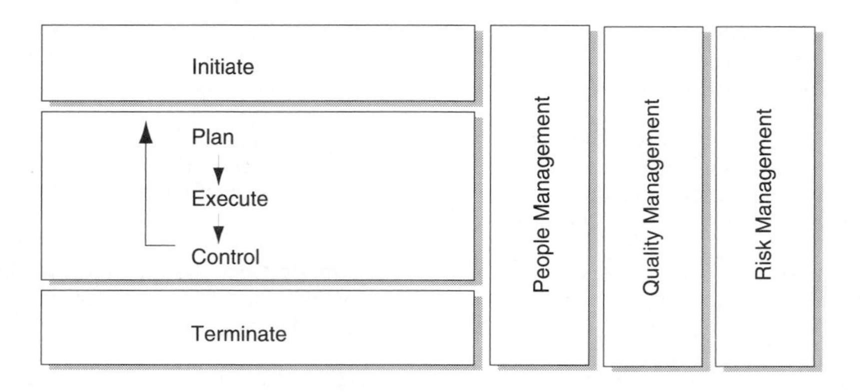

Project Phases *Figure 1.5*

I.S. projects were traditionally managed almost entirely within the I.T. function in the organization, with user involvement only by way of interviews conducted by an analyst.

This picture has changed as systems have become more complex and critical to organizations. First, users were seconded to the project to work directly with the analysts and designers. More recently, it is not uncommon for users to manage the project. A structure (illustrated in figure 1.7) which we have found works very well is as follows:

- The *nominal project manager* is a high-ranking user with enough seniority and *clout* to get the resources necessary for the project and makes decisions when necessary. This person must lend status and priority to the project. He will not normally be able to devote more than about 30 percent of his time to the project, since he will be in a senior line position. He will carry ultimate responsibility to the organization for delivering the project results. Seconded user personnel will have a solid-line reporting relationship to this person and a dotted-line to the project leader

- The *project leader* is normally an I.S. person who will manage the team on a daily basis. He will report to the Project Manager (dotted line) and to the I.T. Management (solid line). This person should have senior status within the I.T. function, as well as some business knowledge. I.T. staff allocated to the project will report to this person

Project Lifecycle

A major concern usually expressed is "If there are so many different kinds of I.S. projects, how do I know how to manage them? Is there a way to manage all of them?" Fortunately the answer is yes!

We can separate out the *Project Lifecycle* from the *System Development Lifecycle* or those of the other project types mentioned. Figure 1.8, which will be used as a reference point throughout, illustrates this. All projects share the phases: initiate, determine feasibility, plan, estimate, execute, and terminate. The execute phase is iterative, with each iteration

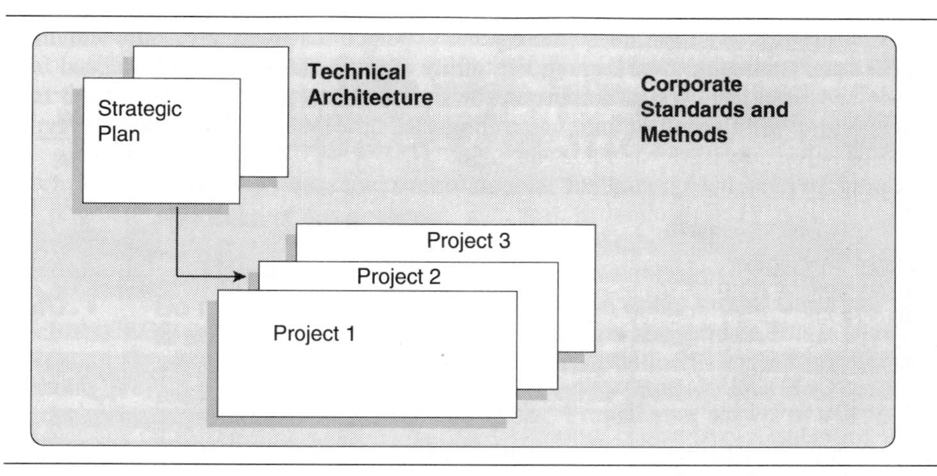

Strategic Fit *Figure 1.6*

representing a phase of the type of project being managed. For a system development project, these may be specification, design, programming, system testing and installation. The systems development (or other) methodology will specify the tasks, deliverables (things that are produced/delivered) and quality standards for each phase. The Project Lifecycle is thus a *container* for the Systems Development Lifecycle (or other type of lifecycle).

In the diagram, we have depicted a generic project lifecycle suitable for all kinds of I.S.

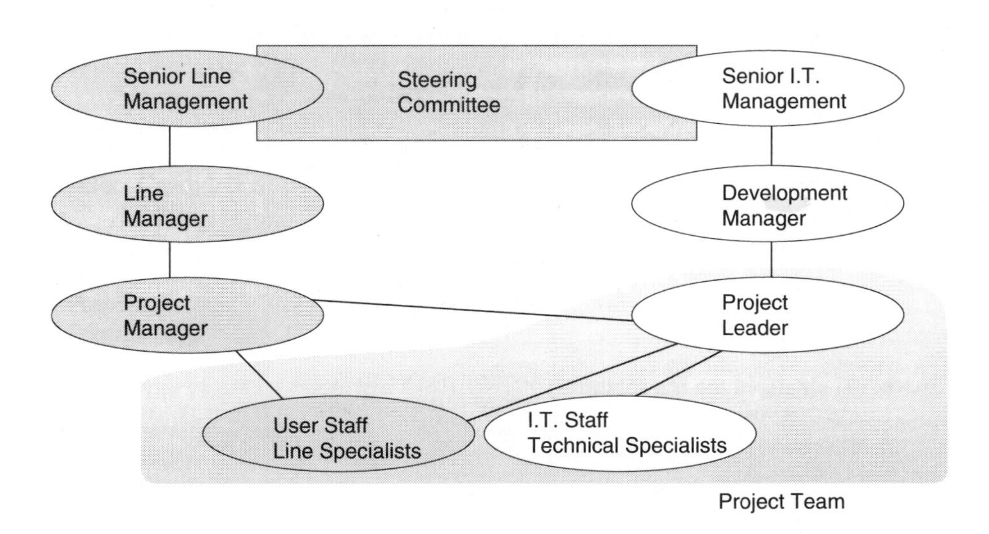

Management Structure *Figure 1.7*

projects. Each project will have all of these phases. Notice that there are some activities which occur once (Initiation, Determining Feasibility, Termination), while others occur for every phase and some per task or activity in the technical lifecycle of the project. These technical activities will depend on the type of project. For a systems development project, they would be defined by the System Development Life Cycle (SDLC) employed. Those of us from a programming background may be comfortable with a *pseudo-code* version of the above:

```
INITIATE PROJECT
DETERMINE FEASIBILITY (USE ESTIMATING)
PLAN PROJECT (USE ESTIMATING)
DO UNTIL ALL PHASES COMPLETE
   DO UNTIL TASKS FOR PHASE ARE COMPLETE
         SCHEDULE TASKS
         OBTAIN RESOURCES
         EXECUTE TASKS
         COLLECT RESULTS
         CHECK QUALITY
         ASSESS PROGRESS
   END DO
   REPORT ON PROGRESS
   PLAN NEXT PHASE (USE ESTIMATING)
   REVIEW
END DO
TERMINATE
```

For those not from a programming background, consider figure 1.9. This depicts the translation of the generic project lifecycle into a time-based bar chart. Activities are listed on the left with a bar representing the time taken for that task in the graph. Activities which are

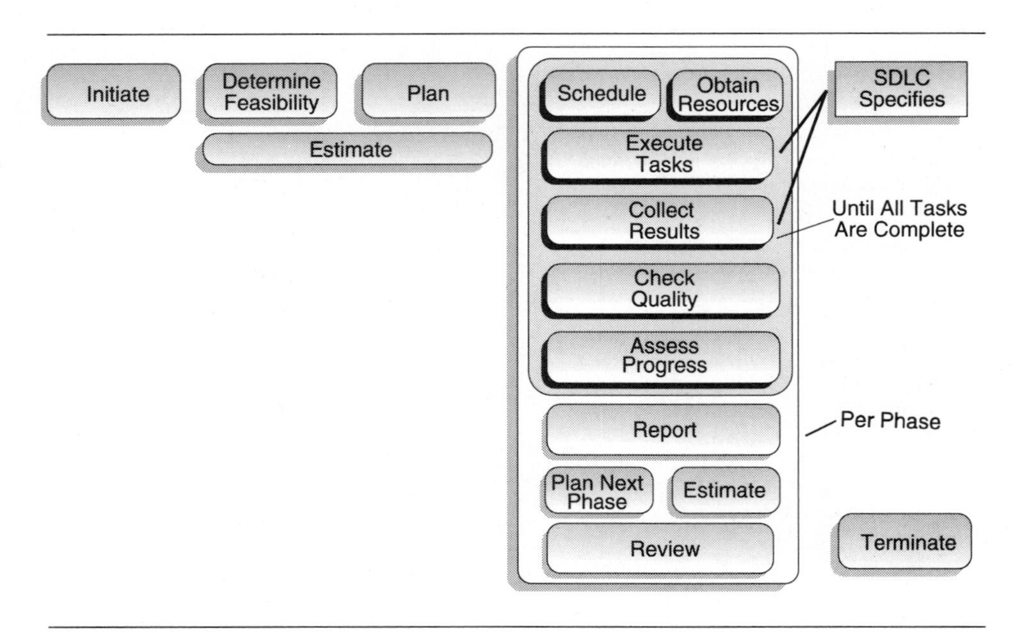

Project Lifecycle *Figure 1.8*

stepped, for example, *Initiate* and *Determine Feasibility,* have dependencies - the latter cannot proceed until the former is completed. Tasks which may occur in parallel (e.g., *Estimating* with *Initiate* and *Determine Feasibility*), can be shown below each other, indicating that they occur during the same time period.

Tasks 1 through *n* represent the activities specific to a phase determined by the systems development or other methodology dependent upon the type of project undertaken. For systems development these would include tasks like *Meet with Users, Develop Data Model* and *Analyze Current System.*

The bracketed set of activities represents a phase within the project. Typical phases for a development project would include: Analysis, Design, Build, Installation, etc. For each phase, the structure will be repeated, but the Tasks 1 through *n* will be unique. See chapter 8 for more details of the various lifecycles.

In the chapters that follow, we will discuss the relevant tasks and techniques for each part of the lifecycle. As we progress, we will highlight different aspects of the project management lifecycle diagram to show the areas to which the topics under discussion are relevant.

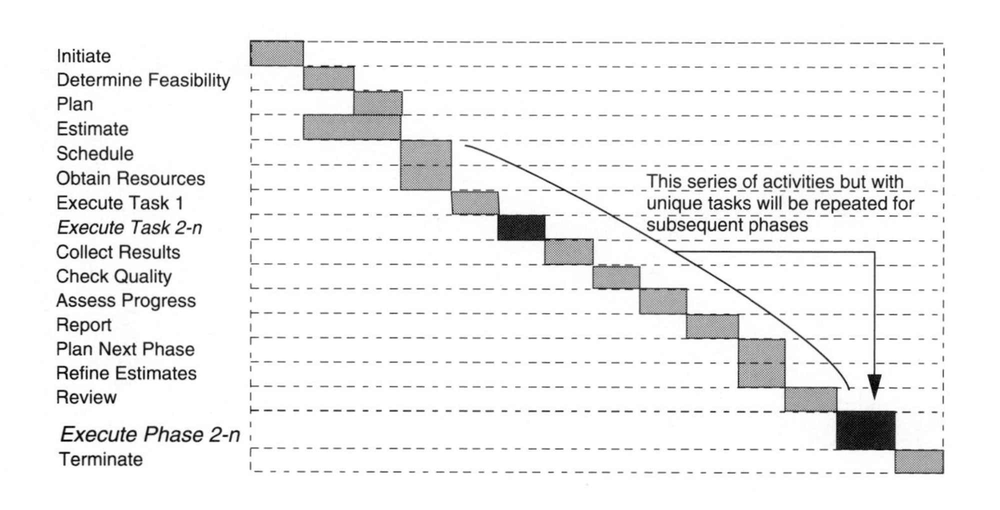

Project Lifecycle as Bar Chart *Figure 1.9*

2 *Project Initiation*

Introduction

In this chapter we will be dealing with the initiation of the project. These are the activities which happen right at the outset, as shown by the highlighted block in figure 2.1.

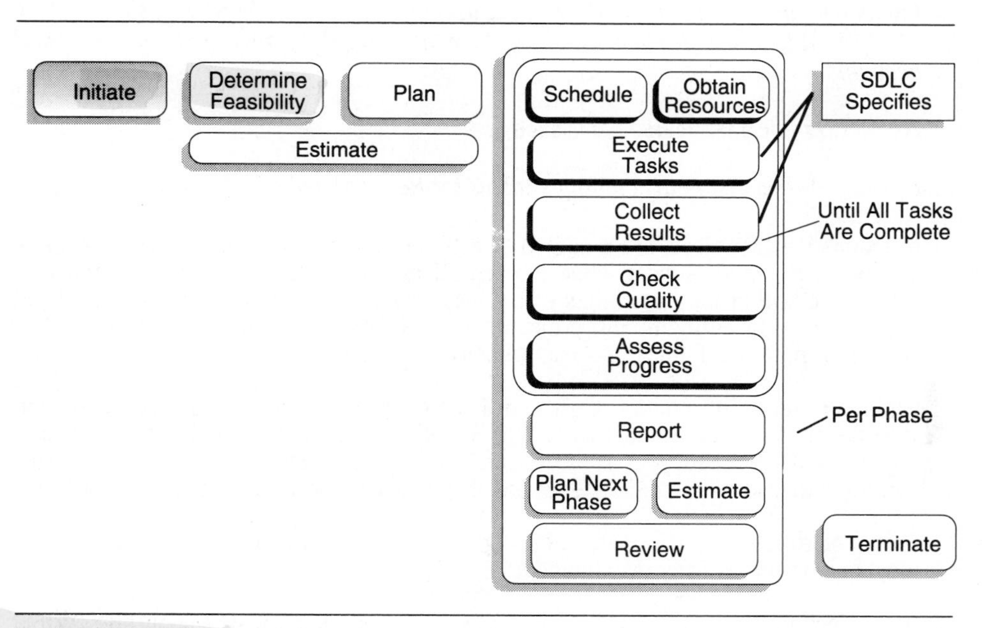

Project Lifecycle *Figure 2.1*

Project Definition

When a project is suggested or derived from high-level planning, one of the first things we need to do is to define it. This allows us to begin thinking sensibly about its goal, scope and feasibility. The project definition could be as small as a one-page form (figure 2.2), or as large as a ten-page typed document. In either case it should include the following items and sections:

- The company or organization name
- The area, division, section or other organizational unit requesting the project
- The individual who originated the project request
- The date that the definition was drafted
- The project sponsor - the person who will take overall *business* responsibility for project funding and success
- A title for the project. This is important! People will have to live with it for the duration of the project. Watch out for titles with nasty acronyms
- A unique project code, which will be used for costing and identifying the project
- The project goal. This must be clearly stated, preferably in one sentence. For example:

  Implement, at all branches of the bank, a user-friendly inquiry system, using existing ATM facilities, which clients can use to inquire on the status of their own bond accounts

  Avoid fuzzy or open-ended goals such as:

  Develop the necessary accounting systems for Retailers Inc.

  Remember that projects are mounted to achieve change. The goal is the *net change* desired in the environment when the project is completed successfully. It must be achievable within the constraints of budget, resources, technology, policy and legal requirements. In setting up our project, we should also ensure that it does not violate ethical or moral guidelines (for example, invasion of privacy)

- Priority in terms of Quality, Cost and Time (Schedule). This should be negotiated with the sponsor, and ranked as 1, 2 or 3, with 1 being most important, and 3 least important. Explain to your sponsor carefully the relationship between these. This ranking will serve as a guide later in the project when trade-offs need to be made
- Terms of Reference. These are enduring constraints or limitations which the project must bear in mind. They may include:
  - Technical issues such as "The project must be implemented using the current hardware configuration and network" or "The application package purchased must conform to corporate standards with respect to DBMS compatibility"
  - Resourcing issues such as "The project must not consume more than three programming resources at any one time"
  - Policy issues such as "Because of the competitive and confidential nature of the project, no external resources are to be used"

PROJECT DEFINITION

Company________________ Project Manager________________

Division________________ Originator____________________

Date___________________ Sponsor_______________________

Project Title______________________ Project Code__________

Project Goal____

________________________________ Priority: Q___ C___ S___

Terms of Reference______________________________________

____

Business Deadline_________________ Budget_______________

Assumptions____

____

Related Projects____

Moral/Ethical or Legal Issues ____________________________

Project Definition Form

Figure 2.2

- Legal issues such as "The system produced must conform to all requirements of the Hire Purchase Act and its amendments at implementation date"

- The Business Deadline is the date by which the project should be complete and have delivered its final results to the organization. This is also the date on which the cost-benefit analysis has been based. Any delay in implementation beyond this date will cause the project benefits to be reduced. For example, there may be a major marketing advantage to being the first insurance company to offer a new type of policy. If we miss the deadline, a competitor may beat us to it, thus negating a large proportion of the anticipated benefits from the system

- It is vital to spell out assumptions which have been made in the project definition and planning. These may be crystal clear to you at the outset, but may be very obscure to someone joining the project later. Explicit assumptions also allow us to check them as the project progresses. If we find that they were not valid, we have an early warning to rethink our approach

- The project budget should be specified in the form of a range and in resource units, e.g., persondays. The latter is to preserve a consistent base in the face of inflation and across currencies. We need a range because, at this stage of the project, we do not know enough about it to determine an accurate estimate. We will return to this topic in chapter 7

- We need to identify Related Projects which:
  - We depend upon for their results, e.g., if we are writing a Sales Analysis system, we will be dependent upon the Point of Sale system which collects the sales data
  - We will integrate with, so that our interfaces are defined consistently. We may, for example, be developing a Personnel system, which will interface with a Salaries Package being implemented by another team
  - Which follow on from our project. We may be creating results which another team will use, e.g., A menu and help system which will be used by all online systems in the installation. We need to keep dependent projects informed of our progress and any issues which they should be aware of in their own analysis and design
- Moral, ethical and legal concerns or issues which must be addressed sensitively should be detailed to sensitize project members to their existence and ensure they are not overlooked. Some of these may also translate into specific activities in the plan to address them. For example, staff whose job function may change as a result will need to be consulted, counseled and possibly retrained

Project Feasibility

Before the project can proceed, we need to establish if it is possible and a smart thing to do. This follows initiation and is done together with estimating, as shown in figure 2.3. There are a number of aspects to feasibility:

- Business issues
- Technical issues
- Time

Feasibility from a business perspective can be judged in several ways, depending upon the organization culture and the type of project.

- We may foresee *cost savings*. This is typical of systems which replace repetitive mechanical or clerical tasks, or which optimize an expensive process. Examples include a Delivery Planning System which would optimize routing of trucks; a Material Requirements Planning (MRP) System in manufacturing which will reduce raw material holdings (and capital costs); a system to calculate the least-cost mix of raw materials to produce a lubricant with required properties; an electronic mail (e-mail) system to reduce postage costs
- We may want to offer *new services* to our clients, thus attracting them to do business with us, rather than with our competitors. An example here is new types of home loan which allow clients to borrow without formalities against the paid-off capital portion to the limit of the original loan amount
- There may be *mandatory changes* required as a result of technology becoming obsolete, or because of legislation. For example, a project to replace a mainframe which will no longer be supported by the vendor or a project to implement the collection of Value Added Tax (VAT)

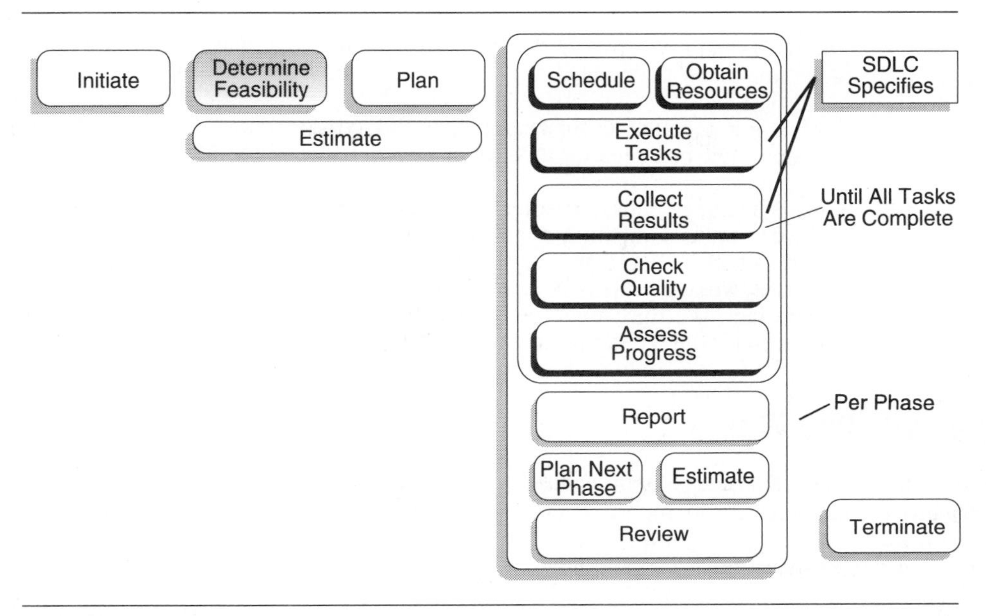

Project Lifecycle *Figure 2.3*

- We could be looking for *strategic advantage* through various avenues:
  - Providing facilities to our clients. Example: Home banking
  - Linking tightly to our suppliers to reduce our inventories and realize discounts. Example: An airline logistics system linking directly to a catering firm which supplies in-flight meals
  - Running more efficiently, thus reducing prices. Example: Retail supermarkets employing scanning to keep precise track of inventory
  - Better decision making - allowing us to use corporate resources more efficiently, to choose which markets to pursue and which products to promote or phase out. Example: The American Airlines *Sabre* reservations system

  With this category of systems, there will normally be no cost saving, and a traditional cost-benefit analysis would reject them. Also, there can be high risks. However, there can be major benefits when they succeed. Timing is usually crucial, and you should examine the effects of delays to the final date before committing to the project

- There can be *technical reasons* for mounting the project, including:
  - Unstable, old technology, which is no longer reliable, for example, a system running on an old network protocol which does not have adequate error control and security

- Performance of the old system may be inadequate to cope with business volumes, and the architecture may not allow an upgrade in the same environment

- The old system was poorly designed and documented, and cannot be reliably or economically maintained

- Skills are not available, for example, an old system was written in assembler, but could now be coded for a PC environment using a high-level language

- Time is a strange consideration at first glance. We include it since management will often want something done in an impossibly short time. There are tasks which cannot be shortened, regardless of how much we spend, or how many resources we apply. This is counterintuitive, but unfortunately an empirically proven fact. We must therefore check that the business deadline is a feasible one, even if all the other project parameters are acceptable.

Feasibility Report

When the various alternative approaches to the project have been investigated, we will normally prepare a feasibility report (figure 2.4) containing the following:

- An executive summary including:
  - Objectives of the study
  - Scope of the study
  - Possible courses of action
  - Pro's and con's
  - Recommendations
- Decision Criteria. The criteria that were laid down in terms of which feasibility was judged. This may include details of how cash flows were calculated, how project estimates were arrived at, assumed cost of resources, etc.
- Source and Reliability of Data. Details where we obtained our information, how reliable this is and what level of error our estimates and calculations contain
- Outline Requirements of the Proposed Solution. What the system or project would have to achieve to be considered successful. This should include the terms of reference, interfacing to other systems and compatibility issues
- Alternatives considered, with details of:
  - Operational Attributes - what is unique about this option in terms of how it would behave in operation?
  - Economic Implications - how does this option compare to others in terms of expenditure and cash flow?

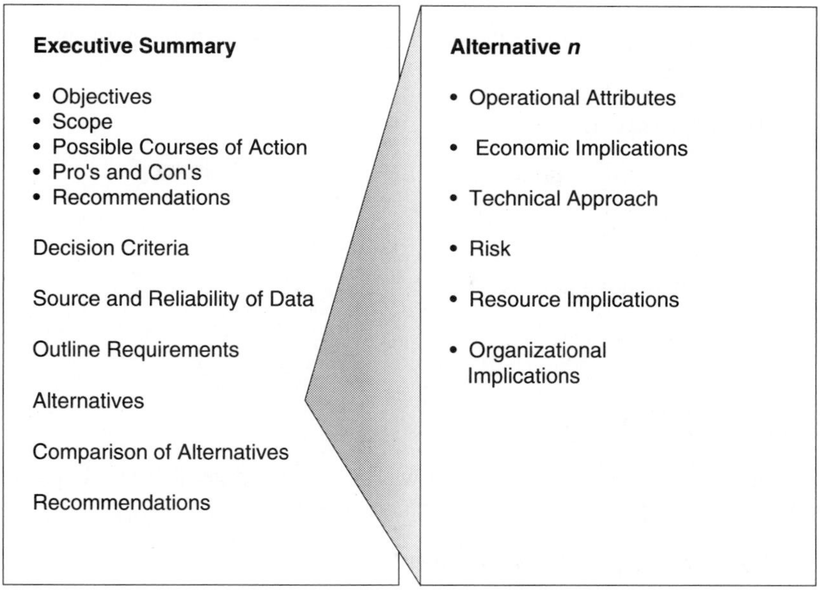

Executive Summary	Alternative *n*
• Objectives • Scope • Possible Courses of Action • Pro's and Con's • Recommendations	• Operational Attributes • Economic Implications • Technical Approach • Risk • Resource Implications • Organizational Implications
Decision Criteria	
Source and Reliability of Data	
Outline Requirements	
Alternatives	
Comparison of Alternatives	
Recommendations	

Feasibility Report *Figure 2.4*

- Technical Approach - what are the unique features of the technical approach; how is it superior to or inferior to the other options? Remember to include support considerations

- How risky is the option; how could this be controlled?

- Resource implications of this option relative to the others. Bear in mind resource seniority, skill level, availability, number required and duration of assignment. Highlight internal and external resource categories

- Organizational implications, with particular respect to changes required in the environment or in the way that users perform their functions. Consider training implications. Pay careful attention to any changes in job function or redundancies which might arise

• Comparison of alternatives, preferably by way of tables and graphics

• Recommendations. This should include a preferred option and a second choice, in the considered opinion of the person/people preparing the report. The alternatives put forward should be well supported by the data presented. Remember that we want to determine feasibility - "not feasible" or "do nothing" are legitimate options in some circumstances. It is also legitimate, where the suggested approach is not feasible, for the study team to suggest an alternative approach, or to recommend further investigation

Project Justification

Projects may be justified regardless of the outcome of a cost-benefit analysis. They may proceed because of:

- Benefits or savings
- Legal necessity
- Technical necessity
- Competitive advantage or
- Purely political reasons

Plan and Design

Once a project has been approved and we are given the go-ahead to begin work, the next step is project design (see figure 2.5). This includes:

- Establishing boundaries and scope
- Identifying standards, methods and techniques to apply
- Identifying the technical environment in which we will be working
- Identifying and customizing the tasks to be performed for inclusion in the project plan

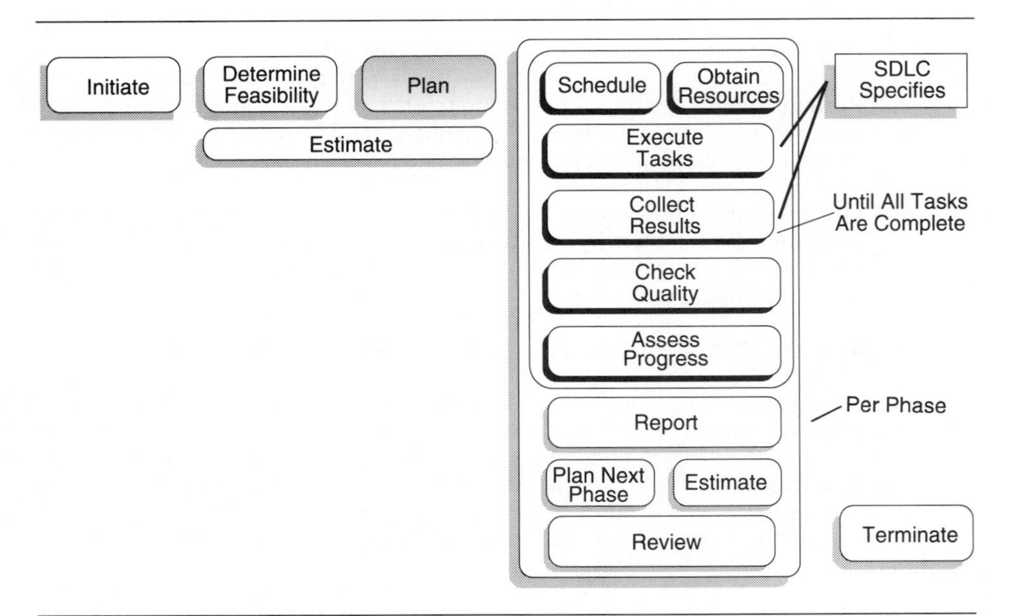

Project Lifecycle *Figure 2.5*

- Determining the skills required
- Estimating effort and durations
- Allocating resources to tasks
- Planning dependencies
- Seeking approval and revising

We will examine each of these in turn.

Scope of Project

Determining the scope of a project and managing it thereafter are vital to project success. If we get the scope wrong, or allow it to balloon, we are doomed. For some types of projects, the scope is fairly clear and can be expressed verbally, e.g.

Convert the existing batch ICL system to an equivalent IBM batch COBOL system without any change in functionality.

Install a Hewlett-Packard 9000 series machine and the AUTODRAFT CAD package in head office to support two design engineers and five draftsmen.

Even these have some loopholes:

- Does the conversion include files, or only the code and system control language?
- What about documentation - is this part of the objective?
- Is training in the CAD package included or not?
- Is there a prepared site for the machine and workstations?

We need to be very explicit in determining the scope. Mistakes can see the project size double, without any change in the available resources or budget.

Where the scope is fuzzy, such as in a system development project, or where new technology is involved, there are some useful diagrammatic techniques which can be employed. We recommend that you routinely use these for all projects.

Context Diagram

A context diagram is a high-level data flow diagram, with the proposed system shown as a single process box. See figure 2.6. Around the perimeter are all those things (external entities) with which the system will interact. These may be individuals, departments or other systems. Between the external entities and the central system are nodes representing collections of data which are involved in the communication. These may represent documents, files, screens, reports, etc. The medium through which the exchange occurs can be indicated in the node symbols. Examples follow:

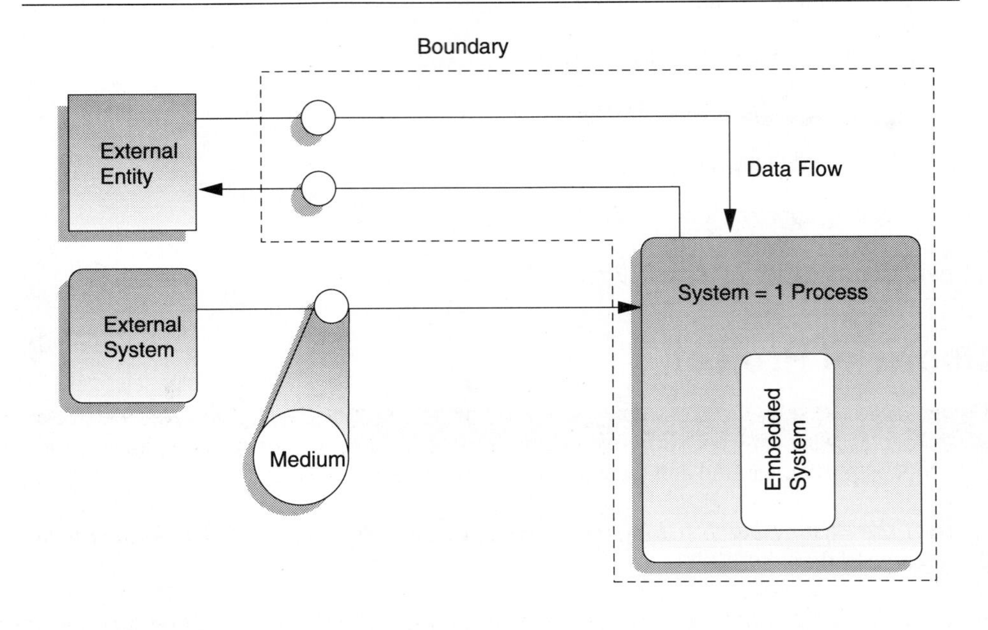

Context Diagram *Figure 2.6*

R	Random access medium
S	Sequential medium
H	Hardcopy, paper forms and reports
C	Communications link
P	Parameter passed in realtime

Arrows are used to indicate the flow of information and are labeled to indicate the collection of information.

A boundary is drawn to indicate the scope of the proposed system. Data groups (nodes) which are within the boundary will be maintained by the system, and the project can define their format. Those outside the boundary are maintained by other systems/sources and the system under consideration would have to adhere to the requirements imposed.

A highly simplified example of a context diagram for a Salaries system is shown in figure 2.7. In this example, the Payroll Clerk will feed the system with monthly changes via an online terminal and receive printed reports. Employee data is read from a database maintained by the Personnel System. Employees receive printed pay slips, and their bank accounts are credited electronically via transmission of transactions to the banks. The envisaged system could dictate the format of the summary reports, the way in which the payroll clerk communicates change data and the format of the pay slip. It would have to

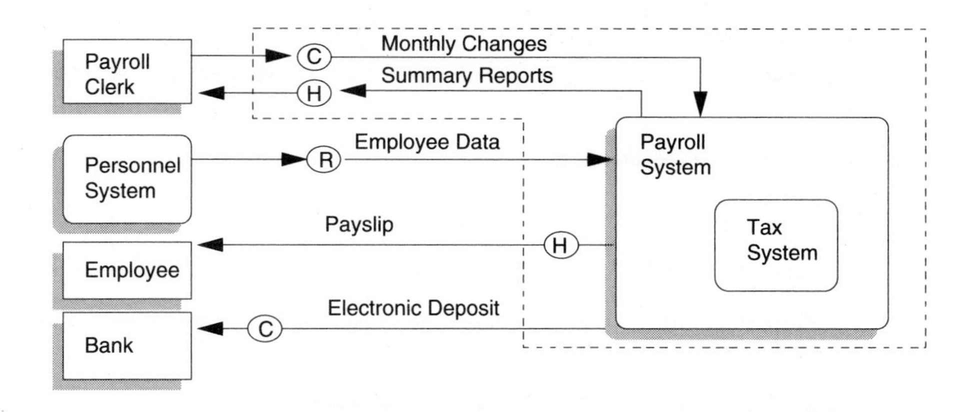

Context Diagram Example *Figure 2.7*

adhere to formats dictated by the Personnel system for employee data and by the banks for electronic deposit transactions. The Payroll system makes use of an embedded Tax system, which might already exist, or be developed as a separate project. We have left out communication with the Tax Authorities, and all detection and reporting of exceptions for simplicity.

Technical Environment

We need to consider the proposed technical environment in some detail, since this will influence the skills which the team needs, the methods that we will employ, the physical location where team members will work, and our task estimates. Some factors for consideration:

- Are we going to use Personal Computers (PCs), Minicomputers or Mainframe? A combination of the above may be appropriate; for example, a client-server application, where the database is centrally held on a server, but intensive graphics processing is performed by workstation software
- Will the system be centralized or distributed? Centralized systems are usually easier to specify and test, particularly in the area of controls. They may be less user friendly, more expensive and less flexible than distributed systems. The choice could influence testing and implementation strategies markedly
- What will the software environment contain?
    - Transaction Processing (TP) Monitor

- Database Management System (DBMS)
- Co-resident software/routines such as menu, logging or help systems
- Network control and management
- User interface: character based or graphical? Full screen or interactive at field level?

- What development tools will we use?
  - Which language
  - What compilers, linkers, debuggers, generators, etc.
  - Are there any collections of library routines, class libraries, components or tool boxes available to simplify our development?
- What types of hardware will the system need to be aware of?
  - Terminals
  - Printers
  - Special equipment, for example, ATMs, Scanners
- What variations will there be between development and production environments?
- Are there particular performance or response time constraints?
- What about security? How critical is the system, and what measures are available to us?
- Who will the direct users of the system be? For example, a system to be operated by members of the public will need a different user interface than one designed for expert data capture operators

A useful tool to help us define the technical environment which we envisage is the Technical Environment Model (figure 2.8). This represents the technical components (hardware and software) with which the system will interact. A central box represents the node in which the software will run. Within this, layers represent co-resident software components, including:

- Network software, for example, Virtual Telecommunications Access Method (VTAM™)
- Teleprocessing software, for example, Customer Information Control System (CICS™)
- Application infrastructure software, for example, a menu system, audit trail logger.
- Co-resident software, for example, a library of functions, or an online help system
- Database software, i.e., the DBMS or file management software
- Operating System e.g. MVS, Unix™, MsDos™

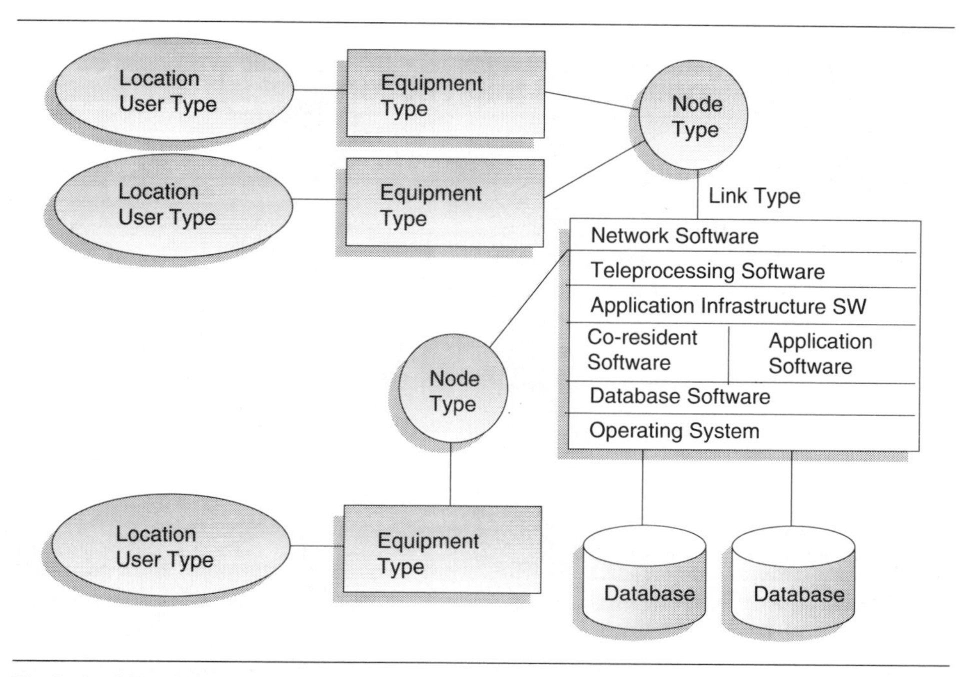

Technical Environment Model *Figure 2.8*

Linked to the central node are communication links, indicating the types of nodes through which communication may pass, the types of communication links employed, the type of equipment at the user locations and the classes of users who will access the system. Also linked to the central node are details of the database(s) the system will access. Database here is used loosely to mean a collection of related files.

Figure 2.9 is an example of a simplified Technical Environment Model for a project which will be implemented in an IBM Mainframe environment, with attached Personal Computer (PC) and dumb terminal workstations (3270 screens). It shows that we are using the CICS teleprocessing monitor, VTAM™ network software, a home-grown help and logging system, the DB2™ database management system and the MVS™ operating system.

Product Breakdown Model

An extremely valuable technique is the Product Breakdown Model (PBM), shown in figure 2.10. This can only be prepared at summary level at this stage of the project, but will be expanded as the project proceeds. It is an essential tool in establishing and controlling the scope of the project and will form the basis for change management and quality assurance.

The PBM is a hierarchical chart indicating the products that the project will deliver. The highest level represents the total result from the project. Finally, the lowest level will indicate individual *deliverables* which will be produced by a person carrying out a project task. At this stage we will probably be able to record only about three levels of detail; nevertheless, the exercise is extremely valuable, because it helps to set the expectations of the team and sponsor correctly.

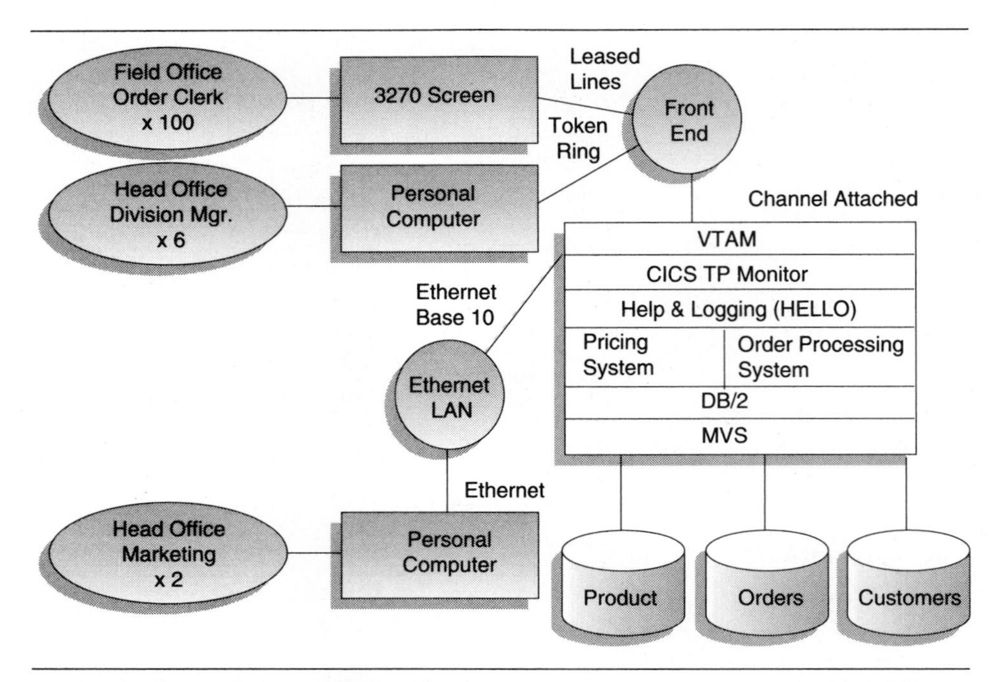

Technical Environment Model Example *Figure 2.9*

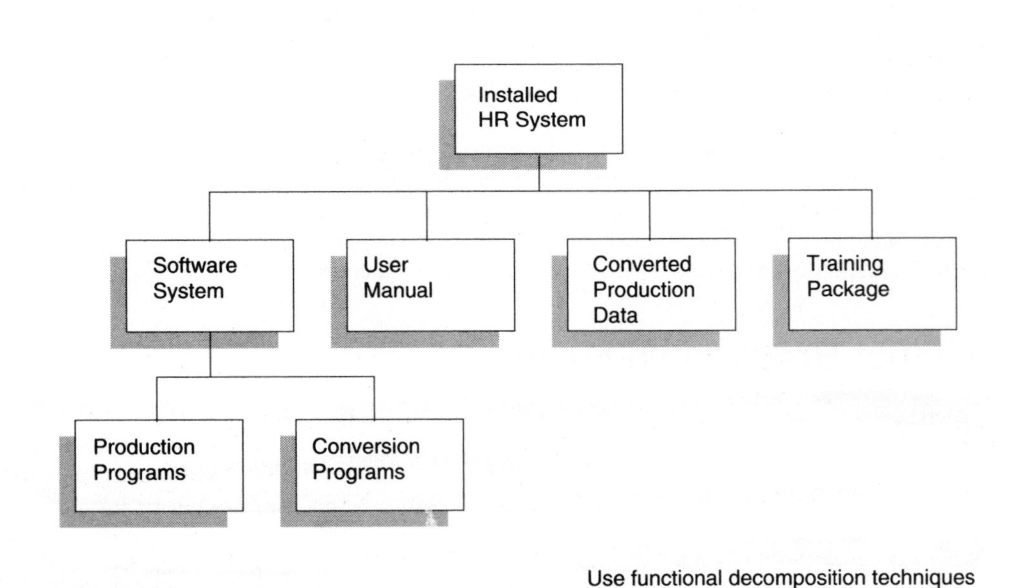

Product Breakdown Model *Figure 2.10*

Standards, Techniques, Methods

We will need to decide how we are going to run our project.

- Are we going to use a particular development methodology?
- Are there corporate standards in place for project activities?
- Are there accepted techniques which our staff have the skills to use?
- Are there quality standards that we can use?
- Are technical and programming standards in place for the environment? If not, do we need to develop these as a subproject?
- What is expected in the way of documentation?
- What techniques will we use for testing and validation?
- Are there automated project management tools available - if so, are there any previously developed or skeleton plans which we can use?

It may be necessary to identify areas where the team is lacking in skill so that we can arrange the use of external resources, or appropriate training.

Determining Tasks to Perform

Tasks to be included in the project can be determined from several sources:

- Further decomposing the PBM to identify individual deliverables, which are indicative of the tasks needed to produce them
- Using the tasks specified in your system development or other methodology
- Adding the necessary tasks indicated in the project lifecycle presented earlier
- Ensuring that you have provided for ongoing tasks such as team management
- Ensuring that you have included necessary training and infrastructural tasks, such as defining missing standards
- Think about each task and whether it has any dependency. Is the prerequisite task included in the schedule?

It may sometimes be useful to construct a matrix, such as the one shown in figure 2.11, mapping deliverables against tasks. A further refinement can be realized by entering O for output, I for input or IO for input and output in the cells. It is possible to use this for determining dependencies, which will be useful later.

Tasks

Deliverables	Collect Forms	Interview Users	Run JAD	Build Dictionary	Normalize Data	Draw Schema	Draft Func. Model	Prototype		
Data Model	X	X	X	X	X	X				
Function Model	X	X	X				X			
User Interface	X	X	X					X		

Include:

- Technical Tasks
- Management Tasks
- Q.A. Tasks
- Dependencies

Determining Tasks to Perform *Figure 2.11*

Determining Skills Required

To determine the skills which the team will need to perform the required tasks, we can again refer to our chosen methodology or construct a similar matrix to the one shown. See figure 2.12. This maps skills on one axis, against tasks and job titles on the other. First list the tasks, then the skills, and then map these to the job titles. Where the team will not have the required skills, consider entering external support groups, e.g., network support, or consultants into the job title area.

Estimating Durations

We will cover the whole area of estimating in detail in chapter 7. At this stage we may need to derive some initial estimates to give us a feel for the overall effort (and hence cost) as well as duration of the project. We should be working in ranges at this stage, since there is a high degree of uncertainty. The most accessible estimating technique at this stage is a Work Breakdown Structure.

		Skills									
		Communications	Business Knowledge	Data Modeling	Database Technical	Development Tools					
Tasks	Interview Users	X	X								
	Normalize Data		X	X	X						
	Prototype	X	X			X					
Job Title	Systems Analyst	X	X			X					
	Technical Analyst			X	X	X					
	Analyst Programmer	X	X			X					

Determining Skills Required *Figure 2.12*

Work Breakdown Structure

We can construct a Work Breakdown Structure (WBS) which represents the project as a whole at the top level, and phases and tasks at the lower levels. The greater detail we can obtain in the WBS, the more accurate our estimates can be. We would achieve an overall estimate by estimating for the lowest level tasks on the WBS, and collating the figures upward. The total of all tasks will be our overall estimate. Figure 2.13 illustrates this.

Determining Dependencies

To determine dependencies we can consult our methodology or refer to a matrix that maps tasks to deliverables. Once we have dependencies, they can be represented in the form of a network. We will cover formal methods for doing this later. We should only include definite dependencies and not worry too much about partial dependencies at this level. Figure 2.14 shows how dependencies can be derived from a matrix.

A Definition of Success

When is a project successful? By our definition when it satisfies all three of the following criteria:

- It meets requirements (of functionality, reliability, maintainability, portability, efficiency, integration and operability)
- It is delivered on time
- It is delivered within budget

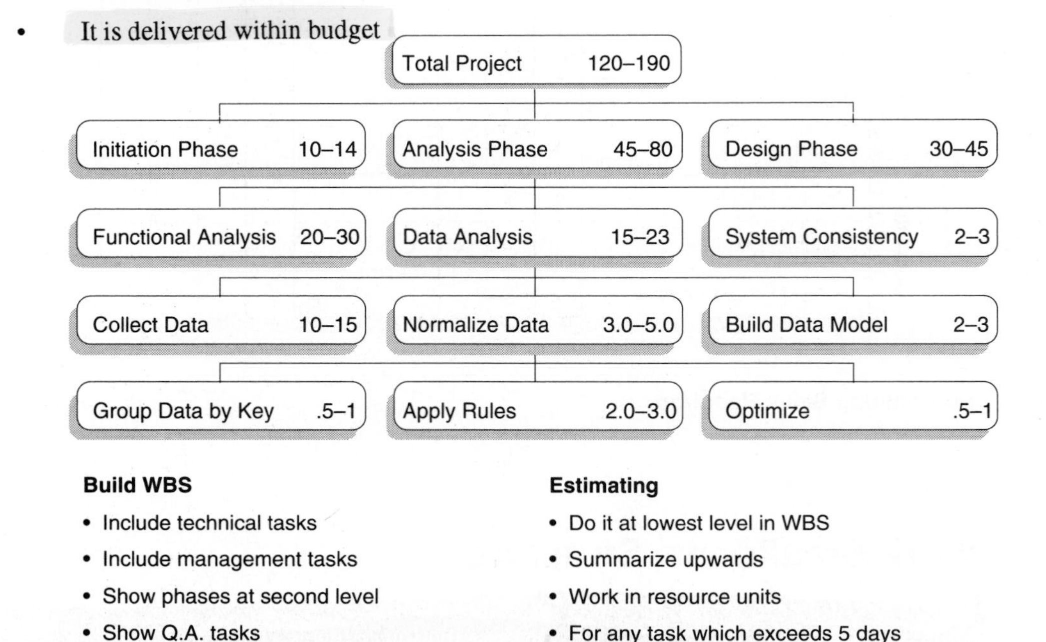

Estimating Durations *Figure 2.13*

In the I.S. industry, we are frequently guilty of meeting only one of these, namely requirements (and then only partially!). Horror stories abound of projects that run several hundred percent over budget, and take twice as long as anticipated. By our definition, less than 20 percent of I.S. projects are successful. This is an extremely worrying figure, and illustrative of the great need for better, more informed, more disciplined project management, such as this book encourages.

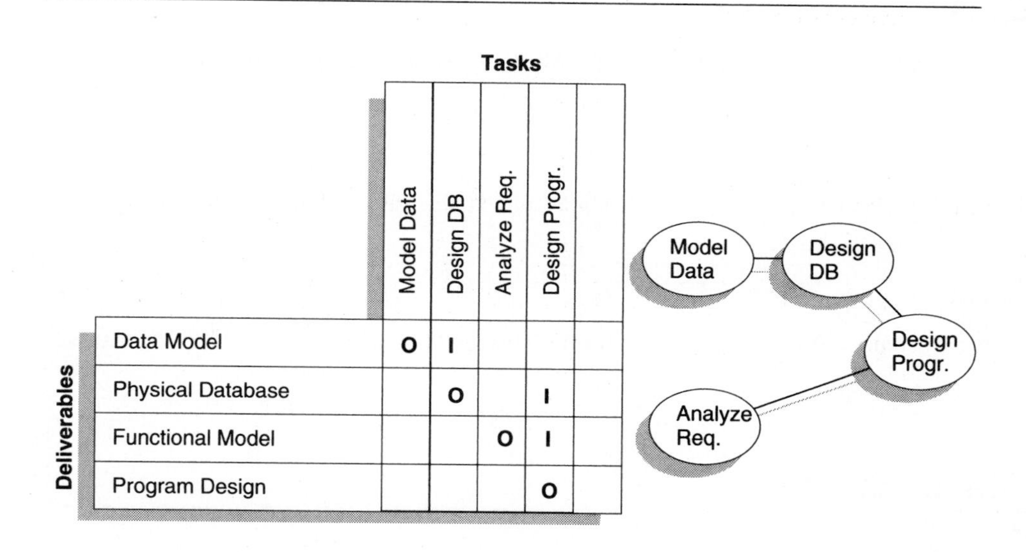

Determining Dependencies *Figure 2.14*

Case Questions

MyWay Organizer

Q2.1

Complete a Project Definition Form for the MyWay Organizer project. (10 minutes)

Q2.2

Scope the MyWay Organizer system using a Context Diagram. (10 to 15 minutes)

Q2.3

Draw a Technical Environment Model for the MyWay Organizer system. (10 to 15 minutes)

Gleam Stores

Q2.4

Prepare a Project Definition Form for a Gleam Stores project to fully document the system and prepare a training package. Assume that both user and system maintenance documentation are required. (20 minutes)

Q2.5

Prepare a Context Diagram for the Gleam Stores pilot system. Include all existing systems with which this will interface, as well as interfaces to the proposed "package" productivity applications. (25 minutes)

Q2.6

Draw a Technical Environment Model for the Gleam Stores pilot system. Include the workstation component and related software on the Head Office mainframe. (25 minutes)

Q2.7

Determine the financial feasibility of the suggested approach of replacing dedicated terminals with "commodity" PC workstations. Figures you have obtained from colleagues and Gleam management include:

Currently Installed Equipment:
460 terminals, of which 200 are masters. Some 15 percent of these require replacement annually.
Hardware cost of installing/replacing a dedicated terminal is $3500.
Equipment is written off over five years with straight-line depreciation.

Software:
Software for dedicated terminals costs $300 for a workstation or $500 for a master. There is a sliding scale discount on volume. For each (additional) ten copies purchased as a bulk order, the software cost reduces by 5 percent. For 12 copies discount is 5 percent, for 20 copies 10 percent.

Plans & Costs:
Forty new stores are planned to open in the next two years. Growth beyond that is projected at 15 percent per annum.
Estimated terminal population in these stores is 90.
Cost of a suitably configured PC to act as a workstation using the new software: $950.
Software cost per PC is $120. Networking software for Server (replacing master) is $500.
Installation of dedicated terminals must be performed by vendor personnel at $75 per hour.
PC installation would be completed by our own staff costed at $30 per hour.
Installation in both cases is estimated to take two hours per workstation, plus half a day per network in new branches. Where PCs replace old machines, the same network cabling can be used.

Pilot:
The pilot software development has to date cost $92 000. Documenting, stress testing and development of training materials is estimated to cost a further $35 000.

In your answer, try to show the position for the alternatives year by year. Would you recommend Gleam go ahead with this approach? Could you suggest a better alternative?
(30 to 40 minutes)

Handover Trust

Q2.8

Think about and detail the ethical issues and organizational issues which may arise as a result of the proposed project. How could you handle these sensitively? (35 minutes)

Q2.9

Scope the system using a Context Diagram and Technical Environment Model. (1 hour)

Q2.10

Determine what skills will be necessary in the team to tackle the project. What job titles would these relate to? (30 minutes)

3 *The Work*

Tasks

When we think of a project, this is usually the first thing that comes to mind - work to be done, tasks to complete.

- There are technical tasks, such as drawing up data models, writing specifications, coding programs
- There are managerial tasks, such as supervising the team, checking quality, controlling risk, and reporting on progress
- There are people oriented tasks, such as selecting a team, motivating the team and counselling a team member

To carry out the project successfully, we need to

- Include all necessary tasks
- Make sure the tasks are of manageable duration
- Estimate the effort and duration for tasks as accurately as possible. This will be dealt with in chapter 7
- Ensure that tasks are allocated to the correct resources and vice-versa. See chapter 8
- Monitor the completion of the tasks as they are carried out to ensure that we are not falling behind and that the quality of delivered work is high. This will be explored in chapters 12 and 15

Source of Tasks

Most of the tasks will be of a technical nature. These are normally dictated by the nature of the project, and can be derived from a formal methodology where one is employed. Examples would include SSADM (mandated by the U.K. Government) or Information Engineering (popular in commerce).

There will also be management tasks - normally to control the project, liaise with management, ensure quality and so on. These are more related to the organization's philosophy of reporting and quality assurance and are likely to be unique to the organization.

A system development project will have tasks such as:

- Establish feasibility
- Plan the project
- Define user requirements
- Specify technical design
- Build the system
- Test the system
- Prepare operational documentation
- Train users
- Install the system
- Post implementation review
- Manage the team
- Report progress to management

These tasks will often be too large to manage as a unit, and will in turn have sub-tasks. For example "Define user requirements" may be broken down to:

- Establish scope of system, major inputs and outputs
- Analyze current system
- Interview users regarding additional requirements
- Prepare functional model of proposed system
- Prepare data model for proposed system
- Review with users
- Prototype behavior of new system
- Finalize requirements document

To be useful, we need to get to a level of task that is of a manageable duration, and which can be assigned to a single person, or small group, for completion.

Work Breakdown Models (WBM)

You probably noticed that there was a sort of hierarchy when we decomposed the "define user requirements" task above. We could have specified sub-tasks for other high-level tasks in a similar manner. We could also carry the decomposition further, for example, by decomposing the task "Prepare data model for proposed system" to:

- Identify major entities
- Collect attribute data
- Identify keys for each major entity
- Construct Entity Relationship Model

A useful way, therefore, to view our tasks, sub-tasks and sub-sub-tasks for a project, in relation to each other, is a hierarchical chart, known as a Work Breakdown Structure (WBS), shown in figure 3.1.

In the WBS, each level is decomposed further into the next level, giving its "children". This can be done using the functional decomposition technique familiar to us from structured analysis:

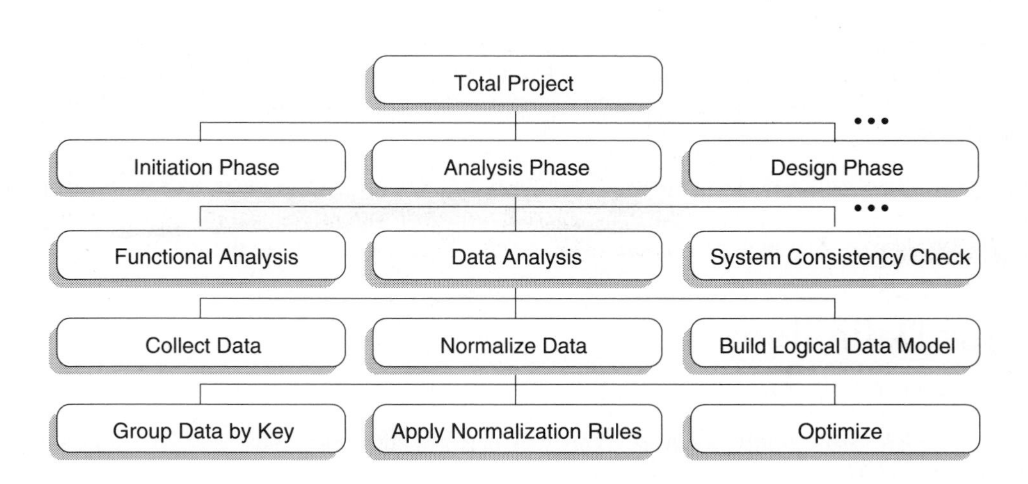

- Include Technical Tasks
- Include Management Tasks
- Show Phases at Second Level
- Show Q.A. Tasks
- Decompose Until "Work Packages" Are Obtained

Work Breakdown Structure *Figure 3.1*

- For each box, ask the question: "What must I do in order to achieve this?" The answers indicate child boxes

- There should be between 2 and 6 child boxes per parent. If there were to be 1, this would, by definition, be equal to the parent box, and is therefore superfluous. If there are more than 6, human short-term memory starts to lose track when we want to consider all the children for a parent together. If you find you have more than 6 children, look for missing intermediate levels that could group some of these

- To check if a child box is in the correct place, ask the question "Why do I do this?" The answer should be the parent box

- Wording in the boxes should have the form:

  [verb] {qualifying clause} [object] {qualifying clause}
  where square brackets [ ] indicate required items, and braces {} indicate optional items, for example:

  Define Functional Model

  Integrate System with Sales Analysis

  Set up Valid Test Data for Transactions

We need to ensure that our Work Breakdown Model (WBM) contains all the tasks, not just the technical ones. Be sure to check that Quality Assurance (QA) steps and management tasks are specified. Tasks may also need to be included to address legal, ethical or human factors issues.

In this book we will refer to items at the lowest level of the WBM as tasks, and to those at higher levels as work clusters. For each *task*, we need to identify the information in the following section. Figure 3.2 shows the relationship of task definitions to the WBM.

Task Definition

- Task *description*

- *People and groups involved.* Who will perform the task? What sort of skills are required?

- *Applicable techniques.* How do we carry out the task? Is this specified in our methodology, or other corporate standard?

- Are there any *prerequisite tasks* which must be complete before this one can be tackled? For example, data collection must be complete before a relational data model can be constructed

- What *deliverables* are required as input to the task? What will the task produce? The functional analysis task, for example, may require as input the current system description, and produce as output the functional model

- Are there any particular *dangers* associated with the task that we should look out for?

- Do we know how to *estimate* how much effort the task will consume and how long it

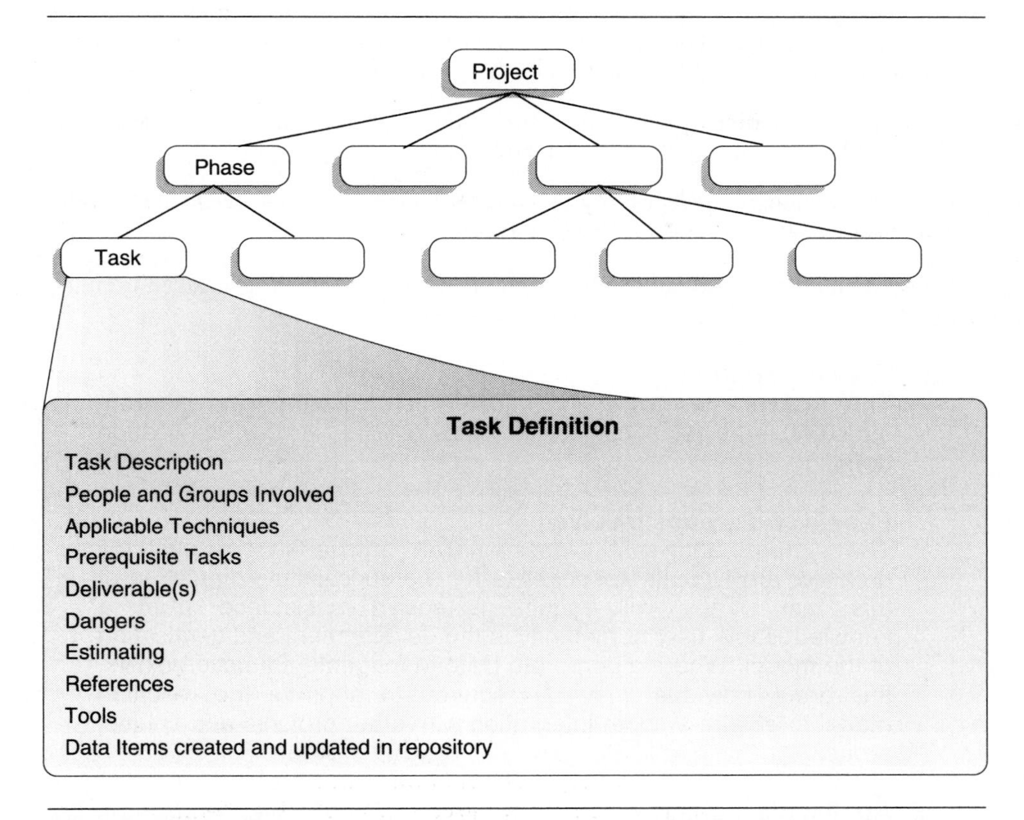

Work Breakdown Model *Figure 3.2*

will take with the resources available?

- Are there any sources of *reference* or *reusable components* which we use in performing the task? For example, a standards manual, programming language manual, methods description, sample library or class library

- Is the task supported by any *tools. For example,* project management software, CASE tools, Data dictionary, compiler, editor, etc.

- If using a dictionary, or repository-based CASE approach, what *data items* in the dictionary or repository are *affected* by this task?

A sample task definition follows for the definition of an entity model. This is taken from the *Inspired* ITIM method used in an IEF CASE environment:

Task description
This task creates an entity model which describes the inherent structure and relationships of data groups in the application domain.

People and groups involved
The systems analyst, business analyst, data analyst or systems architect are typically responsible for the successful completion of this task. They will require extensive assistance from a knowledgeable user community in the development and verification of the model. Where a corporate Data Management function exists, this group/individual may be required to approve the resulting model to ensure smooth integration with other projects and subject areas.

Applicable techniques
The Martin-McClure entity modeling approach should be used. This begins with intuitive naming of candidate entities, followed by collection and allocation of attributes to these candidate entities. Finally, identifiers are chosen for each group, non-entities are discarded, and relationships between remaining entities are determined.

Prerequisite tasks
Data collection. Project Design (including identification of key users).

Deliverables
A neatly presented **entity relationship diagram** making use of the Martin-McClure notation. Many-to-many relationships are permitted and should not be resolved to junction boxes.

Entity definitions listing the name, data items and key components for each entity. Only data items required for primary and foreign keys need be recorded at this level. Other data items serendipitously discovered may be documented, provided this activity does not introduce unnecessary delays. Definitions should be captured in the Data Encyclopædia.

Definitions for all data items discovered or required for keys should be captured in the Data Encyclopædia.

Dangers
Do not go into too much detail - for example, trying to locate all attributes of the entities.
Do not expand the scope to include entities which have no bearing on the application context.
Do not worry about normalization. The entity model does not need to be normalized. Repeating groups are acceptable.
Introducing duplicates of items already in the Encyclopædia without realizing that they already exist in other systems and databases.

Estimate
Allow approximately one week for data collection (longer if user community is geographically dispersed or relatively unavailable). Allow between 1 to 3 days for formal data analysis and capture into CASE tool. Allow 1 to 2 days for review with users and technical auditor. Allow one day to amend model following review.

Reference
See Martin-McClure, "Recommended Diagramming Standards".

Tools
All information and models to be captured into the IEF CASE tool.

Data items affected
Entity Name
Entity Description
Per attribute:
- *Entity Attribute Name*
- *Prime Key Component (true/false)*
- *Foreign Key Component (true/false)*

Existing Database Relations
- *Relation Name*
- *Technical Environment*

Required Retention Period
Archival Frequency

ERD Name
per Entity:
- *Entity Name*
- *Position*

per Relationship
- *From Entity*
- *To Entity*
- *Cardinality*

Phases

In I.T. projects we will normally show our major configuration management phases at the second level of the Work Breakdown Structure, as shown in figure 3.3. We will explore configuration management in more detail in chapters 14 and 15.

At the next level of detail, the first child will normally be a task addressing any issues raised at the previous phase's concluding review. The second last child will be an update of the plan for the next phase, followed by a review of the deliverables completed in this phase.

Note that this is consistent with the generic project management cycle which we presented in chapter 2.

Manageable Unit of Work

Earlier we introduced the idea of a manageable unit of work, or MUW. This is a very useful way of staying out of trouble. If we include large chunks of work in our project plan, say a month long each, and one takes twice as long as expected, we will be a month behind schedule. If we plan at the level of a week, and a task takes twice as long, we will be a week

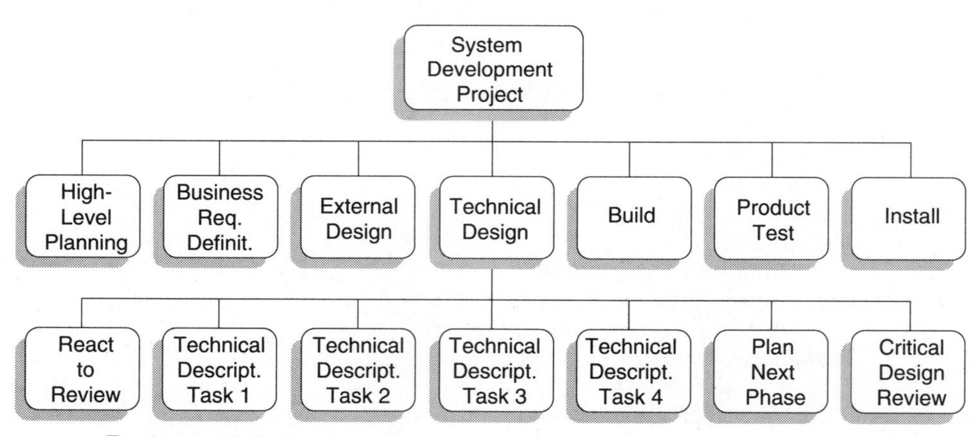

- Top Level = Project
- Next Level = Configuration Management Phases
- Third level starts with task to react to review of previous phase – correct any deviations, formally record any concessions
- The last task on the third level is the Quality Assurance review for that phase

Configuration Management WBM *Figure 3.3*

behind. Obviously there is a trade-off somewhere, since there will be an overhead in the planning, monitoring and recording of each task. You should choose a good MUW for your project. We suggest that you start with a week (calendar time) - this could represent several man-weeks of effort. Next apply the following modifiers to come up with a unit for your project:

Decrease the MUW if

- Your project is using new technology
- Your project is risky
- Your project is extremely critical to the organization
- Your team (or you yourself) is/are inexperienced
- The project is under a lot of time pressure
- The system is very complex
- There is a high degree of integration required with other systems
- You are going to make use of several external resources
- The environment is highly political

Increase the MUW if

- The project is non-critical
- You are using proven technology
- You and the team are experienced
- The project is not very complex, nor tightly integrated
- You have control over all resources
- The environment is apolitical and stable

In any case, the MUW should not be shorter than one day, nor longer than one month.

Once an MUW has been chosen, the WBS should be decomposed to the level where no task exceeds this value. This will have the effect of controlling the degree of formality and control on the project - we are balancing the management overhead against the risk of slipping. We should also find that small projects result in a small number of tasks (maybe 20 to 30) while large projects could have several hundred tasks. This counters the problem we frequently find with inflexible methodologies which can prescribe hundreds of tasks for even a small project.

Case Questions

MyWay Organizer

Q3.1

Develop a Work Breakdown Model for the MyWay Organizer project to 3 or 4 levels. (15 mins)

Q3.2

Assuming your Work Breakdown Model has a task "Unit test modules". Define this task in detail using the guidelines discussed in the chapter. (20 mins)

Q3.3

Determine a manageable unit of work (MUW) for the project. Assume:

- You are an experienced project leader with 10 years experience, the last 4 in the PC environment
- You have a young team of three good analyst programmers
- The specification is well developed and understood
- Only one of your team has developed in the exact environment (compiler, editor, operating system) before. The other two have used the language before, but in other environments
- Success of the project is critical to the organization
- You have tight deadlines, but will receive the full backing of management. Funding is not really a major issue, as this is regarded as a top priority project

(10 mins)

Gleam Stores

Q3.4

Prepare a WBM for Gleam's project to document the pilot system fully and build training materials. (30 mins)

Q3.5

Prepare a WBM for Gleam's planned implementation in 20 regional stores as the first phase of deployment. This should include hardware installation, network installation, data conversion and training. (30 mins)

Handover Trust

Q3.6

You have asked one of your project leaders to prepare a draft standard project plan as a basis for all the project leaders to use in managing the parallel system development projects required. She has produced an indented list representing the tasks and sub-tasks which she envisages in the plan. The indentation represents a hierarchy - similar to a work breakdown. The list is shown below. Please critique it and correct it where necessary.

- Determine feasibility
  - Justify economics
  - Evaluate technical options
- Specify requirements
  - Model data
    - Collect documentation
    - Analyze data
    - Draw entity model
  - Examine existing system
    - Interview users
    - Draw data flow diagrams
  - Define logical system
    - Develop logical DFD's
    - Design reports
    - Design online screens
- Design system
  - Database design
    - Normalize data
    - Draw Relational model
    - Add physical access requirements
    - Calculate sizing
    - Specify physical schemas
  - Program design
    - Define program requirements
  - Design batch flows
    - Specify batch process
    - Define Job Control
- Write system
  - Code programs
  - Test programs
- Parallel test with old system
  - Prepare data
  - Move software to production

Run test
Verify results
Take system live

(45 mins)

4 *The Product*

Deliverables

Projects are mounted to deliver products and achieve results. Hopefully, what we deliver will be tangible and useful. To this end we introduce the concept of *deliverables.* A deliverable is a tangible result delivered when a task is complete. It may take the form of a paper model, a report, program code, a computer file, etc. Some examples include:

- A Data Flow Diagram depicting the scope and boundaries of the system
- An Object Model showing the attributes, derivation and behaviors of classes
- A report on the feasibility of the project
- A library of coded programs
- The User Operating Manual for a piece of software
- The installed network components for an online system
- A set of test results for a program

Deliverables are important, because they are tangible, and thus serve as evidence of progress. They also allow us to assess quality of work performed as we proceed. We will address these issues further when we cover project control, reporting and quality assurance in chapters 11, 12, 13 and 15.

Each deliverable should have an unequivocal definition. This is essential to allow us to measure completeness and quality, as well as to build expertise in the production of deliverables. A formal definition also helps to capture the experience gained in projects for future use. The definition of deliverables can usually be obtained from your methodology. For each deliverable we should know the:

- *Purpose*: Why do we need this deliverable - what does it do for us?
- *Structure of the deliverable:* Is it a document, file, chart, etc. What is the presentation format and medium?

- *Manner in which the deliverable is used practically* (illustrated by an example)
- *Validation rules* which should be applied in collecting information and constructing the deliverable. For example, "all data items used on screen and report prototypes must be present in the data dictionary"
- *Notation conventions* which will be followed in presentation
- *Quality standards* which can be used to verify that the deliverable has been correctly produced and meets the necessary standards
- *Associated tasks* which are involved in producing the deliverable. These effectively form a cross-reference to the WBM discussed in chapter 3
- *Prerequisite deliverables* which are needed in order to build this one. These can be determined by the data which is required as input to this deliverable
- *Tools* which support the development of this deliverable, or contain the result
- *Data content* of the deliverable to assist in the determination of dependencies

A sample deliverable definition (for an entity model) from the ITIM methodology is shown below:

Deliverable Type
Entity Model

Purpose
To understand the inherent data structures used in the application area, the logical business grouping of these, and the inter-relationships between these groupings. The entity model also allows us to verify with users that the analysis group has correctly understood many business relationships and issues. Later it will serve as input to the design of physical databases, migration planning for data conversion and possibly a structure around which to begin prototyping.

Structure
The entity model contains entities (things about which we wish to store data) and relationships between those entities. An entity could be a concrete object (or category of objects) e.g. FURNITURE, VEHICLE; a type of person e.g. CUSTOMER, EMPLOYEE; a conceptual grouping for analysis e.g. PRODUCT CATEGORY, REGION; or a record of a transaction of the enterprise with external parties e.g. DEPOSIT, SALE, DELIVERY.

Occurrences or instances of entities must be identifiable by one (or a small number) of identifying attributes (key). For example, I can tell one CUSTOMER from another by the CUSTOMER-NO attribute and one EMPLOYEE from another by the STAFF-NO attribute. If more than one attribute is required to identify occurrences of the group, examine it carefully, because it may be a relationship, not an entity.

Relationships exist between entities. They are labeled to indicate their nature. An example would be a relationship between CUSTOMER and ORDER labeled "Places". It is possible for entities to have relationships with themselves. For example, the entity EMPLOYEE may have a relationship labeled "Manages" to itself. This would link together the employee who is a manager with those who are managed. Several relationships can exist between two entities. For example, between PERSON and POLICY, in an insurance model, we might have relationships of "Life Assured", "Owner" and "Beneficiary". These would link different PERSON instances (records) to a given POLICY instance.

Validation Rules
Entities to be shown as square-cornered boxes. All boxes to be labeled with entity name in the singular - Client not Clients. All relationships to be shown with cardinality indicated and label present.

Notation Conventions
We use the Martin-McClure standards for representation of entity models. This means that entities are shown as square-cornered boxes, with the name of the entity written in the box. Relationships are shown as labeled lines. The label describes the nature of the relationship. Ends of lines indicate cardinalities or ratios between the entities connected. A minimum and maximum is indicated in each case. A zero is shown as an open circle. One is shown as a line at 90 degrees to the link. Many is shown as a crow's foot symbol. Labels are above or to the right reading the relationship from left to right or top to bottom. Labels to left or above are read in opposite direction.

Example
A simple entity model is shown in figure 4.1.

Note that cardinalities can also be expressed numerically, where this is significant. For example, a transfer transaction must be related to exactly two accounts.

Quality Standards
Boxes correct shape. Entity names singular and meaningful to business people. All meaningful relationships shown and labeled correctly. Cardinalities to be indicated. All entities must have specified keys and foreign keys to create relationships. Martin-McClure conventions followed. No "clones" of entities already present in current databases.

Associated Tasks
Produce Entity Model.

Prerequisite Deliverables
Context Diagram. Project Definition.

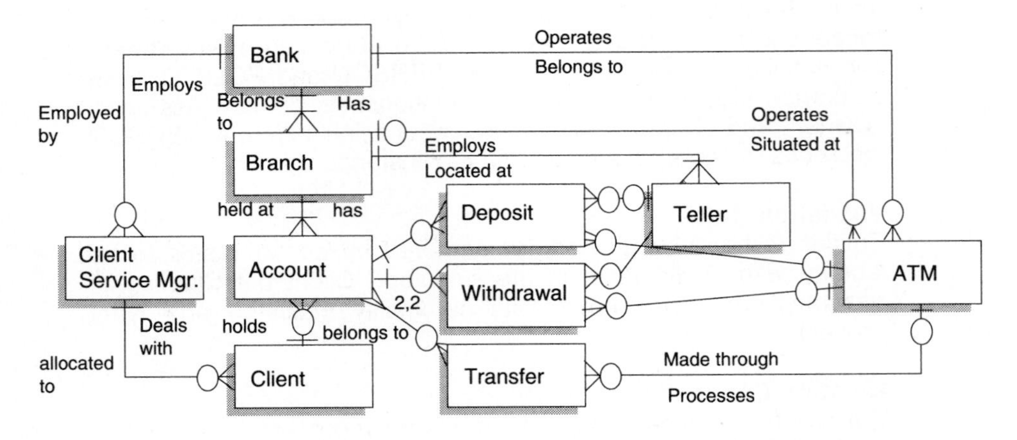

Sample Banking ERD *Figure 4.1*

Tools
Use IEF entity model diagrammer and attribute maintenance tool.

Data Content
Per Entity:
Entity Name
Entity Description
Per attribute:
- *Entity Attribute Name*
- *Prime Key Component (true/false)*
- *Foreign Key Component (true/false)*

Existing Database Relations
- *Relation Name*
- *Technical Environment*

Required Retention Period
Archival Frequency

Whole ERD:
ERD Name
per Entity:
Entity Name
Position
per Relationship
From Entity
To Entity
Cardinality

Products

A concept less common in project management than the Work Breakdown Model, but equally powerful, is that of Products. A product can be defined as a collation of deliverables. Our project could produce several products. A system development project could produce:

- The working software system including programs, system control language and files
- User documentation

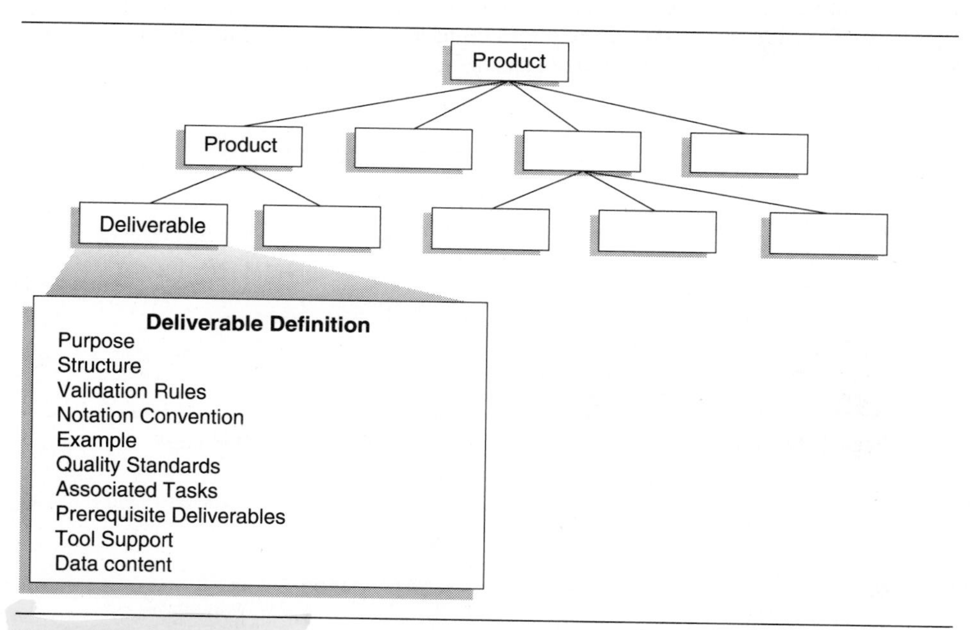

Product Structure Model *Figure 4.2*

- Technical documentation for maintenance and tuning
- A license agreement for users of the software, setting out legal conditions pertaining to the use of the software

It is critical that we include *all* expected products in our planning, since this will radically affect the scope (and hence required effort and cost) of our project.

Product Structure Model

Just as we showed the relationship of detail and summary tasks in the Work Breakdown Model, we can show the relationship of products and deliverables in a Product Structure Model (PSM) - figure 4.2. This shows the project goal product as the top box, and then decomposes each product in turn, until detailed deliverables are shown at the bottom. The PSM represents a *configuration,* since it shows the components which comprise each part of the final product. It will be used during planning, but also later for estimating, project control, change management, quality management and right through to maintenance, once our solution is in production. Most project management tools provide facilities for capturing Work Breakdown Structures, but very few have corresponding facilities for Product Structure Models. These are sometimes found in CASE tools and Configuration Management tools. If these are available, they should be used to capture and track the changes to the configuration. Particularly on complex projects, this can prevent a disaster where the scope of a change to specifications is not assessed correctly.

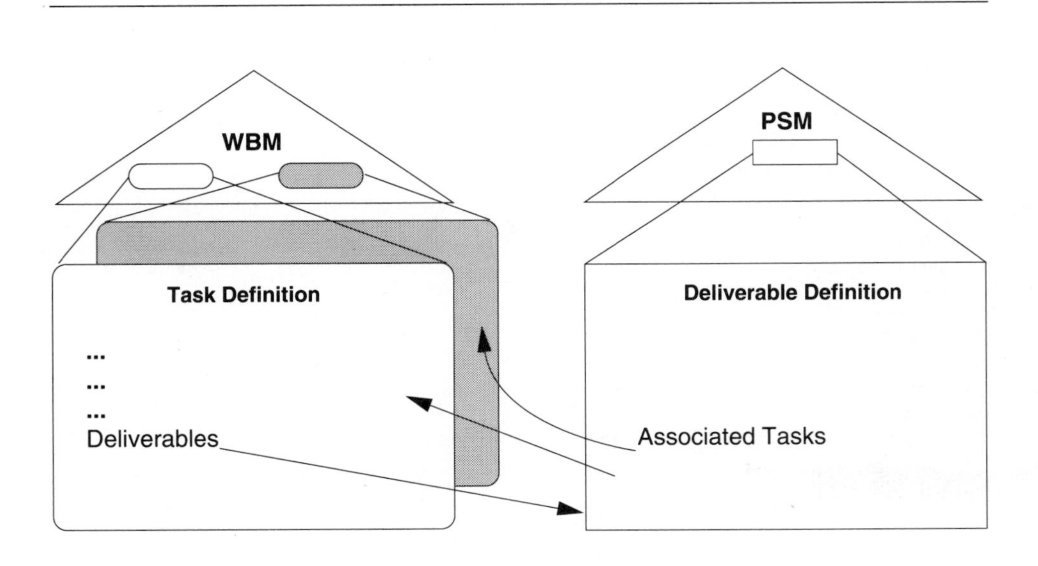

WBM to PSM Relations *Figure 4.3*

Relationship to tasks

Obviously, there is a relationship between the tasks in the WBM and the deliverables in the PSM. For example, the task *Develop Data Model* will result in the production of the deliverable *Relational Data Model.* The correspondence is not always one-to-one though. For example, the single task *Perform Functional Analysis* may result in the deliverables *Functional Decomposition Chart* and *Function Narratives.* Likewise, the tasks *Develop Prototype, Review with Users* may result in a single deliverable *User Interface Definition.* The relationships between the WBM and PSM are shown in figure 4.3.

Case Questions

MyWay Organizer

Q4.1

Prepare a Product Structure Map for the MyWay organizer project. You should include all deliverables that will go in the shrink-wrap box. Ignore other deliverables for now. (10 mins)

Q4.2

Prepare a Product Structure Map for the deliverables which the company will need to support the Organizer product in the field and which will be necessary to maintain and enhance the product. (10 mins)

Q4.3

Add to the models from Q4.1 and Q4.2 all the necessary deliverables to track and control the project, report to management, and ensure quality. Think about others you might add to help us improve performance on future projects as well. (10 mins)

Gleam Stores

Q4.4

Prepare a Product Structure Map for all the deliverables (technical and management) required of the project to fully document the pilot system and produce a training package. (30 mins)

Q4.5

Assuming you have a deliverable called "Test Plan" in the PSM, develop a detailed definition of this deliverable using the guidelines in the chapter. (30 mins)

Handover Trust

Q4.6

Using the list of deliverables for the network installation project below, develop a matrix showing the corresponding tasks. Structure these tasks into a WBS. (1 hour)

- Network Requirements Definition
  - Applications Characteristics Summary
  - Transaction Volume Summary
  - Technical and Compatibility Constraints
- Network Plan
  - Candidate Hardware Technologies
    - Cabling
    - Routers and Hubs
    - Modems and Network Interface Adaptors
  - Candidate Software Technologies
    - Link Protocol
    - End-to-End Protocol
    - Network Management
  - Compatibility Matrix
  - Cost Summary
  - Performance Summary
- Pilot Plan
  - Cabling
  - Network Equipment
  - Installation
  - Test Plan
  - Test Results
- Pilot Installation
  - Hardware
  - Software
  - Public Carrier Contracts
  - Management Process
- Installed Network
  - Prepared Sites
  - Installed Cabling
  - Installed Network Equipment
  - Public Carrier Contracts
  - Service Level Agreements
  - Network Management Document
  - Network Test Results

5 *Resources*

People

By far the most important resource on the majority of I.T. projects is personnel. People will consume as much as 80 percent of the budget in a systems development project. Also, they are highly specialized, and seldom interchangeable. Great care must be taken to match people to tasks to ensure that:

- The person has the knowledge, skill and experience to perform the task successfully and productively
- Staff members are developed and challenged sufficiently so as not to become bored (more on this in chapter 18)
- The time of highly skilled (and expensive) staff is not wasted on tasks which could have been performed by lesser qualified staff

Infrastructure

In order to function, the team will need infrastructure. This includes:

- Office space
- Administrative support
- Access to hardware and software tools
- Standards and guidelines
- Access to training in required skills
- Re-usable components from previous projects (or obtained externally)
- Resources external to the team: vendors, support groups and contract staff

Tools

Tools used by the team will normally take the form of software. These could include:

- Project management software to assist in the planning, control, tracking and management of the project
- System development tools, including programming language support (editors, compilers, linkers, libraries, code generators); utilities (file management, sorting, system management); end-user tools (Query/Report writers, spreadsheets, graphics packages, etc.) and Computer Assisted Software Engineering (CASE) tools which automate modeling and management of a repository of planning, analysis and design information
- Configuration management software to manage the complex set of related components which will comprise our final product. The same tools may be employed in change management for production systems

Standards

Standards are vital to ensure consistency and quality. They relate to the way in which procedures are performed, the representation of deliverables, and the way technologies are utilized. Some examples are:

- System Development Methodology which determines the tasks and deliverables for a development project
- Deliverable definitions which determine exactly what information a deliverable should contain and how it will be portrayed
- Programming standards which define how the particular language and environment are to be utilized, thus encouraging good practices and discouraging bad ones

Resource Requirements

Figure 5.1 indicates the typical profile of resource consumption during a development project. The profile for other types of projects will obviously differ. Notice that resources do not drop to zero upon implementation. There is usually a period of monitoring and fine adjustment once the system is in production. The different types of resources required also peak at different times. Analysts are heavily involved up front and during requirements definition, less during programming, and heavily again during system testing and implementation. Programmers may only be needed when the technical design firms up. Project management occurs throughout the lifecycle, usually peaking in effort with the point of maximum resource involvement, and when final testing and installation take place.

Resource Profiles

The team will make use of a wide variety of resources both internal and external. Some of the more common roles are as follows:

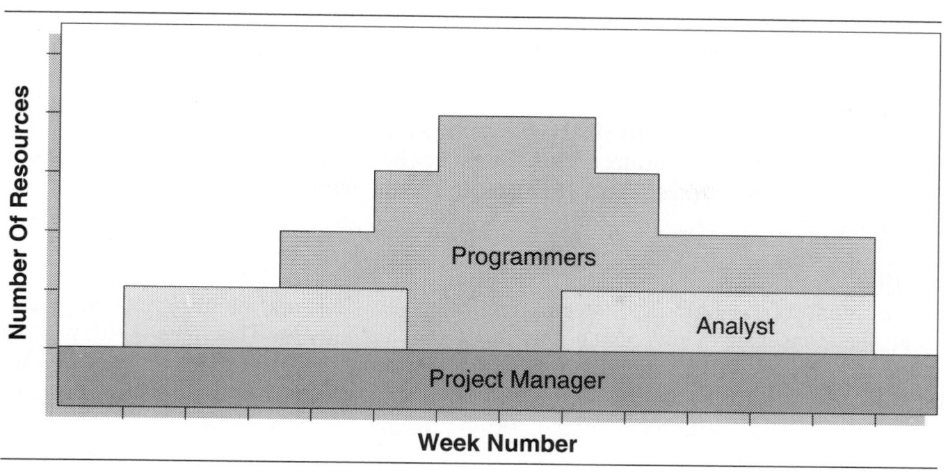

Resource Histogram *Figure 5.1*

Executive Sponsor
This is the most senior user/line manager sponsoring the project. He is ultimately responsible for committing resources and funding and ensuring that business benefits are delivered. This person may have limited time to devote to the project, but is essential to lend status and confirm organizational commitment to the project. A further role is observing the project progress at the macro level to ensure achievement of objectives and minimize risk.

User System Manager
This is a member of the client organization who owns the products from the project. This person is the primary specifier of requirements and must provide leadership from concept through implementation, into production and thereafter.

Project Leader
This is normally an information systems professional who is responsible for the successful completion of the project through the effective and efficient application of the assigned resources. He will manage the project team on a daily basis, assign tasks, collect deliverables, record progress and ensure that quality is consistently delivered.

Activity Manager
An individual reporting to the project leader who is assigned a particular set of tasks or deliverables to supervise and manage to completion.

User Managers/Supervisors/Staff
These are persons selected by the User Systems Manager to contribute to the definition of requirements, installation planning, testing activity, etc. In short, they, together with the User Systems Manager, represent the interests of the user community on the project. They are normally chosen because of their knowledge of the business processes and because they will use the system or facilities provided.

Information Systems Management
These persons represent the levels of I.S. management to which the Project Leader reports. They should be involved in project planning, feasibility analysis, resource allocation and quality assurance. They should be kept informed of project progress and any factors which

may affect the achievement of project objectives.

JAD Facilitator

This is a specially trained individual who acts as a consultant and facilitator in the application of JAD within the project. This should not be the project leader, as the facilitator should have no vested interest. The JAD leader should work under the direction of the project leader.

JAD Scribe

The scribe supports the JAD facilitator by recording decisions and building models which represent the consensus view, using the modeling techniques. This person may be a facilitator in training. Where automation (e.g. Integrated CASE) is used within the JAD sessions in an interactive way, the scribe will need to be expert in the operation of these tools.

Business Analyst

These individuals are charged with specifying the business requirements of the new system or technology. They should have an intimate knowledge of the industry, the organization and the application area. They need not be technical specialists, although they should have an appreciation for what is technologically feasible.

The business analyst should be capable of conceiving the new system in its entirety, including the business changes and implications around the computer system. He should concentrate on achieving a business system which will optimize the benefits for the organization - the computer system should only be a component.

Where the individuals concerned also have technical skills, they may act in the role of Systems Analysts as well.

Systems Analysts

These are normally senior individuals who have both a knowledge of business processes and information technology. They frequently progress to this position through the technical ranks of programming and systems design. They should be skilled in the techniques of eliciting and synthesizing user requirements so as to produce complete models of system requirements capable of being communicated accurately to system designers.

Analyst/Programmers

Analyst/programmers normally have a blend of analysis and programming skills. They are frequently found where high-level languages are employed, especially where prototyping is used in the development lifecycle. They generally work with broad user requirements, prototype with operational level users to refine these, and then go on to implement the production programs which meet these requirements. They normally need to work under the supervision of a senior analyst or project leader who coordinates progress within the overall system architecture.

Programmers

Programmers may have a wide range of experience levels. Junior programmers may be "coding clerks" who have a limited set of skills usually related to one particular programming language/environment. These coders would be responsible for accurately translating detailed specifications from an analyst or designer into a working program.

Senior programmers may have very highly developed skills and be capable of working in a variety of different technical environments. They may well perform significant design activity and may require specifications at a logical level only, since they will perform detailed program design. Really good programmers will demonstrate very much higher productivity, and the temptation to bog these individuals down in administrative or managerial tasks should be resisted.

Development Support
This team or individual supports the development personnel in respect of standards, guidelines, environment, methods and tool usage. There is normally a technical role in the maintenance of a productive development environment, as well as a consultative role to the project teams.

Data Manager/Architect
This area or individual is responsible for the management of data as a corporate resource. It is a business rather than a technical role. Data, like other business resources, costs money to plan, design, acquire, and maintain in good condition. Like other business resources, we should gain business benefits from the investment in data. Ensuring that adequate management takes place to achieve this is the role of Data Management. Duties normally include: Corporate Data Modeling, Custody of the Corporate Data Dictionary, Maintaining a directory of which data resides in which media and locations, coordination of new data added through projects to minimize conflicts and redundancy, ensuring that proper procedures are in place for the design, capture, validation, updating and maintenance of data.

Database Administration (DBA)
This function or person is responsible for the smooth operation, integrity, and efficiency of production databases. Development databases may also be managed, but usually less formally. The DBA role is highly technical and demands a very good understanding of the particular DBMS technology and surrounding/supporting tools, including the Data Dictionary. The DBA function may provide consulting/audit services to project teams in the area of physical database design. Duties would normally include tuning the databases for optimum performance, and performing recoveries when necessary. The DBA function is not responsible for the accuracy of data captured into the database, which is a line function, but acts as a custodian.

Facilities Management
This area or individual is responsible for the smooth operation of the computer environment. This could include equipment, maintenance, operations, network, system software, hotline support to users, archival and recovery, capacity management, and security.

Network Management
This area/individual normaly reports to Facilities, and is responsible for the smooth operation of the telecommunications network. This would normally include telecommunications equipment (such as terminals, lines, modems, etc.), network operation and management software, network configuration, and problem resolution. The function will normally work closely with the I.T. Architect to ensure that adequate telecommunications facilities continue to be provided.

Capacity Management
This area or individual is responsible for the provision of adequate hardware capacity to meet the needs of the business applications running in the organization. It is normally only formalized for mainframe and network components of the infrastructure. Growing business

dependence on minicomputers, workstations, and local area networks is causing these to become new areas of responsibility. Duties normally include: Strategic Capacity Planning as a parallel activity to Strategic Application Planning; Evaluation of Upgrade and Tuning Scenarios; Recommendations with respect to upgrades, technology replacement, and likely performance implications of technology choices. The group will sometimes provide consultancy services to project teams in the area of performance engineering and prediction modeling.

Information Systems Architect
This person or area is responsible for ensuring the applications portfolio meets business requirements. Application options are normally planned through derivation of business requirements by business modeling and examining options for realizing strategic advantage through application systems. A key responsibility is ensuring the integration and compatibility of applications across the enterprise. Assistance must also be provided to business management in selecting high-yield projects and prioritizing these to maximize business benefit while minimizing technical compromises.

Information Technology Architect
This person or group is responsible for the investigation, evaluation, selection, integration and suitable application of technology in solving business problems. They will normally architect the overall technical environment within which applications will be developed, maintained and operated.

Quality Assurance Auditor
This person or group is responsible for identifying non-conformances to established quality standards during the execution of projects. He is not responsible for producing a quality deliverable - this is the responsibility of the staff member who produces the product. The auditor's role is to detect when this has not been done, so as to limit the impact of non-conformances outside the project. This person typically conducts the reviews at key points in the project lifecycle.

Internal Business Auditor
This group is responsible for ensuring that adequate controls exist and are applied within business processes, including those implemented in computer systems. They must be involved in requirements definition to ensure that these requirements are incorporated into application systems.

Software Support
This area/individual is responsible for the installation, configuration, tuning, upgrading and advice on system software products. They need to work with Facilities Management, Network Management, and the DBA to ensure compatibility, efficiency and reliability across all system software components.

Project Secretary
This person serves as an administrative resource to the project team. He may be responsible for scheduling meetings, taking minutes, typing, filing, organizing documentation, doing project accounting, etc. He may also double in the role of scribe for JAD sessions, with appropriate training. Companies traditionally under-resource this area. Even normally sensible companies seem to think that it makes sense for project members to act as very expensive (and often inefficient) typists, filing clerks and receptionists.

Project Librarian
Responsible for the cataloguing, safekeeping and retrieval of project deliverables, documentation, and objects (such as program files, test databases, etc.).

Information Center
The I.C. normally provides services and support to End Users. It will assist users to access data held in systems and databases via Query tools, report writers, and sometimes custom-written code. It may also assist users to work with professional productivity software such as spreadsheets, personal database managers, graphics packages, word processors, etc. These are normally employed on Personal Computers (PCs), workstations, or the corporate computing facilities.

The I.C. should not be developing applications which have the potential to become departmental or corporate, since these should be handled through the development group where professional standards and quality assurance can be applied. Doing these through I.C. facilities can lead to lack of integration, unmaintainable systems, capacity/performance problems, and business risk.

Operations
This area is responsible for the daily operation of the hardware configuration, taking securities, running batch work, distributing output, managing media storage, etc. Operations departments are normally found in mainframe sites - most smaller systems have little need for these activities, and the role is normally fulfilled by a user or I.S. technical staff member.

Organization Model

This is a structure familiar to us as an organization chart. If we add definitions for the roles of the participants in the various positions, in a similar manner to which we defined the tasks in the Work Breakdown Model, and the deliverables in the Product Structure Map, then we have an Organization Model. We should be aware of both formal and informal reporting channels in this structure, as well as enduring and temporary ones.

Relationships to Work Model and Product Model

Each task at the lowest level of the WBM must ultimately be assigned to an individual or group for completion. Likewise, each deliverable at the lowest level of the PBM will be the responsibility of an individual or group. These responsibilities should be recorded on the respective models.

Task Assignment

Great care is needed in assigning tasks to individuals. I.T. people thrive on growth (learning new things) and challenge (doing difficult things, or performing at a level not attained before). This should not be confused with putting people under too much pressure, or asking them to perform the impossible. Do not challenge to the point of failure. This will only demotivate staff and result in a drop in productivity.

We should look for tasks which the individual has the potential to perform well. The task should be within the scope of the person's skills, aptitudes and capabilities. It may require some "stretching", but be careful not to overdo it. Experienced senior staff should not be given tasks which, to them, are merely routine and offer no challenge. Remember that a task which is routine for one person can present a challenge for another. This is where good management comes in. If we can assign the task to a person who will:

- Be able to perform it (perhaps with some assistance)
- Learn something in the process
- Enjoy the challenge of the task, and feel a sense of achievement on completing it

then we have succeeded.

The Role of a Mentor

The importance of mentors cannot be overstated. It is an old-fashioned concept in this age of specialists and canned training courses, but one which we believe has massive benefits. Where a task would be routine for your senior person, but too challenging for a junior, assign it to the junior, but under the mentorship of the senior. The junior will feel a great sense of challenge and responsibility, but will not panic since help is at hand. The senior person would have been bored by the task and is relieved not to have it assigned to him. At the same time, he is challenged to see how well he can convey his experience and knowledge to the junior and to see how well the latter can perform under his guidance. Productivity, self-esteem, motivation and quality all benefit.

Case Questions

MyWay Organizer

Q5.1

Prepare a list of the resources that you anticipate you will need on the MyWay Organizer Project. Include roles and non-people resources. (10 mins)

Q5.2

Assuming you have an Analyst Programmer role in your team, prepare a job description for this person in the context of the organization and the project. (15 mins)

Gleam Stores

Q5.3

Using your (or a provided model) answer for the WBS of the Gleam Stores project to document and build training material based upon the pilot system (see question 3.4), prepare a matrix to determine what resources you will need to carry out the project successfully. Express your resource requirements as a hierarchy showing legs for people and other things. Show reporting channels (solid line) in the people hierarchy. "Dotted line" relationships can be shown as well (as dotted lines!). (30 mins)

Handover Trust

Q5.4

Examine the Handover Trust case. Document what external resources you might need to accomplish the overall objective set by Mr. Renfrew. How would you obtain these resources? How could you make Handover self-sufficient in the medium term? (40 mins)

Q5.5

You have been appointed project manager on the ThoughtWell Books project. You have available to you within MacroSoft the following resources:

Joe Blains, an experienced analyst who has worked in networked minicomputer environments for the past 8 years. He has 12 year's experience overall, having worked initially in a mainframe bureau environment. Joe's particular strength lies in his excellent communication skills and good user relations.

Mary Long, an analyst with some 4 year's experience. Mary has a college degree and is bright. She can get a bit detached from the pragmatic everyday issues and detail required to make a working system though. She is a good conceptual and lateral thinker. She is a bookworm and loves the idea of working with ThoughtWell - she has specifically asked to be assigned to the project.

James Mbangwe is an analyst programmer with extensive Clipper and dBase III (not IV) experience. He is very technical and is the person many colleagues go to with their language, tools and compatibility queries. He has 9 year's experience, all on personal computers, having started on the original IBM PC in ROM Basic. James can be difficult to manage and is not always a team player, but performs very well if motivated.

Chermaine Phillips is a programmer with a computer science background. She has an Honors degree with a major in communication protocols. She has been with MacroSoft for four years. She previously worked for another system integration house for about three years, leaving them because the work was "not technically challenging". Chermaine is experienced in c and Pascal programming, and is busy learning Clipper, which she finds is "like c in places, but with weird arrays" and "frustrating because you can't get at the machine as easily."

Peter Wilson is an end-user computing specialist. He assists clients in using PC and LAN technology to improve the productivity of their staff, particularly managers and knowledge workers. He has a very good knowledge of all the popular PC productivity software, including spreadsheets, graphics packages, word processing, and end-user databases (including dBase IV). He has just completed an internal project at MacroSoft to connect the company's LAN to the Internet, thus providing all our staff with e-mail, Net-News and World Wide Web access.

Penny Ohlsen is a mature programmer who has worked her way up from being a mainframe operator. She has extensive mainframe background, mostly in COBOL. She also has a good knowledge of several TP monitors, and a variety of Relational databases. She has 20 year's experience, of which 15 has been in programming. She is an excellent mentor and is well liked by the younger staff. Her particular strengths are thoroughness, excellent testing and debugging skills, and wisdom gained through hard experience.

Lars Bontsen is an analyst programmer who recently joined MacroSoft. He has a bachelor's degree in business computing and accountancy. He is a people person and very outgoing. He has 18 month's experience in industry. His record shows that he works hard and is very keen to learn and develop his career. His major exposure to date has been in spreadsheets, the Paradox database and some Lotus Notes programming.

Denise Frentsen is a programmer with about 2 years experience. She has worked mainly in c and c++ with some exposure to databases. She is thorough and works well, but is shy and does not mix easily with people she does not know well. She has recently completed a LAN-based order processing system.

Management has indicated that you can assemble a team of four, plus yourself as project manager. You can use a maximum of three senior people.

Choose your team based upon your understanding of the project, the technical environment, the type of application and the backgrounds of the people involved. Give details of the role you anticipate for each person in the team. Justify your selections. Give reasons which you can use to explain to those *not* selected why you have not chosen them. (40 mins)

6 *The Project Lifecycle*

PLC versus SDLC

Most of us are familiar with a system development lifecycle (SDLC) which prescribes various phases and tasks. These normally progress something like this:

Initiate Project
Feasibility Study
User Requirements Definition
External Design
Technical Design
Build (Programming)
System Testing
System Installation

While these are familiar and useful, they are not necessarily comprehensive, or applicable to every kind of project. The SDLC specifies the necessary technical tasks for a development project. There are many other tasks which are needed to ensure a successful project. These include the managerial tasks such as organizing work, scheduling resources, reporting to management, liasing with external groups and so on, as well as the tasks required to monitor risk and assure quality in all tasks performed. These activities normally occur in parallel to the technical tasks, but could also be interleaved with them. This means that we need a Project Management Lifecycle (PLC) which can be considered an *umbrella* incorporating the SDLC.

There are also many other types of projects which you will encounter in I.T. These include:

- Implementing a package
- An end-user computing development
- Implementing a technology, for example, a new Database Management System
- Installation of a computer or network
- Converting data from one system to another
- and so on

Each of these will have unique technical tasks and a unique series of events in the lifecycle. We need a common framework within which we can handle all the various types of project.

Generic Lifecycle

The generic lifecycle presented in figure 6.1 has the advantage that it is consistent for virtually all types of I.T. projects. This has major advantages, including the following:

- Project Managers do not have to re-invent the wheel for each new project
- The senior management and steering group(s) to which projects report will be able to compare projects meaningfully
- Project reporting and terminology can be consistent in terms of phases and review points
- Expertise can be built up with respect to estimating techniques and past performance
- Standard project plans can be built up in tools, needing only slight modification to provide a solid, comprehensive plan for a new project

The lifecycle presented is adapted from one proposed for Software Engineering projects by the Institute for Electrical and Electronic Engineers (IEEE), built on the concept of Configuration Management. This approach has proven successful in handling some very large and complex projects. In system engineering and aerospace, it is not uncommon to be

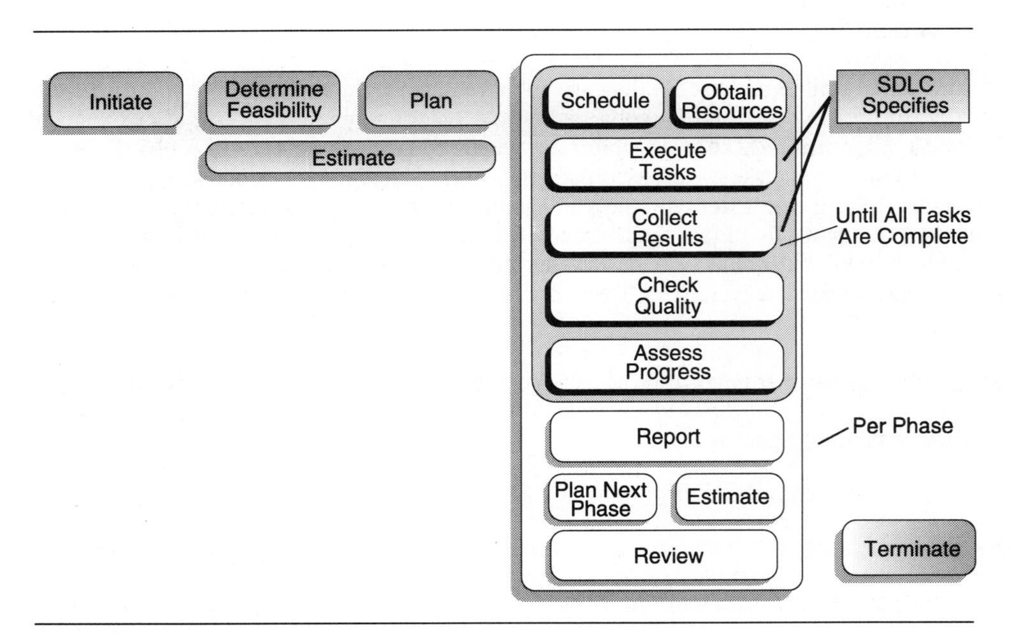

Project Lifecycle *Figure 6.1*

creating software and hardware at the same time. The systems are thus being designed to run on hardware which is itself at the design stage! When one considers the complexity of the software which might be several hundred thousand lines of assembler code, one gains an appreciation for the difficulty of the task.

But wait, you say, isn't this all getting too formal and too rigorous, and too expensive? I don't want to launch missiles, or navigate airliners, all I want to do is install a business application. In the past, we have got by without much in the way of formal project management, haven't we? Yes, and what a sorry record we have to show for it! Look at the number of projects which have not met requirements, were delivered late, or ran over budget. Look at the current levels of system maintenance in our organizations - typically between 70 to 80 percent. Now think about the increasing complexity of the systems which we are tackling, and their increasing criticality to the organization. Maybe a little formality will save a lot of tears later. A recent survey we conducted including more than 29 business system projects found a surprisingly high average size of 439 man months. At an average cost of $3000 per man month, this is an average cost of over $1.3 million per project - these are hardly tiny or cheap projects.

Realistically, though, can we have a formalized lifecycle and still accommodate modern approaches like prototyping, Joint Application Development (JAD) and Rapid Application Development (RAD)? Yes, we can. What we need to do is to contain the techniques and iterations to *within* the phases, rather than across the lifecycle. It is also possible to allow iterations through a few phases if we wish, but this requires very careful management, and increases risk.

Discussion of the Lifecycle

The project begins with the initiation phase already discussed in chapter 2. Here the scope is defined, goals are set out and the participants identified.

Following this, there comes the feasibility study and planning (see previous page). Parallel to these, and central to their success is the estimating activity. Once the feasibility is established, approval obtained, and the initial plan set up, the project can begin in earnest.

For each phase that follows, there is a common structure incorporating the following steps:

Repeated for all tasks designated in the Phase:

- Schedule the tasks in detail. This will normally involve consulting the methodology in use for the type of project being tackled, and adjusting this for the specific project's unique characteristics. Technical dependencies between tasks need to be understood. For example, we cannot install computer hardware until the premises and power are ready.

- Next, we obtain any new resources required to carry out the tasks. These may be people or other resources such as equipment, tools, etc.

- We then allocate and execute the tasks, collecting the work results (deliverables) as these are produced. Each of these should be quality assured before being accepted as complete.

- By counting completed, quality-checked deliverables received, we can monitor our progress on an ongoing basis.

We will repeat the above cycle until one of two things happens: Either all tasks for the phase are complete or we reach a mandatory reporting deadline (e.g., we are required to report to the steering committee every two months). When either of these occurs, we then prepare and present the necessary report information in as concise a way as possible.

At the end of the phase, we plan the next phase in detail, including re-estimating tasks, since we now have a much better understanding of what they will involve. We then conduct a formal review with our sponsors and an outside auditor/facilitator. This is to ensure the technical quality of the work produced as well as to ensure that the project is still meeting business goals, and indeed, that we are aware if these may have changed.

We can then move on to the next phase. When all phases are complete, the project terminates. This could also occur prematurely at any interim review if insurmountable problems are encountered.

In succeeding chapters we will cover each aspect of the framework and elaborate on the concepts introduced above.

Case Questions

MyWay Organizer

Q6.1

Integrate the activities for the MyWay organizer system development project into the generic project management lifecycle. You can assume a MUW of one week. Your management require a monthly report of progress. Present your answer as a list of activities in time sequence. (20 mins)

Handover Trust

Q6.2

Using the list of activities below for the installation of the Handover Trust network, merge these with the generic project lifecycle. Include all necessary management and quality assurance tasks in addition to the technical tasks. Your answer should be in the form of a Work Breakdown Structure. (45 mins)

- Define Network Requirements
  - Summarize Application Characteristics
  - Determine Transaction Volumes
  - Determine Technical and Compatibility Constraints
- Plan Network
  - Select Candidate Hardware Technologies
    - Determine Cabling Options
    - Determine Router and Hub Options
    - Determine Modem and Network Interface Adapter Options
  - Select Candidate Software Technologies
    - Determine Link Protocol Options
    - Determine End-To-End Protocol Options
    - Determine Network Management Options
  - Prepare Compatibility Matrix
  - Prepare Cost Summary
  - Do Performance Summary
- Devise Pilot Plan
  - Plan Cabling
  - Plan Network Equipment
  - Plan Installation
  - Plan Test Plan
  - Plan Expected Test Results
- Install Pilot
  - Install Hardware
  - Install Software

- Conclude Public Carrier Contracts
- Implement Management Process

Install Network
- Prepare Sites
- Install Cabling
- Install Network Equipment
- Conclude Public Carrier Contracts
- Draft Service Level Agreements
- Write Network Management Document
- Collect Network Test Results

7 *Estimating*

Project Success

In chapter 2 we defined project success as meeting requirements, delivering on time, and remaining within budget. We also talked about establishing the feasibility of the project, and choosing the right projects to support the business strategy. None of these can be achieved unless we develop realistic estimates of cost and delivery date. As can be seen in figure 7.1, estimating is essential at the early stages of the project, but is revisited at the end of each phase.

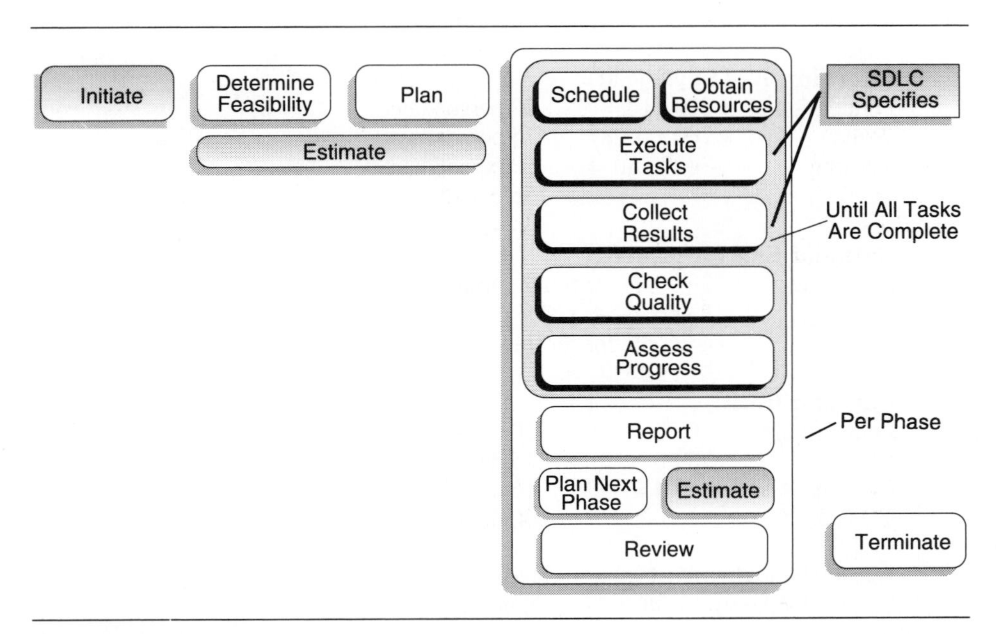

Project Lifecycle *Figure 7.1*

Unfortunately, estimating information systems projects is a very thorny issue, and currently as much of an art as a science. To estimate accurately, we would need the following factors:

- The size of the job being tackled
- The productivity that can be expected (i.e., how quickly can we perform the work?)
- The resources available to us
- The environment and constraints under which we will work

An Analogy

A construction manager planning a building project might do a calculation like this:

Requirements
200 meters of foundation
200 meters of wall, 3 meters high
3000 square meters of roofing

Productivity factors:

Foundations:	10 meters per personday
Walls:	50 bricks per square meter
	Bricklaying at 400 bricks per personday
	8 square meters per personday
Roofing:	50 square meters per personday

Effort involved in project:

Foundations:	200 / 10 =	20 persondays
Walls:	200 x 3 / 8 =	75 persondays
Roofing:	3000 / 50 =	60 persondays
TOTAL		155 persondays

Calendar time for project:

Resources:	2 people for foundations
	5 people for walls
	4 people for roofing

Calendar times:

Foundations	20 / 2 =	10 days
Walls	75 / 5 =	15 days
Roofing	60 / 4 =	15 days
TOTAL		40 days

Assume 5 productive days per week
Total calendar time = 40 / 5 = 8 Weeks

Cost of project (excluding materials)
Assuming labor at 600 per personday
Cost: 155 x 600 = 93 000

Unfortunately, as software engineers, we do not have the same luxury. Let's look at each aspect in turn:

- The size of the project is largely unknown until we have completed the analysis phase, by which time we may have done a third of the total work required. Even then, we cannot be totally sure, as there may be unforeseen issues in the technical design, or the requirements may change during development
- Productivity norms are sadly lacking. Some measurements are available, but different studies conflict, and the range of values is enormous. Furthermore, many factors unique to the project; e.g., the person performing the job, and the technology used, can have drastic effects on productivity achieved
- We seldom know at the outset what resources will be available to us. We may be told that "you will have four programmers when you need them", but this may or may not be the case. The actual resources could have vastly different productivity from that expected
- The environment under which we work is constantly changing with new tools, technology, techniques and approaches. We never stabilize things long enough to be able to collect the measures that we need for better estimating next time! It is like the construction engineer using a new type and shape of brick each time

Estimating Dilemma

We thus have the picture shown in figure 7.2 - garbage information in, and useless answers out.

How can we improve this situation? A better scenario would be to collect the information

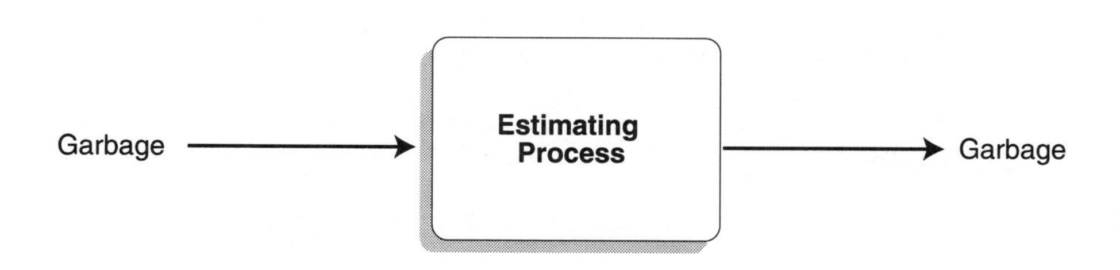

Estimating Dilemma *Figure 7.2*

that we need, building a database of norms which we could use when faced with future estimating exercises (see figure 7.3). But, you say, this would take us a long time, and we can't keep things stable just to collect some figures - especially when the new techniques promise higher productivity and we have a huge backlog of requests.

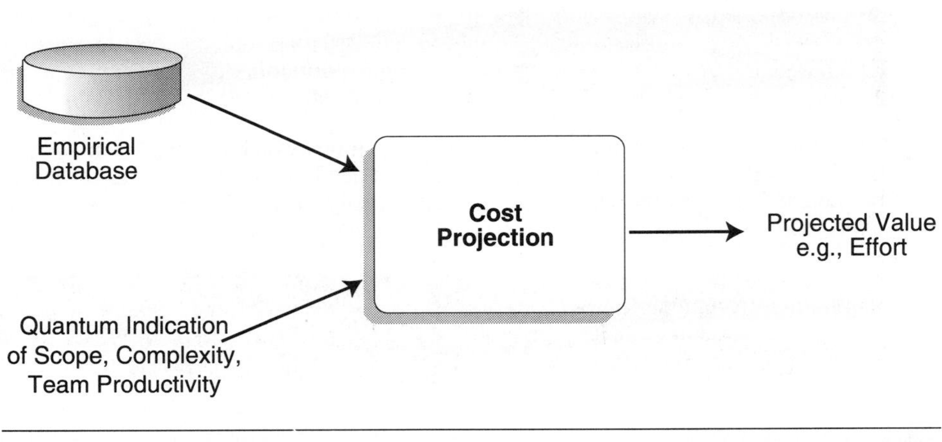

Estimating – A Better Scenario *Figure 7.3*

The truth is that empirical evidence shows that very few of the techniques which promised vastly greater productivity over the years actually delivered on that promise. Some of the most productive shops around are still using COBOL '74. Consider figure 7.4 from the respected consultants Nolan and Norton.

According to their study, the biggest increase in software productivity was the change from machine code to assembler. Everything else since then has had only a marginal effect. Maybe we *can* afford to stabilize a little and collect some data. Still in a hurry? Fortunately, a lot of useful data has already been collected by various researchers. We will take a look at some of these figures a little further on. First we need to establish some foundations.

An Estimate Is

The popular view of an estimate can be gauged from the following scenario:

Arriving at the office one morning, you bump into the AGM Marketing in the lift. On the way up, she asks you, "How long it would take 'you guys in systems' to develop a Sales Analysis system?" You think a bit, and off the top of your head, say, "About six months." What have you done? You do not know the size of the system, what the requirements are, what resources we can expect or anything else useful. But most things take about six months in our optimistic world view, unless we have tried them before and know that they are more complex than we thought.

By lunch time, you have given it some more thought: There will need to be an interface to

1950's	Input Output Control Systems/Assembler/Machine Code	
1960's	3GL Programming Languages	**+ 8% per annum**
1970's	Modular Programming & Database	**+ 3 - 6% per annum**
1980's	4GL Programming & CASE	**+ 1- 2% per annum**
1990's	00, Re-usability, 5GL	**No figure yet available**

Source: Nolan/Norton

Improvements in Software Productivity due to Technology *Figure 7.4*

the Point of Sale system which is in a nasty old technology; there will be some tricky communications stuff to get the data into head office; and the performance of the analysis system will have to be very good since you have very little time left in the overnight batch slot. Maybe nine months is more like it? You see her at lunch and tell her. She goes white - she told the board this morning that they could have it in six months, and now, before we've even begun, you say it's going to be 50 percent late.

You were guilty of giving the answer the person would *like*, not a real one. The estimate you gave was *the smallest number with a non-zero probability of coming true*. What we need to do is become more professional. If you consult a structural engineer and ask for a quote for a bridge the answer will depend on the nature of the job, the terrain, the weight the bridge must carry and other factors, not on what you would like to hear. We have to give *real* estimates. An estimate is *a prediction which is equally likely to be above or below the actual result* (see figure 7.5). If you had quoted a wide range based upon the level of information that you had (say 6 to 12 months) the AGM would have understood the term "estimate". She may have asked you for something more definitive, and you should then have said: "I will need more information to develop a more accurate estimate. Would you like me to invest some effort in this?"

Certainty versus Project Stage

As mentioned earlier, we have to go some way into the project to understand its scope fully. It is thus impossible with systems projects to give exact estimates at the start - we simply do not have the information available. Our early estimates are likely to have a wide margin of error, and we must quote a *range*. Only as we gain real data on which to base later estimates can we firm these up. This is illustrated in figure 7.6.

We thus need to do estimates at several levels:

- At the strategy level - Macro estimates which allow us to compare potential projects one against the other. A high degree of accuracy is not required, but the estimates

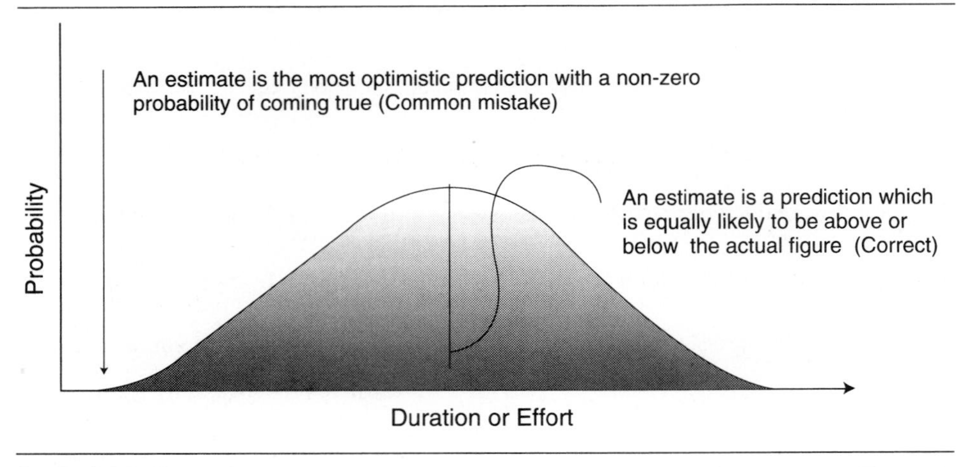

An Estimate Is . . . *Figure 7.5*

must give a *relative* cost and duration to allow sensible selection of projects

- At the feasibility stage - A somewhat more accurate macro estimate after the project objectives and scope have been firmed up and we have an idea of potential technical environments and resourcing. This needs to be accurate enough to allow a cost/benefit decision to be taken as to whether to proceed with the project

- Once feasibility is established, and we begin the project, we need an estimate per phase of the project. These estimates will be firm for the next phase, but soft for later phases. As we complete a phase, we will re-estimate in detail for the next phase, and revisit our soft estimates for later phases

- In addition, we may need to do micro estimating for individual tasks and deliverables within the current phase. For example, we may need to estimate the development time for each program when assigning these to individual programmers

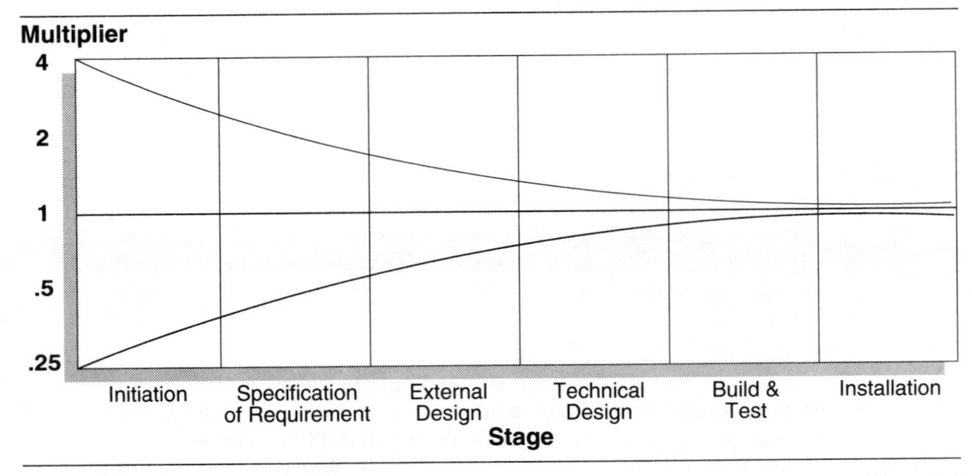

Certainty Versus Project Stage *Figure 7.6*

At each of these levels we may have different information available to us and may thus need to use different techniques. We will explore these after establishing a suitable background.

Factors Affecting Effort and Duration

There are a great many factors which influence the effort and elapsed time for a software project of a given size. These include:

- *Complexity*. We intuitively understand that something simple will take less time to do than something complex. A report program from one file will be less difficult to develop than a real-time online multi-file update. McCabe, Halstead and other authors have proposed metrics for complexity. Unfortunately many of these can only be calculated after the design (or sometimes even after the code) is complete. Nevertheless, it is interesting to see what the range of values can be, i.e., what influence complexity can have on the development effort. Complexity measures have been found to be closely correlated with the number of source statements in the product produced. The measures can assume a wide range of values, with no apparent maximum. It is important, however, to remember that increasing complexity is always associated with more effort, and higher fault rates

- *Skill of team members* has a major influence on the effectiveness of effort applied. One programmer may write in one day what will take another several days. Studies indicate a 1:7 ratio between best and worst performers. Some studies indicate a 1:20 ratio. These figures are startling. There seems to be at least *an order of magnitude difference* between inexperienced novices and top performers. What is also extremely interesting is that the variation seems to be less among individuals than among environments. This may be because certain organizations have superior environments allowing the people there to perform at a higher level, or it may be that high-achievement environments attract more high achievers

- *Elapsed time of project*/degree of specification change. The longer a project runs before delivering its final results, the more likely it is that the environment or business conditions will change during the lifecycle. This leads to changes in objectives, requirements and constraints. All of these need replanning, respecification, redesign, redevelopment, retesting etc. Everything with a "re" in front of it means it is costing us time and money to redo something which we had already completed or begun. Short projects are much less likely to encounter large specification changes. Practically, we should try to keep each chunk that we develop to no more than 9 elapsed months from specification to delivery.

- *Staff turnover*. Loss of key staff members can significantly impact project schedules and costs. There is an inevitable learning process for the new staff member, and a loss of experience that goes out with the old staff member. There is also the effect on the team which has adapted to working as a group and will now have to go through that process again. The lost productivity is reflected in longer schedules and higher costs - see figure 7.7. The project may also incur additional costs by carrying both staff members during a handover period to try to minimize the problems mentioned. Risk is increased due to the possibility of loss of information and expertise

  Productivity measurement figures show that a small amount of staff turnover (up to 10 percent per annum) is actually healthy as new skills and perspectives will be brought into the group at a rate which is not disruptive. Higher levels of staff turnover can result in a loss of productivity as shown in the figure

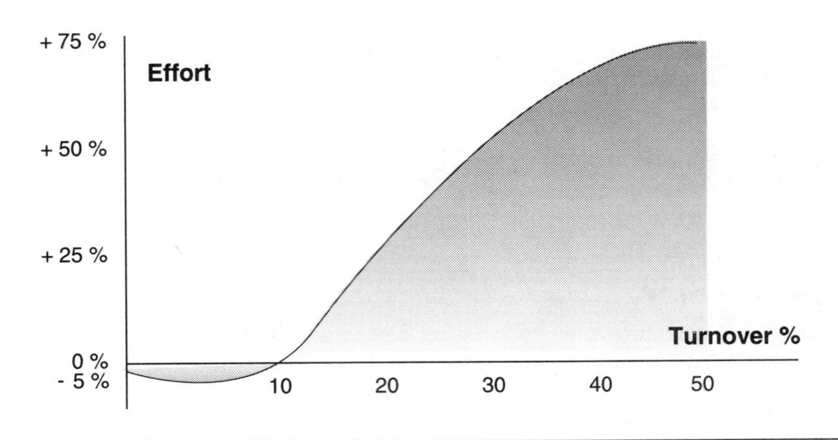

Change in Effort versus Staff Turnover *Figure 7.7*

- *New methods/techniques/technology.* While we frequently adopt these in the hope of increased productivity, the truth is that they often negatively impact productivity and increase risk on the first projects where they are applied. The relative productivity figures quoted earlier indicate that we should be skeptical of claims made by vendors for dramatic productivity improvements through technology

  For example, 4GLs were claimed to "cut development time to a third" and the like. Even if the claim were true, the improvement would apply only to the programming component of the lifecycle, which is about 20 percent of the total project time. So, if we saved two-thirds of this, we would still only save 13.4 percent on the whole project. The analysis, design, training, documenting, implementation and other aspects of the project remain unaffected. These benefits would also only be achieved once the team were fully conversant with the new technology

  Two technologies now being adopted hold real promise of increased productivity and shorter project durations. These are Integrated CASE (I-CASE) in support of Information Engineering (IE) and Object Oriented Technologies (OOT). We should be aware, however, that to attain these benefits, we will need to have in place a disciplined, managed development process and methodology as well as the skills to manage and utilize the technology correctly. Just as putting a piano in a room will not create a concert, so plugging a tool into an organization incapable of utilizing it correctly will not produce faster development

- *Size of team.* Common wisdom from the construction and engineering disciplines is that we can shorten projects by adding more resources. A typical question is "how many people do you need to get it done by the New Year?" Unfortunately, this relies on the assumption that the work is easily divisible and that resources are easily interchangeable and roughly equally productive. These assumptions are not true for system development projects. Firstly, the resources on a team need to communicate with each other. This adds an overhead for each additional member. Second, the tasks are not always easily divisible. Third, we have seen that resources can display very great variation in productivity

  As can be seen from figure 7.8, each additional person adds another interface and corresponding communication overhead to every member of the team. As a rough rule of thumb, when we were managing package development projects, we used to

	Team Size	Number of Interfaces	Effort Multiplier
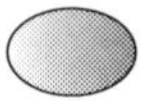	1	0	1
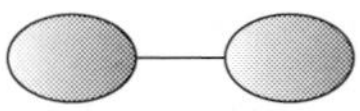	2	1 ea 2 total	1.26
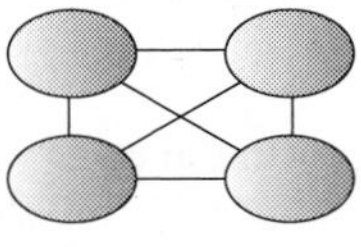	4	3 ea 12 total	1.44
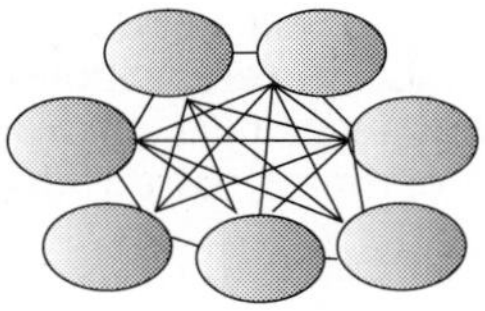	7	6 ea 42 total	2.0

Source: Tetrarch

Effect of Team Size *Figure 7.8*

deduct 10 percent productivity from everyone on the team for each new member. This means that by the time the team reaches 10 strong, no one is doing anything else but talking to each other! This may be cynical, but you can almost watch it happen.

A mathematical prediction from the proprietary Tetrarch methodology calculates the overhead at n to the power 1/3, as shown in figure 7.8.

This communication overhead can be reduced in several important ways:

- Using small teams with high levels of skill
- Breaking the task up into manageable chunks with minimal interaction between them - this is the role of a systems architect
- Using highly structured, graphical specification and design techniques which reduce the communication problems between team members and between teams
- Using development methods, such as Object Oriented Analysis and Design, which reduce the number of translations between different models and representations

• *Development environment/language* (despite our skepticism) can have a major influence, particularly on the build phase of development projects, but also on the speed with which system maintenance can be safely carried out. What we need to avoid is applying the productivity improvements to the whole project estimate - typically we will not see a major effect there. In our experience, it is the total environment, rather than the specific language syntax which makes the most difference. A COBOL environment with good supporting tools, a library of useful functions, and standard program structure skeletons can be just as productive as a good 4GL environment. Graphical, object oriented development environments with purpose-built editors and browsers, and a very rapid modify/test cycle, such as Smalltalk, can display very high productivity, even though the language itself may seem somewhat cryptic.

Do not interpret the comparative numbers from table 7.1 as time per function point, however, since it may take just as long to write 20 lines of Smalltalk as it does to write 300 lines of COBOL. Benefits do occur in reduced complexity, and testing and maintenance effort, however.

We should also realize that maturity and level of experience with the environment, as well as non-technological factors, have a major effect on productivity, rather than the particular tool or product. In our experience, we have had teams which were producing over 200 lines of debugged working COBOL code per person per day. These figures are far above the industry norms of around 20, but were achieved with a highly motivated, highly skilled team in a supportive management environment and using a lot of home-grown utilities to customize and optimize our development environment

- *Motivation of team members* can have a major influence on the productivity achieved. We will examine the whole subject of motivation in chapter 18. Suffice it to say that attitude and motivation level alone can have a significant impact on productivity achieved, all other factors being equal

- *Time pressure*. This is an area where systems projects differ radically from construction projects. Common sense tells us that the more resources we add to a

Assembler	320
C	150
Algol	106
COBOL	106
Fortran	106
Pascal	91
RPG	80
PL/1	80
Modula/2	71
Prolog	64
Lisp	64
Basic	64
4GL/DBMS	40
APL	32
SmallTalk	21
Query Languages	16
Spreadsheets	6

Source Statements per Function Point - Fenton *Table 7.1*

project, the faster we can complete it. Unfortunately, this does not occur with system projects. To quote Fred Brooks, a manager of one of the largest system projects of all time, the development of the IBM System 360 Operating System (which eventually evolved into MVS):

"Adding resources to a late software project invariably makes it later."

Is this true? Does it really happen, and if so why? Norden of IBM collected data on over 2000 projects and built a database with a view to better understanding the behavior of the software development process. Work by Norden and Larry Putnam shows that the relationship is indeed true. It appears that the minimum effort expended on a project occurs when we have only one resource, no communication difficulties, and no rework. As we add resources, we increase communication overhead, communication failures and consequently rework. There is actually a point at which adding more resources will negatively impact the total delivery rate. There is also a minimum possible elapsed time for a project. Any number of resources will not make the project achievable in less time. This holds very important messages for how we design, estimate and staff our projects as well as how we do our cost/benefit calculations. We will discuss the Norden/Putnam model in more detail later

- *Quality and documentation requirements* affect the time and effort of projects dramatically. This can be easily seen in systems engineering projects building safety critical systems such as air traffic control support, or nuclear power plant management software. The validation and verification effort can sometimes cost more than the development effort.

  In commercial systems, we have in the past not been too worried about quality and reliability, since our systems were not *mission critical.* However, many have now become so: What would happen to a modern bank if the national ATM network posted things to the wrong accounts, or just decided to hold client cards at random?

Techniques for Estimating

Determining the Size of the Job

Our first step in estimating is to determine the size of the job we are tackling. This involves several things:

- Having a clear definition of what it is we are trying to achieve
- Establishing a clear boundary (or scope) which details what is and what is not included
- Understanding our terms of reference and constraints
- Trying to obtain a measure of the size of the product we are expected to deliver

The first three steps deal with the project goal and objectives; the scope and context of the project, and the terms of reference. We have dealt with these already in the chapter on Project Definition.

There are several techniques which we can use to try to gauge the size of the project. Before we do that, let us step back and look at the measurements available to express the size of a computer system.

Three major measures are used:

- Lines of Source Code (SLOC)
- DeMarco's BANG
- Function Points

Lines of Code

The easiest measure after the event is lines of code - all we have to do is run an editor against our source library and let it count them for us. This is a popular measure, probably simply because it is so easily available. Unfortunately, it suffers from many shortcomings:

- It is highly language and environment specific. 1000 lines of COBOL is not the same as 1000 lines of Clipper or Fortran, and may not even be the same as 1000 lines of COBOL on another machine or operating system
- Lack of counting standards. What do we count - source lines in the editor, or source instructions? What about comments? What about included routines or data definitions which appear more than once? Do we count them every time, or only once? Some texts refer to KNOCSL (kilo non-comment source lines) as a measure. Attempts have been made to standardize on a measure called Effective Source Lines of Code (ESLOC). This usually works as follows:
    - Write one instruction per line where possible. If an instruction (e.g., an IF-THEN-ELSE or CASE) needs several lines, then place each condition or action on a separate line
    - Include full-line comments in the count, but not where these only create spacing
    - Ignore partial-line comments
    - Count library code and data definitions which will be re-used in many programs once. Count the lines which include or invoke this code

Utilities are available to scan source libraries and calculate counts based on a consistent approach.

- SLOC's are only available after the fact. Source lines only exist to be counted after we have written the code. This is very late in the project, and useless for a meaningful cost/benefit analysis
- The measurement of source lines encourages verbosity. "Be careful what you measure, some damn fool will try to manage it . . ." A long program which accomplishes the same task as a short one is probably not as well designed, efficient or maintainable. If we measure source lines, we may regard the better programmer as less productive

BANG

This is a measure developed some time ago by Tom de Marco. It attempts to assign a functional "weight" to the system, independent of the implementation technology or language used. The name probably stems from the amount of "Bang for the Buck". De Marco gives two forms of the measure, one for "function rich" and one for "data rich" systems. Function rich systems are those which perform a lot of processing, or data transformation. Data rich systems are those which are mainly concerned with management and use of complex data structures, e.g., a Decision Support system. The former is based on a dataflow (i.e., process) model, and the latter on a data model. While these measures were temporarily out of vogue, we may see a resurgence of their use in I-CASE environments, since their computation can be automated where specifications are in machine-readable form.

Function Points

This technique was first developed at IBM by Albrecht. It was initially used as a productivity measure, and subsequently for sizing and estimating. It is similar in concept to BANG, in that we develop a functional weight for the system, regardless of implementation technology. We try to calculate what functionality the system is delivering to the end user. Various aspects such as number of inputs and outputs, number of files maintained, number of updates, etc., as well as their complexity, are taken into account. Using these and weightings derived from measurements of previously developed systems, a raw Function Point count for the system is calculated. This is then adjusted with weighting factors such as complexity, deadline pressure, performance constraints and so on. These adjustments produce an adjusted function point score which is then used to determine productivity (FP/Manmonth of effort) or estimate elapsed time. The latter assumes we know how many function points per manmonth of effort we can deliver. This presumes some prior measurement in our own situation, or that we have obtained suitable starting figures from a site (or sites) similar to our own.

As a measurement tool, Function points has a number of advantages:

- It is technology and language independent
- It is much easier for users and nontechnical managers to relate to than lines of source code. They can meaningfully compare the size of two projects without requiring programming background
- It can be calculated earlier in the project - typically once the requirements definition or prototype is complete
- There are standards for how function points should be calculated, published by the International Function Point Users Group (IFPUG)
- There is a wide body of knowledge available concerning the range of productivity measured in Function Points for languages, environments, application areas and industries. This helps organizations who do not have their own collected data to get meaningful figures to start with more easily and quickly

Unfortunately, there are still some limitations we should be aware of:

- Function Points tend to have a Mainframe/ Character Terminal/ Centralized systems view of the world and are difficult to apply to Graphical User Interfaces, LAN systems and Client Server applications. The IFPUG does, however, try to adapt the approach to cater for new technologies

- Function Points are impossible to calculate, and difficult to estimate until the full Requirements Definition, and preferably External Design, is complete

- Like SLOC, we tend to get what we measure. With function points this will not be more lines of code to implement a given function, but rather extraneous functions which do not necessarily have business value

  With the advent of business process re-engineering, we should have as our goal the simplest systems which will support the effective running of the business. Using FPA we may give credit for reports which merely entrench old and inefficient ways of doing things

Even with the limitations mentioned, Function Point Analysis is still probably the best measure we have for comparing the size of systems meaningfully. There are also guidelines to allow us to use the technique in a systems maintenance context.

Calculating Function Points

If this is a development or maintenance project, you should calculate function points as soon as you have the following information:

- The inputs and outputs (Screens, reports, files) to the system are defined
- Interfaces to other systems are defined
- The logical data base (conceptual schema) for the system has been defined
- The target environment has been chosen

The form in figure 7.9 shows the basis for calculating function points. Proceed as follows:

- Record the number of inputs to the system. These include: Screens, parameters accepted, and any inputs from special devices such as a bar code scanner. Rate these as simple, average or complex
- Record the number of outputs from the system. These includes reports, outputs to special devices such as a bar code printer, etc. Rate them as simple, average or complex
- Record the inquiries to which the system will respond. These may be made in batch or online mode. Rate as simple, average or complex
- Record the number of files (or database relations/tables) maintained by the system. Rate them as above
- Record the interfaces with other systems, e.g., parameters passed to service modules, records read from a temporary file, etc. Rate as before
- Multiply each category by the prescribed weighting, and add together giving a total unadjusted function point score
- Next, rate each of the adjustment criteria on a scale of 0 (no influence) to 5 (critical/major influence). Add all of these scores giving a total degree of influence

Function Point Calculation

Project ______________________ Estimator ______________

Comments ____

____

____

Date ______________________ Adjusted FP Count ___________ D

Raw Function Point Count

	Complexity			Total
	Simple	Average	Complex	
Inputs	___x 3	___x 4	___x 6	_____
Outputs	___x 4	___x 5	___x 7	_____
Inquiries	___x 7	___x10	___x15	_____
Files	___x 5	___x 7	___x10	_____
Interfaces	___x 3	___x 4	___x 6	_____
Total Unadjusted Count				_____ A

Adjustment Factors

Factor	Influence (0-5)	Factor	Influence (0-5)
Data Communications	___	Online Update	____
Distributed Functions	___	Complex Processing	____
Performance	___	Reusability	____
Heavily Used Config.	___	Installation	____
Transaction Rate	___	Operational Ease	____
Online Data Entry	___	Multiple Sites	____
End User Efficiency	___	Facilitate Change	____
Total Degree of Influence			____ B

Adjustment Factor = B/100 + .65 ____ C

Adjusted Function Point Count = A x C ____ D

Function Point Calculation

Figure 7.9

- Calculate the adjustment factor as follows:

  .65 + (0.01 • degree of influence) = Complexity Adjustment Factor

- Calculate the final Function Point Measure as follows:

  Unadjusted FP count * Complexity Adj Factor = Final FP Count

How you decide whether an input is complex, average or simple can get fairly involved and is beyond the scope of this text. For full details consult Albrecht or the IFPUG documentation.

Unless management and non-technical tasks have been included in the FP project history figures, function point counts will help us to estimate only the technical tasks in the project. Other techniques (e.g. Delphi, analogy, history) can assist in estimating other tasks.

Other Techniques

There are a number of techniques which we can use to help us obtain estimates of size, especially early in the project.

The Wideband Delphi Technique

Developed originally at the Rand Corporation, this technique draws upon expert opinion and assists in reaching a consensus view. A summary of the process is shown in figure 7.10. Applied to estimating, it works like this:

- Several experts who can meaningfully estimate the task at hand are located
- As much information as is available is provided to the experts to peruse
- The experts meet under control of a chairman. Each expert writes down, without discussion, his or her estimate. These are collected by the chairman
- Usually, there will be a majority whose opinions differ only slightly. There may be a small number who disagree and provide estimates which are much higher or lower than the majority opinion
- The chairman asks those who differ to explain their viewpoints to the group
- All members write down new estimates, and these are collected again. Depending upon the strength of the arguments presented by those who differed initially, the majority opinion may shift up or down
- The process is repeated (usually three to four iterations are required) until the estimates are no longer being changed - i.e. the majority opinion range is not getting smaller
- For situations where there is a high degree of certainty a very narrow range of estimates should be achieved. Where there is more uncertainty, a larger range will be present in the final answers, and more iterations will not cause them to converge. This is healthy and can be used as a confidence-level indicator. For example, if the range is unacceptably large to management, this is an indication that too little is

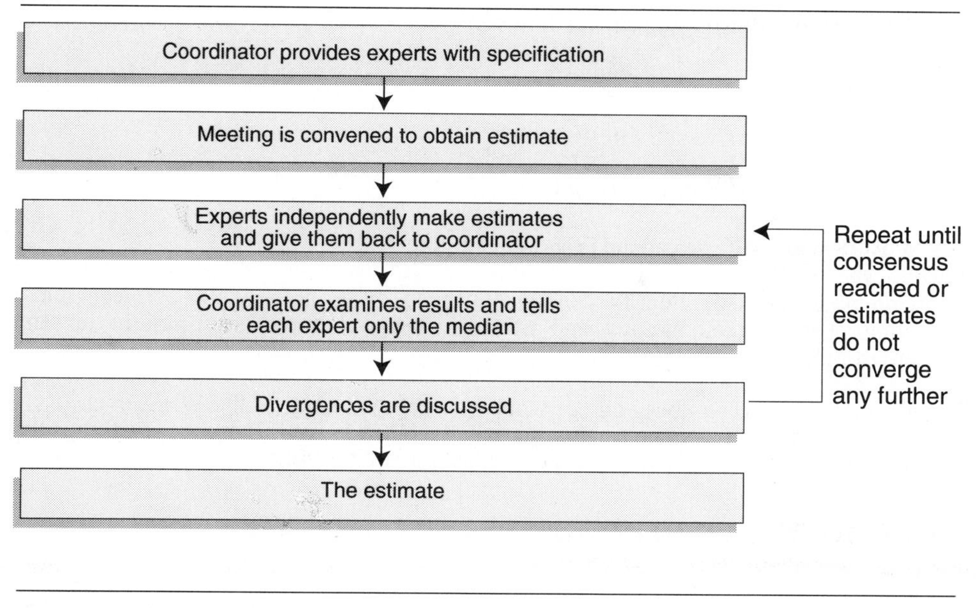

Wideband Delphi Technique *Figure 7.10*

known about the problem, and further information should be gathered before committing to an estimate or decision

Work Breakdown Structure

We saw earlier that the WBS is a valuable tool in designing our project. It can be very valuable in estimating too. The idea is that there will be much less subjectivity in asking a detailed question like "How long will prototyping user reports take?" than in asking a macro question like "How long will the entire project take?" A thorough WBS also ensures that we do not omit major tasks and activities from our estimating process, and that we include time for management, administrative and quality assurance activities. To use a WBS for estimating, we proceed as follows:

- Develop the WBS using the guidelines in chapter 3
- Estimate each task using the best techniques available to you. If there is an empirically based technique (such as past measurement) use this in preference. If not, use expert opinion and Delphi
- When you have the detailed estimates, summarize these upwards in the WBS structure
- Where estimates exceed the Manageable Unit of Work (MUW), decompose the chart further
- Ensure that the WBS contains all:
    - Technical tasks

- Management activities

- User liaison

- Administrative activities

- Quality Assurance activities

A WBS with estimates added is shown in figure 7.11.

Completeness and integrity can be enhanced by using a WBS derived from a comprehensive methodology, and refined by usage in previous projects in the same environment

- Don't revise the number you get when you get to the top of the WBS, even if you don't like it. If it has to be smaller, look in the WBS to see if there are any things that you don't need, or reduce the scope of what you are tackling

Management Overhead

Our experience indicates that you should allow about 10 percent overhead for management

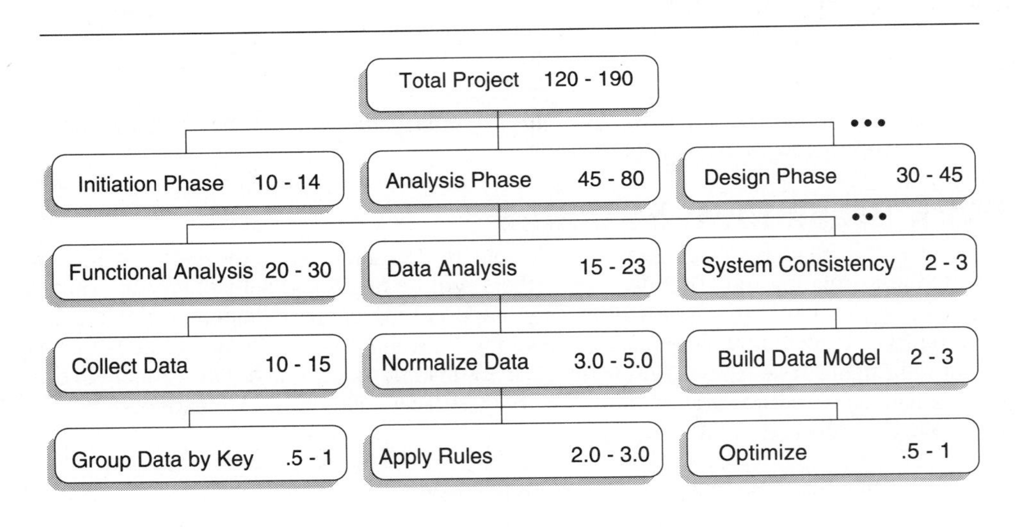

Build WBS

- Include Technical Tasks
- Include Management Tasks
- Show Phases at second level
- Show Q.A. Tasks
- Decompose until "Work Packages" are obtained

Estimating

- Do it at lowest level in WBS
- Summarize Upward
- Work in Resource Units
- For any task which exceeds 5 days
  - Decompose further
  - Estimate Components

Estimating Duration *Figure 7.11*

activities on commercial development projects. This will be lower with small, well-knit teams, and higher with large, unfamiliar teams, or where using subcontractors.

Quality Assurance

We prefer not to see this as an overhead or "add on", but rather as a necessary part of every single activity. Nevertheless, we should ensure that some Q.A. only activities are included in the plan and estimates. These are primarily the review points at the end of each major phase. Experience indicates that these will typically involve about two days' effort each on an average size project. A variable amount of time will be required following each review to address problems identified in the review. This is hard to predict and will depend on the history of the team and organization with respect to delivering quality work.

Converting to Calendar Time

Once we have a quantum of effort which the project is likely to require, we can begin to schedule activities, apply resources and estimate an elapsed duration. There are many factors which come into play here to influence the result. As before there is also a range of techniques which we can apply.

Analogy and Experience

This is probably the easiest and most widely used technique. Unfortunately, it is also frequently the least accurate. Estimating by experience is essentially "gut feel". Someone who has been in the environment for some time, and has a feel for the application and technology quotes a figure. This could be fairly useful, if we have at least established a relative quantum of effort as above. Often this has not been done, and the results are disastrous. This is probably attributable to the natural optimism of systems people.

Using analogies involves trying to find a similar sized application in a comparable environment and extrapolating from this experience. If a suitable project can be found for comparison purposes, and if all other factors remain equal (e.g., team size, team skills, staff turnover, technology, motivation, management, degree of specification change, etc.) we may get good results. More likely we will encounter problems. Even where good data is available, mistakes can be made.

Analytical Models

These have been proposed in various methodologies and by a number of independent authors. They attempt to cater, in some kind of algorithm or formula, for all the factors which influence the duration of the project. They normally take as their basic input the quantum of work to be accomplished in lines of code, run units, or function points. Various other factors which have an influence are then assessed and included in the formula. Finally the computation yields a calendar time estimate.

In our experience these are highly environment and methodology dependent and seldom work when taken out of context. They can be usefully employed, however, in a stable technical environment using a well-defined methodology routinely and in a disciplined fashion. In these rare situations, the techniques provide a useful framework catering for the various factors and their relative degree of influence. You will still have to calibrate the range of each factor for your own environment. Until you have done this and validated the results, the approach should be used with caution.

> I once audited a project plan for a life assurance company. They had set up a reasonable plan and project design and had even done a reasonable job of scoping and sizing the project. When I looked at the projected deadlines, however, I was concerned that the time looked hopelessly too short. "How did you arrive at this elapsed time?" I asked. The project manager replied: "Well, we just finished another project about half this size, and it took half as long, so we doubled the time that took."
>
> "How hard did you work on the previous project?" I asked. "It nearly killed us," he responded. "We were working twelve hours a day, six days a week for the last four months of the project."
>
> They had not realized that they had just planned to do the same thing again, only for twice as long!
>
> GM

There is a danger too, that as soon as you give someone a formula, particularly if it is implemented in a tool, they stop thinking and accept the results as gospel. We have seen project managers happily plug one set of assumptions into a formula and triumphantly announce: "*MethodName* says the project will take 230 elapsed days." This is absolute nonsense. The methodology designer has no idea of the exact conditions and environment under which the project will run; the answer is only an estimate; and it is only likely to be close to right if all the assumptions which the project planner has fed in are right! This kind of model can be much more usefully employed to assess a range of alternatives, i.e., to play *what if?* We should plug in a variety of assumptions, or scenarios and examine the likely results of our decisions, then use this as a guide to sensible ways to structure our projects. As a minimum, we should put in our most optimistic and then our most pessimistic assumptions and obtain a range within which the project is likely to fall.

Empirical Models

These are by far the best approach. Models derived from data collected from actual projects are used to discover the relative degree of influence of various factors. This information can then be used to estimate the duration for a new project under consideration. Examples include the Norden/Putnam model based upon the PADS database, and the CSC Index (formerly Butler Cox) P>E>P model based upon the P>E>P database which is structured similarly to PADS. The PADS database contains a wide variety of project types, including microcode, embedded systems, telecommunications, military and commercial systems development. It is mainly comprised of large projects (>70 000 SLOC). The P>E>P database contains mainly commercial projects. Both of these are derived from original work by Norden of IBM and Putnam from Quantitative Systems. In each of these, a significant number of projects were surveyed and the data captured in a consistent format, creating a database. Statistical techniques were then applied to find models which would describe the behavior of the data, and hence the behavior of systems projects. The data fits a curve known as a Rayleigh curve, well known in engineering disciplines (figure 7.12). Putnam's software equation relates size (measured in LOC) to a constant technology factor (C), total project effort including maintenance in manyears (K) and elapsed time to delivery in years (t_d). The formula is as follows:

$$Size = CK^{1/3} t_d^{4/3}$$

The Rayleigh curve dictates that t_d will occur where the curve reaches a maximum. To apply

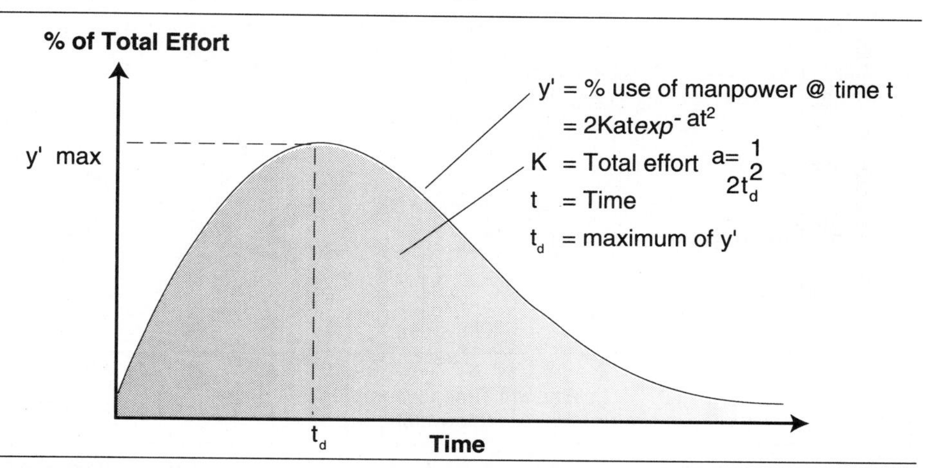

Norden-Putnam Curve: An Adaptation of the Rayleigh Curve *Figure 7.12*

the formula, we must be able to estimate source statements; determine the value of C which is a composite of factors such as technology, environment, skill level, etc., and can assume up to 20 different values; and we hold either K or t_d constant.

A value of C for your environment can be calculated by using data from previous projects and making C the subject of the formula:

$$C = \frac{S}{K^{1/3} t_d^{4/3}}$$

This, of course, assumes that you have data available for previous projects. If you do not, then you will have to try to use figures for an environment as similar to yours as possible (e.g., same technology, same industry).

One very interesting implication of the Putnam model is the relationship between delivery time and effort. From the basic equation, we can derive one which describes the relationship thus:

$$K = \frac{C}{t_d^4}$$

This says the effort is inversely proportional to the *fourth power* of delivery time. The economic impact of the foregoing is astounding, as shown in figure 7.13. A project that would cost $168 000 if completed in 11 months, will cost $690 000 if completed in 8 months, all other factors being equal. It could not be done in 6 months, regardless of how many resources we committed. Even if we do not have access to the proprietary databases, knowing how software projects behave, and being able to present the likely cost impact of unrealistic deadlines to senior management will put us in a much stronger position to negotiate realistic project plans and expectations.

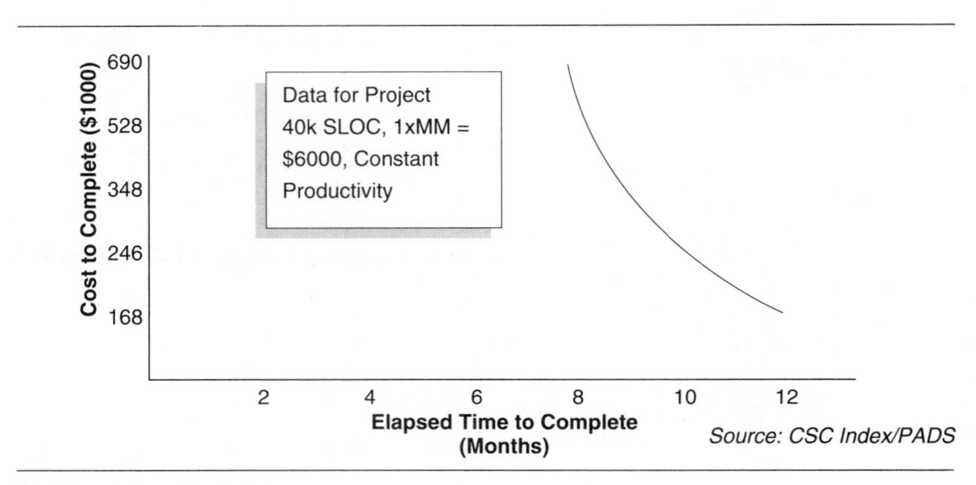

Project Cost versus Time *Figure 7.13*

The equations presented so far assume that we know either K (total effort) or t_d (time of delivery). These are obviously not known until the end of the project. To allow prediction of time or effort, Putnam introduces another factor D_0 which is the *manpower acceleration*, i.e., the rate at which we add resources to the project. This is essentially the slope of the curve from the origin to t_d.

$$D_0 = \frac{K}{t_d^3}$$

Values from the database are provided for specific project types, e.g., 7.3 for new software with many interfaces; 27 for re-implementation of an existing system. Using the previous equations and the one above, we can now get a formula which excludes either K or t_d (and allows us to calculate the other):

$$K = (S/C)^{9/7}(D_0)^{4/7}$$

$$t_d = (S/C)^{3/7}(D_0)^{11/21}$$

Putnam found that, for large projects, t_d occurred at the point of maximum manpower loading, with approximately 39 percent of the effort expended. De Marco examined smaller, commercial projects and found that t_d for these occurs significantly further to the right. For a sample of 24 projects between three and five man-years' effort each, the average point for t_d to occur was 2.5 times time to peak.

Related work by Barry Boehm provides a way to calculate the optimum manpower loading to deliver a product in a realistically short time. This takes the form:

$$t_d \text{ nominal} = 2.5 \times (M)^{.33}$$

where t_d is the most likely delivery time for a project requiring M manmonths of effort to complete. M is the effort to the left of t_d, i.e., the effort to delivery date, excluding maintenance. Thus, for a project of 100 manmonths the typical delivery time will be

$$2.5 \times (100)^{.33} = 11.43 \text{ months}$$

This assumes an average manpower acceleration factor. Your environment may differ, but completing the project in less time will typically require extraordinary productivity, or greatly increased effort. There is strong empirical evidence to suggest that there is a limit to the acceleration factors that can be applied. Boehm provides a formula for calculating the boundary to the "impossible zone". Projects which try to do M manmonths of effort in less than

$$1.9 \times (M)^{.33}$$

months are in the impossible region - see figure 7.14. This does not mean that it cannot ever be done, merely that no project included in the analyses to date has ever done it! You have to be very confident or stupid to ignore evidence like that.

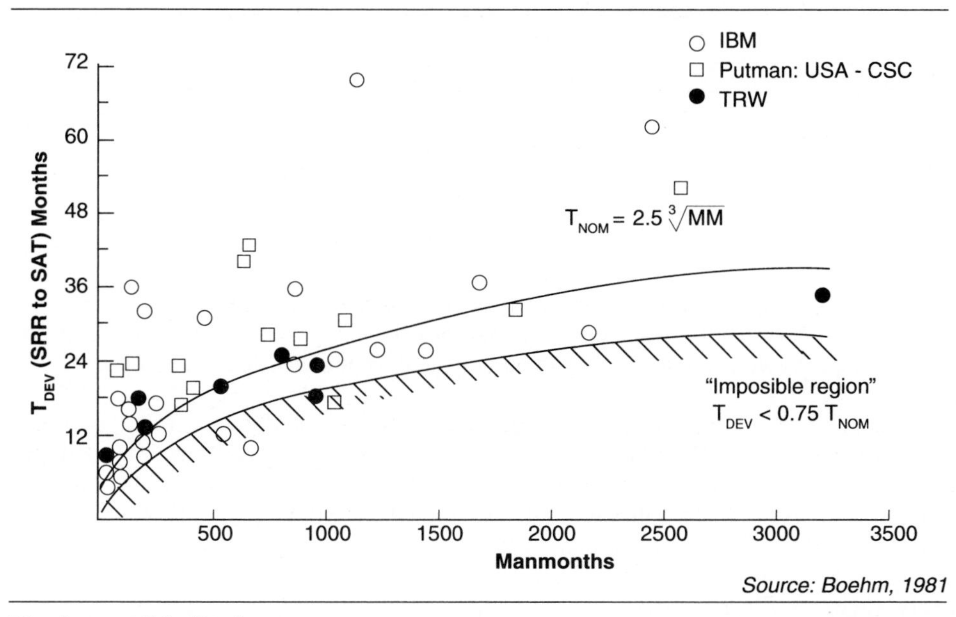

Source: Boehm, 1981

The Impossible Region *Figure 7.14*

Boehm, Barry, W., Software Engineering Economics, © 1981, p.182. Adapted by permission of Prentice Hall, Upper Sadle River, New Jersey.

Another candidate is the Parr model (figure 7.15), an adaptation of the Putnam model which better fits smaller projects (< 15 manyears) where the team is already partly established at the start of the project. Parr's equation is as follows:

$$\text{Manpower (t)} = 1/4 \ \text{sech}^2 \left[\frac{(at+c)}{2} \right]$$

- Similar to Putnam-Norden but
  - Assumes some resources are in place at start of project
  - Better fit to a smaller project (e.g.,<15 manyears)
- Manpower (t) =1/4 sech $^2 \left[\frac{at+c}{2} \right]$

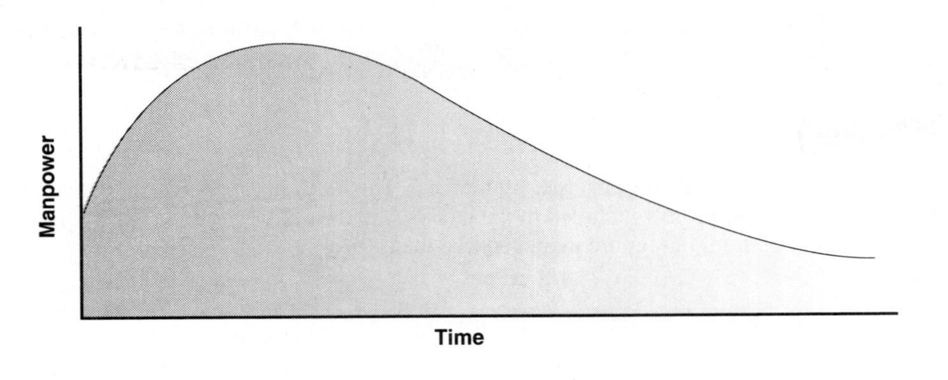

Parr Curve *Figure 7.15*

Sech^2 is the hyperbolic secant. These models have some exotic mathematics, which is simply useful to describe the shape of the curves and the behavior of the variables. Once you have a correctly drawn curve, you can rely on graphical techniques alone, if you prefer.

It is interesting to note that commercial projects display high productivity relative to other types of projects (table 7.2). This is probably due to the lower complexity and less stringent quality, reliability and performance requirements. The need for higher quality in commercial systems as they become more mission-critical may reduce productivity unless we can also achieve re-usability.

Additional Factors

Various other factors can affect the calendar time and costs. You should consider the following to the extent that the information is available:

- *Skill level.* Is the team more or less skilled than average for your environment? Are they more or less skilled than the teams from which your estimating model or figures have been derived?

  At the outset, we will use an average for the installation, or the team, if this is known. At later stages, when specific resource assignments have been made, we may wish to alter the estimates based upon the skill level, or historical productivity, of the particular individual. Remember that there can be an order of magnitude difference

- *Percent allocation to task.* To what extent will people be committed to the project? We all too often blithely say that someone is committed 100 percent to the project, failing to recall that he is also required to do maintenance on two other systems, runs the social club and occasionally gets called upon to support users with complex

Business	**16**	**Each index point means about .25 more/less effort**
Scientific	**12**	
Systems Software	**12**	
Telecommunications	**11**	
Process Control	**9**	
Command & Control	**8**	
Radar	**7**	
Avionics & Space	**5**	
Real time/embedded	**5**	
Firmware/ROM	**4**	
Microcode	**1**	

Source: CSC Index/PADS

Productivity Indices by Project Type - PADS

Table 7.2

problems. We can also not expect 8 hours of productivity in an 8-hour day. The norm from experience is about 6.8 for a contractor or consultant dedicated to a project, and about 5 for an in-house employee

- *Training required.* New techniques or application areas will often require that we train team members before they can successfully perform tasks that the project will require. This training and the attendant loss of productivity and costs must be incorporated into our estimates and plan. A good average is to budget for two weeks of full-time training per person per year. As a project manager, you are responsible for developing your staff - you must plan for growth

- *Leave.* People take leave if you let them. This will normally average one month per individual per year. Although it is a problem for project scheduling, project costs and continuity, it is essential that leave is planned for on a long-term basis. People do need time away from the mill to gain perspective, recharge the batteries and renew their personal relationships. These are all vitally important in the long term. Do not cut back on leave routinely, it will eventually cost you some of your star performers

- *Illness*. If you don't let people take leave, they will get ill and take it anyway. There are also unavoidable illnesses, accidents, etc. These are of course, very difficult to predict, but we should nevertheless put in some contingency for their occurrence. Work on 10 days per person per year. This should be much less in a highly motivated team, but better to have the cushion if you need it

- *Management/Administration overhead*. There is inevitably a management and administration overhead that each individual in a professional team carries. Some of this is necessary, but we should avoid unnecessary meetings as far as possible, and ensure that the paperwork required is the minimum that will get the job done. We are constantly amazed at the lack of secretarial support for project teams. Organizations will commit seven or ten very expensive systems professionals to a team and allow the project manager (probably the most expensive) to act as a typist, filing clerk, spreadsheet jockey, etc. This is ludicrous. Organizations need to provide clerical support to project teams on an economic basis. Two or three days of the project manager's time saved per month will probably pay for a full-time project secretary. Motivation of team members is also vastly improved if they are not bogged down with menial tasks

- *Quality Assurance/Rework*. There is an overhead associated with quality assurance activity. This should be offset by achieving higher quality, and consequent reduction in wasted effort, future maintenance, etc. We will return to this topic in detail in a future chapter. For now, let us remember that doing things properly sometimes takes a little longer. The old adage goes: "*We never have time to do it right; but we always have time to do it again!*" There will also be particular Q.A. activities that should be included in our project planning, such as time for review of deliverables, and time to redress any errors discovered during these reviews. We frequently see plans including reviews, but no time to actually respond to their findings. If we are going to ignore what the review discovers, we may as well not bother

- *Unavoidable delays*. We will often encounter unavoidable delays, particularly when we are dependent upon outside resources, or when we are gathering information from high-level users. We may need to interview board members. If we could schedule this at our convenience, we might need only two days. However, it may take three weeks or more to fit the interviews into the board member's busy schedules

- *External dependencies*. As mentioned above, we will frequently be using resources outside the team (e.g., DBA) or outside the organization (e.g., External Auditors). These dependencies need to be carefully coordinated with the organization supplying the resources or service to ensure they are available to the project with minimum delay. Of course, you need to keep them informed of any slippages or changes in your schedule!

- *Holidays*. Public, company and political holidays may introduce delays. You can expect to lose two weeks of productive time over the Christmas/New Year "silly season"

- *Company cycles*. Plan to avoid critical company cycles as far as possible. These include financial year end, tax year end, management strategy sessions, critical trading peaks (e.g., a retailer over the festive season). Unfortunately, some systems projects will require you to synchronize with these occurrences. In these cases, you should anticipate that there will be other demands on your team, sponsors and users and that extra resources may be required

- *Task Dependencies*. The old adage goes:

  "If a woman can produce a baby in nine months, surely two women can do it in four and a half"

  As we know, it simply does not work that way. People and time are not interchangeable. Some tasks simply must wait for others to finish. There is no point, for example, designing physical database structures until our logical data model is complete. We must respect all of these dependencies in our planning. We can usually find guidance in terms of these dependencies from our systems development methodology.

Estimating Principles

Emerging from our discussion to this point are some important principles that we should apply to our estimating:

- *Separate the estimators from the doers*. De Marco and other researchers have shown that developers who estimate their own performance are consistently optimistic. These biases appear to disappear when the same people estimate for other resources. The message in this is that we should separate the estimators from the doers on a given project. De Marco suggests creating a separate estimating group to assist teams in developing estimates. This has the following advantages:

  - Provides an independent, unbiased opinion

  - Increases consistency of estimates across projects in the installation - at least if they are wrong, they are likely to err in the same way. The relative size of projects is still maintained and correction of planning once the deviation is discovered is greatly simplified

  - Builds estimating expertise. Having a team with this focus allows the individuals in it to gain a much greater exposure to the estimating process and more practice in applying the techniques than the typical project manager would get

  - Provides a central point for collecting and analyzing estimates, actuals and performance

- *Estimate at several levels*. Estimates must be adjusted as we validate or debunk assumptions, and as our knowledge of the project, the resources and the system under consideration improves. For example, we may use an average factor for team member productivity at the outset when the team has not yet been assembled. This may be adjusted later when we know who the individual team members are and what their past performance has been

- *Quote a range*. Never quote a single figure until the project is finished. Generally speaking, the less you know, the broader you go. Initial estimates will have a high margin of error and a correspondingly wide range. Later estimates can become more certain and specify a smaller range

- *Use techniques appropriate to the phase*. Try to use the best techniques available to you with the information available to you at that point

- *Use several techniques.* Different techniques make different assumptions and are sensitive to different factors. It is thus wise to employ several techniques and compare their results. If they correspond reasonably well, then your estimates are probably sound. If there is a wide discrepancy, you should look first to see if any errors have been made in applying the techniques. If this is not the case, then look to see what each technique is sensitive to that the other one might not be. For example, one technique may take into account technological complexity, where another one doesn't. If this is a significant aspect of your project, you may need to adjust the estimate from the technique which is insensitive to this variable.

- *Qualify your estimates.* Try to give management a confidence level in your estimates. This can be done as follows:

  - Determine the range of your estimate (Minimum and maximum)

  - Find a "most likely" estimate which may not necessarily be the midpoint of the range

  - Adjust the estimate by weighting the most likely as follows:

    $$NewEstimate = \frac{Min + 4xMostLikely + Max}{6}$$

  - Approximate one standard deviation from the new estimate by calculating

    $$\frac{Max - Min}{6}$$

  - Quote your range as the NewEstimate + or - this figure

  - The confidence level in this estimate is 68 percent (one standard deviation)

  - If a 99% confidence level is required, approximate 3 standard deviations by using

    $$\frac{Max - Min}{2}$$

    and quoting the NewEstimate + or - this figure

- *Revise your estimates* based on the actual performance achieved, or any change in the underlying assumptions or information upon which our estimates are based. For example, if the specification phase ran 50 percent over the estimated duration, we can expect a similar overrun on later phases. If the initial ten programs take half the expected time to develop, we can probably reduce the estimates for the remaining thirty programs (provided the initial sample is representative!)

  Our standard project lifecycle caters for the concept of a Creeping Window (figure 7.16), where we proceed as follows:

  - Estimate in detail for the first phase (i.e., for each task). Prepare soft estimates for later phases

  - As tasks are completed, record the actual performance

  - At the end of the phase, summarize the actuals for the previous phase. Using the knowledge gained to that point, estimate the next phase in detail to task level. Redo the soft estimates for later phases

- Adjust project plans to reflect the new estimates and commence work on the next phase

- Repeat this process to the end of the project

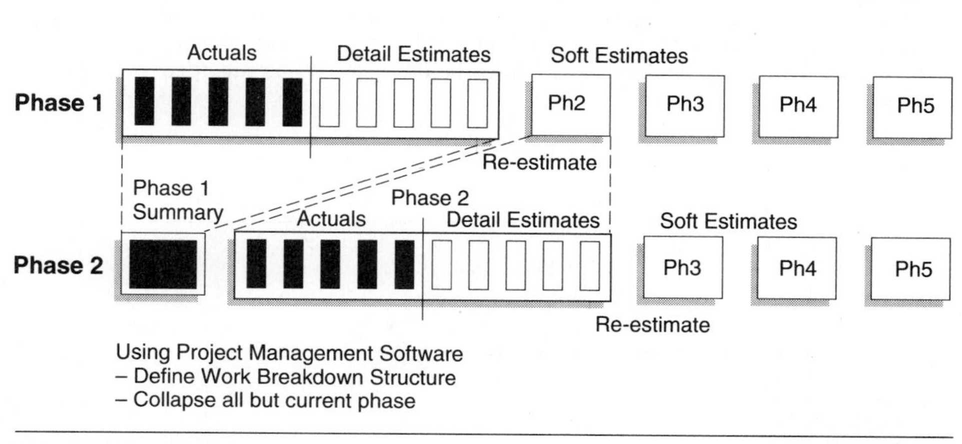

The Creeping Window *Figure 7.16*

An automated project management tool greatly facilitates this process. We can also focus our attention on just the current phase, while viewing it in the context of previous completed phases and later phases presented as summaries only. In this way, we may reduce the number of tasks we need to deal with at any one time to say 20 to 30, rather than the 200 to 300 which the whole project may require

- *Collect data.* If we are ever going to improve our estimating performance, we need to collect actuals, to know how we did on previous estimates. This can best be done by a central group (the estimating team?). They will obviously be dependent on input from the project managers, and they in turn on input from their team members. We will address the kinds of information that need collection and how they can be analyzed and reported in more detail in a later chapter

- *Check reasonableness.* When we apply science, we should not throw common sense out of the window. It is all too easy to drop a digit or put a decimal point in the wrong place when doing a series of calculations. If the numbers that come out of our models or formulae don't look right to your experienced eye, then at least check them. Also, try another estimating technique as advised earlier

- *Add an overall contingency* (about 20% is normal) for unforeseen circumstances. If we do this, it is important that this contingency not be seen as part of the estimate by the project team. "Work expands to fill the time available." Quote the contingency in your estimates to management, and use the figure in your cost-benefit calculations

- *Monitor your effectiveness.* Given the poor state of estimating performance in most organizations, and the large sums of money spent on projects, it is startling how seldom estimating effectiveness is actually checked. An excellent measure is the Estimating Quality Factor (EQF) proposed by de Marco.

  A graph is drawn (figure 7.17) with time on the horizontal axis, and cost (or effort) on the vertical axis. As estimates are made or revised, the projected cost is plotted as a horizontal line from that point in time, to the next revision point. At the end of the project, the actual cost or effort for the project is plotted as a horizontal line parallel to the X axis, at the appropriate cost level. The EQF is determined by dividing the area under the actual cost line, by the sum of the areas where the estimate deviates from the actual cost line. (The shaded area in the figure.) The higher the result, the better the estimating performance. Scores below 8 should be regarded as unacceptable

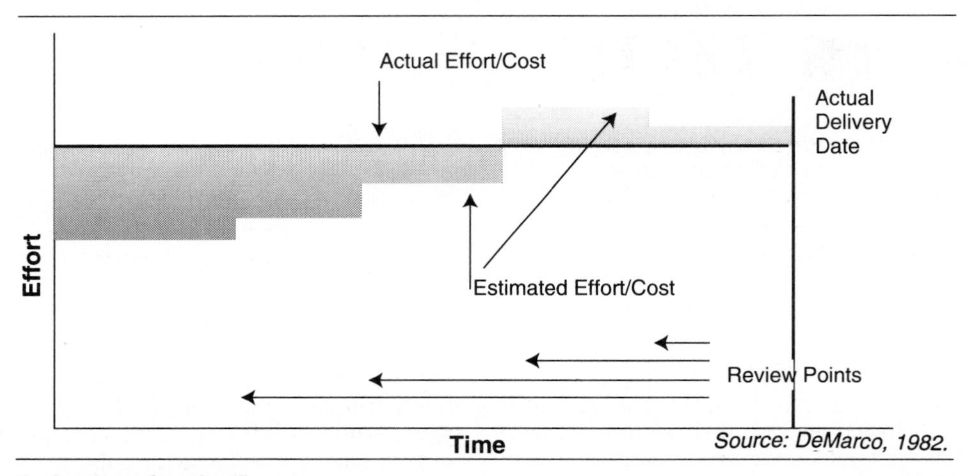

Estimating Quality Factor *Figure 7.17*

Applicability of Techniques versus Phase

From the foregoing, it should be obvious that various techniques offer different advantages and require different inputs. These may or may not be available. We are thus forced to use a variety of techniques at different points in the project lifecycle. We should obviously try to use the most accurate techniques we can, given the information available to us at that point. Figure 7.18 is a summary of which techniques can be usefully employed at what stage.

Putting It All Together

We have discussed a wide variety of techniques and approaches, concluding with the principles you should apply. You are probably wondering how to put this all together in practice. Below is a list of activities in the sequence you would perform them on a typical

	Initiation	Feasibility	Ext Des	Tech Des	Build/Test	Install
Wideband Delphi	• •	• •	•			
Analogy	• •	• •	•	•	• •	
Mathematical Model (using Function Points)		•	• •	• •		
Mathematical Model (using SLOC)			•	• •		
Mathematical Model (using Design Metrics)			•	•		
Constraint Models		• •	• •	• •		
Top-Down	•	• •	•	•	•	
Bottom Up		•	•	• •	• •	

Source: Saker 1990

Applicability of Techniques versus Phase

Figure 7.18

project. The initials in brackets indicate primary responsibility for the task (PM = Project manager or delegate; EG = Estimating Group). The accompanying "Estimating Engine" diagram, figure 7.19, should also help to put the various techniques in context.

- Define project including goal, objectives, constraints (PM)
- Scope the project (PM)
- Prepare initial PBM for results from project (PM)
- Estimate size of deliverables. Use more than one technique (EG)
- Estimate effort (EG)
- Develop WBS to detail available (PM)
- Add the effort known (PM)
- Estimate effort/duration for tasks where estimates are not available (EG/PM)
- Convert effort to elapsed time. Use more than one technique. Adjust for individual identified resources. (EG/PM)
- Develop an optimistic/pessimistic scenario (EG)
- Quote the range adjusting for most likely and required level of confidence (EG)
- Record your estimates (EG)

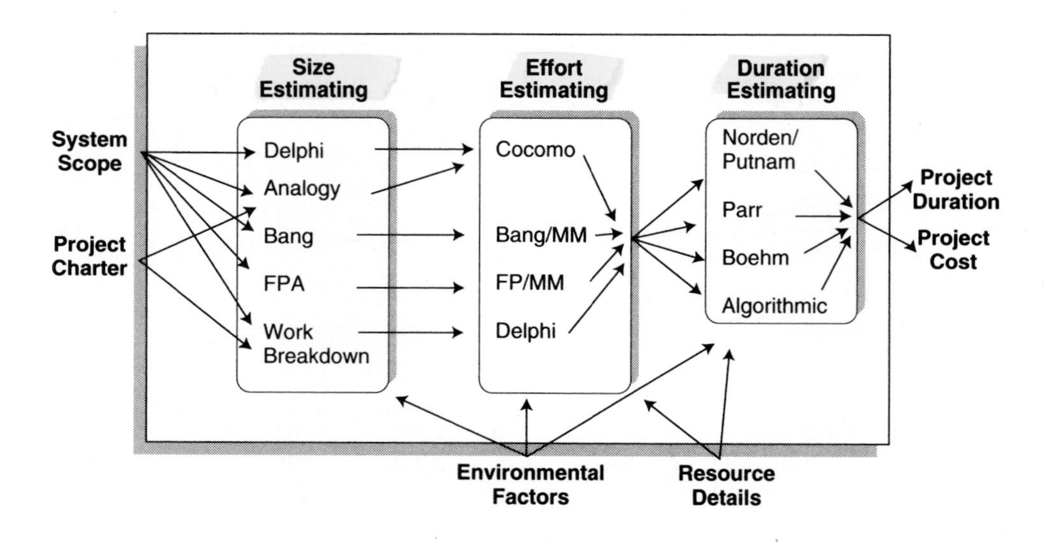

The Estimating Engine *Figure 7.19*

- Repeat as more detail becomes available (PM initiates, EG performs)
- Collect actuals to improve performance (PM provides, EG records and analyzes)
- Monitor estimating performance (EG)
- Improve techniques (EG)
- Distribute techniques (EG)

Case Questions

MyWay Organizer

Q7.1

Your management has asked you to give an opinion on the feasibility of completing the project in four months. Prepare an estimate and justify your answer, giving a degree of certainty. Proceed as follows: Add to the Work Breakdown Structure for the MyWay Organizer development (your own or a sample answer to Q3.1) your estimates for duration of each task at the lowest level. Assume you have yourself as project manager, an experienced analyst, an experienced analyst programmer, and a more junior programmer on the team. Summarize these upward to determine an overall project duration. (15 mins)

Q7.2

Add to the Product Structure Model for the MyWay project (your or sample answers to Q4.1 through 4.3) your estimates for the effort to complete each deliverable at the lowest level. Assume adequately qualified staff will prepare each deliverable. Summarize these upward to determine an overall project effort. (15 mins)

Q7.3

Using the context diagram for the project (answer to Q2.2), determine a raw (unadjusted) function point count for the MyWay Organizer. (15 mins)

Q7.4

Using the description of the technical environment and your knowledge of the project to date, determine the adjustment weightings for use in the FP count for the Organizer. Determine the adjustment factor. (15 mins)

Q7.5

Determine the final FP count for the organizer. If your organization typically displays a productivity of 16 fp/mm for a team of four, what is the likely elapsed duration of the project? (10 minutes)

Q7.6

Form a group of six. Appoint a chairman. Make sure that you have at least four people with good programming background. Conduct a Delphi session to estimate the ESLOC in the MyWay organizer. (15 mins)

Q7.7

Taking the answer from Q7.6, or a number provided by the instructor, determine the most likely and minimum completion times using Boehm's guidelines. Assume a delivery rate of 900 ESLOC per manmonth. What factors in this project might cause you to take longer than these figures? (15 mins)

Gleam Stores

Q7.8

Using the context diagram developed in Q2.5 (your own or a sample answer) and the information provided regarding the technology environment, determine the adjusted function point count for the Gleam stores pilot system. (30 minutes)

Q7.9

Using the adjusted FP count from Q7.8, and the knowledge that the pilot system was developed, documented and tested by a team of 4 people in 9 months, determine the productivity level which the team exhibited. Management has asked how many people you will need to tackle a new project dealing with stock optimization and interbranch transfers between stores. The proposed system is estimated to be 1.5 times the size of the pilot system and should be completed in 7 months. (30 mins)

Handover Trust

Q7.10

An estimate of 2300 FP has been produced for the rewrite of the New Business (NB) system for Handover Trust. This processes applications for new policy contracts from receipt to their final conversion to policy contracts or rejection. The system will capture applications, verify completeness, capture new client details if necessary, capture medical details, assist medical assessors and underwriters in reaching decisions, interface to an industry database for checking and to advise on rejections, validate banking details, project benefits and generate a contract or a variety of rejection notices. It is also possible that applications may be accepted with modifications, which the system, or its users, will suggest. It will also interface to a reassurance system in cases of large cover or high risk.

The system will be developed in a client server environment using a relational database server and high level GUI development tools in the client workstations. Client server projects in the environment to date have produced productivity levels of about 13 FP/MM with small focused teams (3 to 4 people). Due to the urgency of the New Business project, and its criticality to the business, you have a free hand with resources. Management has indicated that you can use up to forty selected people to achieve the most rapid implementation possible.

Estimate the productivity you can expect for different team sizes and prepare a recommendation to management for the team size you would recommend and your reasons. You must provide a well-supported estimate of the overall duration of the project.

How could delivery of key functionality be brought forward?
(1 hour)

Q7.11

An analysis of the early client server project estimates has yielded some interesting results. The quality assurance and planning support group has asked you to perform some further analysis to determine the accuracy of estimating. They have provided you with two project histories. They have not identified them to ensure that the answers are objective:

Phase	**Estimate for full project** (Calendar weeks)	**Actual time for phase** (Calendar weeks)
Project A		
Initiate	15	1
Requirements	15	3
Prototype & Design	17	3
Build & Tuning	17	2
Integration/stress test	16	4
Installation	16	4
Project B		
Initiate	25	2
Requirements	27	5
Prototype & Design	30	12
Build & Tuning	30	10
Integration/stress test	35	5
Installation	35	2

Determine the Estimating Quality Factor for each project. What contingency should be allowed on plans with the current level of estimating expertise? (30 minutes)

ThoughtWell Books

Q7.12

Using the previously developed Context Diagram, determine the likely size of the required software system using Function Points. (20 minutes)

Q7.13

Assume your organization achieves a productivity level of 17 fp/mm per person on average LAN projects. A typical project team would have four members. Your management has indicated that you could assign up to 8 people to your project. Resources are charged at $400 per day per person and $650 per day for the project manager. Assuming it is now June 1, 1995, what is the earliest date you could promise delivery of a working system to the client? What will the total cost of the project be, excluding any hardware purchases? Show all workings and support your deductions in a manner defensible to management. (40 minutes)

8 *Project Design*

Lifecycle Choice

We have already introduced a generic project lifecycle (PLC) in earlier chapters (figure 8.1). This serves as an umbrella and a framework within which the work can be accomplished. We also alluded to the existence of other lifecycles (such as the Systems Development Lifecycle [SDLC]) which may be embedded in the project lifecycle to tackle a specific kind of project. We now need to consider some of the alternatives for the SDLC, and to discuss the pro's and con's of each of these. There are no wrongs or rights here - each is appropriate to a particular type of project and set of circumstances. The intent of the discussion is to inform you of some of the options and their strengths and weaknesses so that you can better

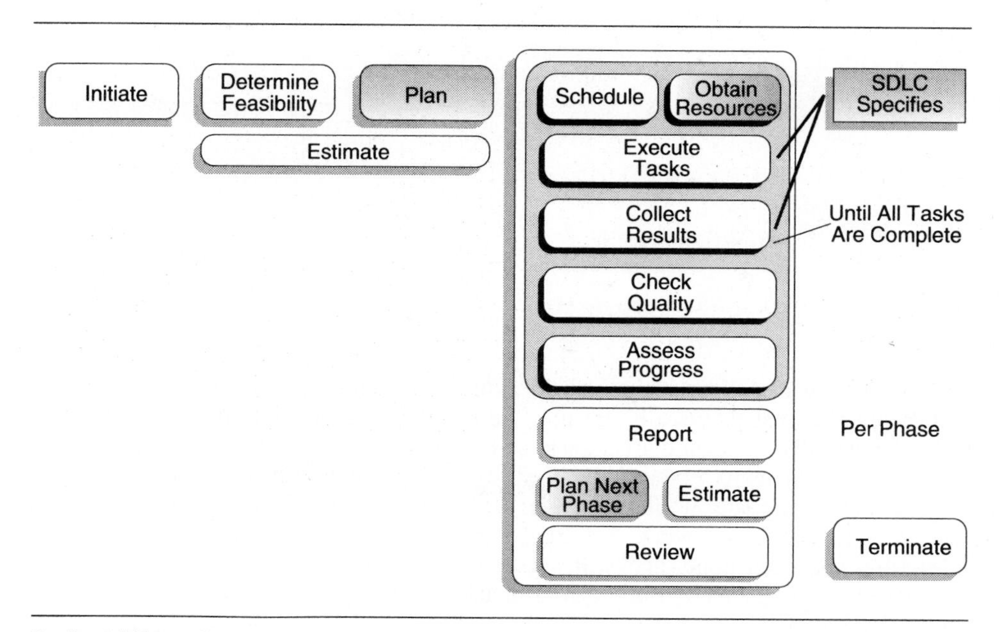

Project Lifecycle *Figure 8.1*

choose and adapt these to your own situation. In the section which follows, the discussion will concentrate on variations in the SDLC. We will then go on to discuss other lifecycles appropriate to other kinds of projects (e.g., system maintenance).

Waterfall Model

This is the classical SDLC model first described by Royce in 1970, illustrated in figure 8.2. It begins with *Requirements Definition*, followed by *Specifications* (what must be done/delivered), then *Design* (how the specifications will be met); next *Implementation* (building the product); *Integration* (making sure all components work together) and finally *Operations* (when the product is deployed into its working environment). These stages are assumed to be discrete, with one completing before the next stage commences. Thus the *Specification* would be signed off before *Design* would begin. In practice, it is almost impossible to get each stage's output 100 percent right before proceeding, so there is provision for feedback, to allow errors detected in later stages to cause corrections to output produced in earlier phases. To this extent, the broken line arrows in the diagram can be considered to represent maintenance activity.

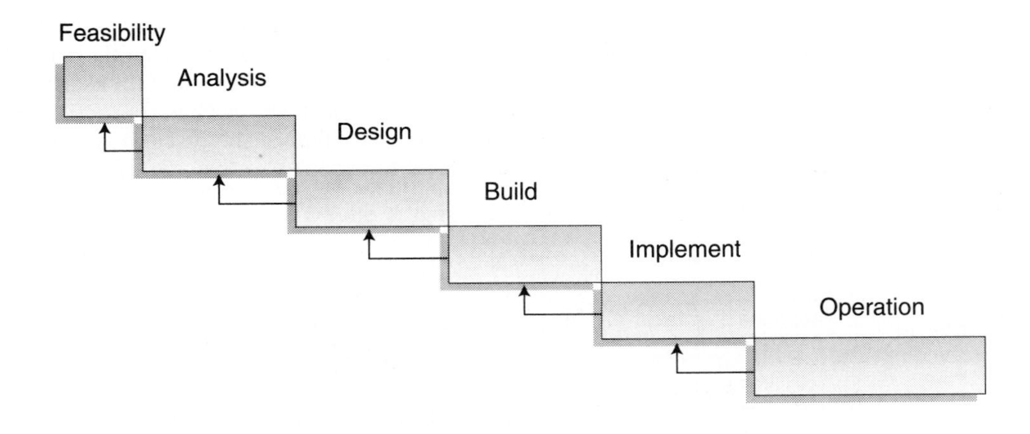

Waterfall Model *Figure 8.2*

Problems with the waterfall model include the assumption that the phases are discrete, which we find in practice that they really are not, the maintenance involved in going back and correcting supposedly finished deliverables, and the overall time taken (little can occur in parallel). A major difficulty is that the specification may contain serious errors which will only be discovered very late, at the integration or operations phase. This is very expensive to correct, as a lot of detail work, e.g., program and file design, coding, testing and documentation has been done, and will need to be redone. Estimates of the size of the problem have been produced by Boehm, Martin and others. Consider the summary produced by Schach in figure 8.3.

The huge maintenance effort (and hence cost) is attributed to poor requirements definition

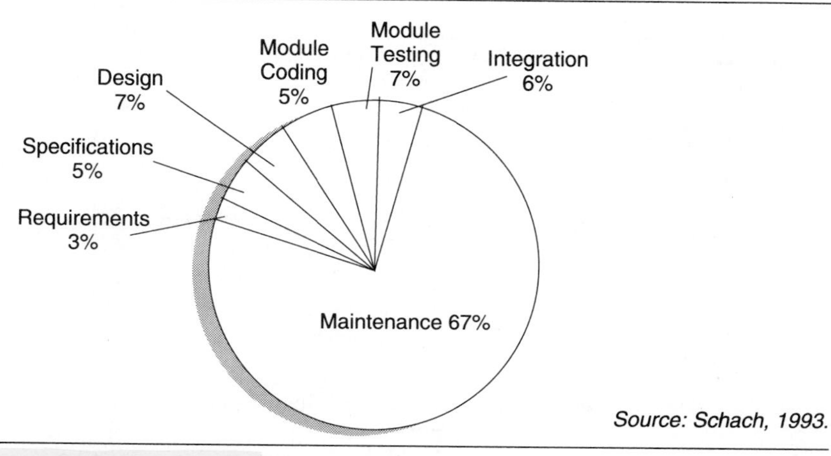

Relative Costs of Lifecycle Phases

Figure 8.3

and specifications. Martin came up with the estimates of effort expended to correct errors shown in figure 8.4.

Phase	Percent
Requirements	82%
Design	13%
Coding	1%
Other	4%

Source: Adapted from Martin 1993.

Effort on Correction of Errors

Figure 8.4

Even more alarming is the *relative cost* to correct the same error, depending upon where it is detected. This is summarized in the graph shown as figure 8.5.

These facts have serious implications for how we should manage projects. Firstly, we must be rigorous in producing quality through every stage, and second, we must try to choose lifecycles and techniques which will reduce errors of requirements definition and specification. Most of these errors occur because of a lack of understanding (detail) or

Source: Adapted from Stephen R. Schach, *Software Engineering*, (Burr Ridge, IL: Richard D. Irwin Publishing and Asken Associates, Inc., 1993), P.11.

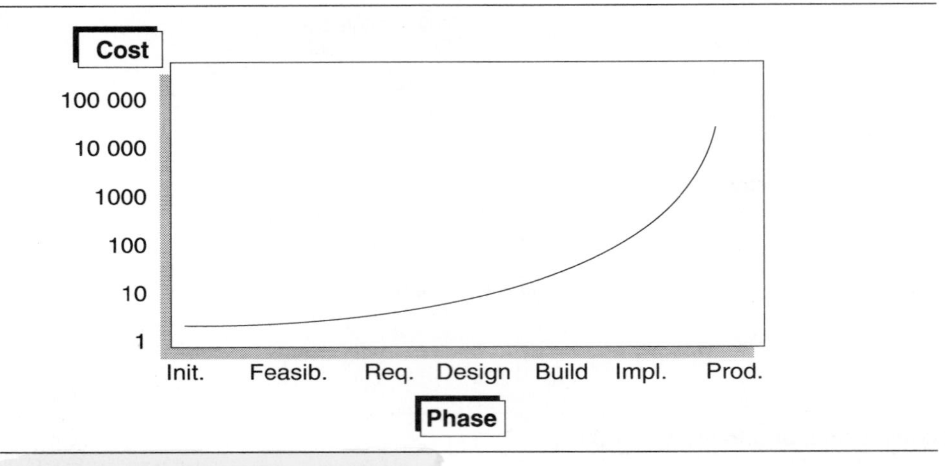

Relative Cost to Correct an Error *Figure 8.5*

failure of communication between the developers and user community. Various lifecycle variations try to address these in a variety of ways.

Systems Engineering Lifecycle

This is based upon and closely related to the classical waterfall model. It is typically used in system engineering projects where the resulting products must have very high reliability and integrity. An example would be a missile guidance system where errors would have extremely serious consequences. Other examples would be software controlling a steel mill in real-time, or an airliner navigation system. In all of these, the cost of errors (in human and monetary terms) is unacceptable. The IEEE have over time standardized a Systems Engineering Lifecycle. This is formally defined in IEEE 1058, 1987. The plan is fairly flexible, and will, with tailoring, handle virtually any kind of software/systems engineering project. In the latter, engineers are frequently concerned with the concurrent development of both hardware and software that will run in the as yet unavailable hardware.

The lifecycle, as shown in figure 8.6, is a variation on the waterfall model, with standard named phases terminating in formal reviews. Borrowing from the engineering discipline of Configuration Management, they introduce the concept of *baselines,* where a baseline is defined as the set of deliverables that will be complete (and frozen) at a particular review point. Two other important concepts are those of *concessions* and *deviations*. Concessions are things which were present in the specifications at an earlier baseline which the sponsor concedes will not be delivered in the ultimate product. An example will be the original specification calling for voice input of terminal commands. This may prove to be too difficult technically, or too expensive to justify. The sponsor then formally concedes that this will not be present in the design specification and delivered product. Note we said formally - it must be in writing and carry the sponsor's approval.

Deviations are requirements from a previous baseline which have not been met when a review is done. The sponsor does not concede that these can be left out. They must therefore be addressed and included/corrected before the project continues. An example here would be

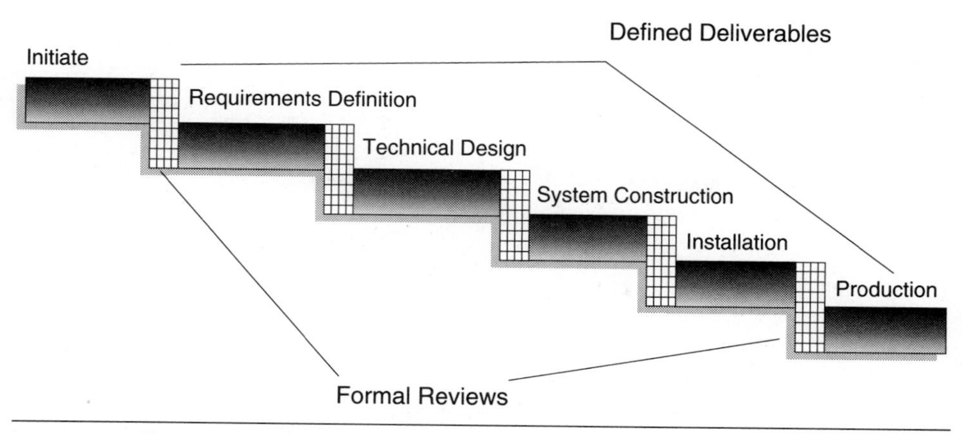

Software Engineering Lifecycle *Figure 8.6*

that a response time of less than three seconds has been called for in the design specifications. When the product is tested, response times greater than three seconds are experienced. The project must halt and correct the problem before commencing with the next phase. Again, these are formally recorded, and the correction must be formally approved by the sponsor.

The lifecycle provides very tight control over scope changes and supports quality management well. We will discuss configuration management and quality assurance in more detail in later chapters.

Drawbacks of this approach include the formality, extra documentation, extra effort and discipline, which can discourage innovation and demotivate team members. Careful marketing by the project manager is necessary to sell the benefits to the organization and team. These are primarily cost containment through scope control, and high quality through formal reviews and correction of deviations.

Most commercial organizations will balk at the degree of formality and potential overhead, but we believe that a modified version of this approach is appropriate in business, for mission-critical applications, given the expense and criticality of systems now being tackled.

Overlapping Phases

In commercial use, we normally see a variation of the waterfall lifecycle with the phases overlapping somewhat. See figure 8.7. The chief advantage is reduced project time. However, we must realize that increasing overlap generates increasing risk, since the chances that we are doing detail work on incorrect specifications or assumptions increases. Resultant rework will increase correspondingly. We probably do not want this to happen, given the figures we saw earlier showing where our effort already goes in the lifecycle. We may find our overall effectiveness and productivity would be increased if we followed the old adage to "hasten slowly".

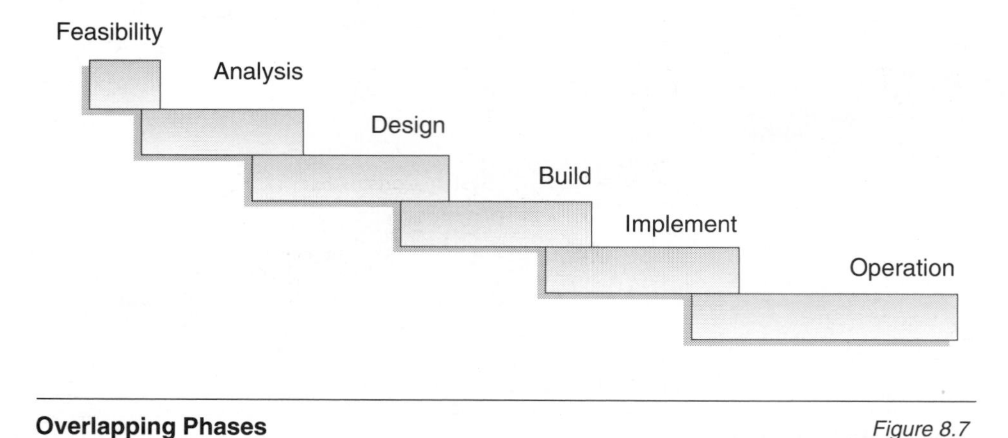

Overlapping Phases *Figure 8.7*

Prototyping

Prototyping as a feature of the lifecycle has been adopted as a result of two factors:

- The ability to do it inherent in online tools (such as 4GL languages, screen painters, Communications Control packages, etc.)
- The need to show the user something and to verify the requirements and user interface suitability early in the project before a large amount of detail work is done

A prototype is a model of the real thing which will exhibit some (but not all) of the same behavior as the final product. It is built with a view to testing concepts and verifying viability before major commitments are made. In this sense an architect's paper model can be considered a prototype. It will allow the assessment of the overall concept, appearance, relationship to the environment; how the structure will relate to the direction of ambient light, and so on. We would certainly not want to live in it, however, even if if were life-size. It would not have any insulation, it does not have electricity, plumbing, or sewerage. These will only be present in the final product, i.e., the building.

The model is extremely useful, however, to correct any misconceptions of the designer before we proceed. The client may have asked for a particular style of roof, which in practice does not suit the rest of the design. This could be apparent from the paper model. Various options can then be tried at minimal cost, and a good alternative chosen before construction begins. This is the proper use of a prototype: to verify requirements, or feasibility.

An example of verifying feasibility would be as follows: A supermarket decides to implement a system whereby clients can directly debit purchases to their bank accounts. This will entail the simulation of an ATM transaction by the Point of Sale (POS) equipment. If this link in the system is not viable, then the whole concept falls down. We may decide to build a technical prototype in collaboration with the banks, to hook up a POS terminal and test its communication ability acting as an ATM. When this is successful, we can then proceed with the rest of the project. Used in this way, prototyping can significantly reduce project risk.

There are several ways to use specification prototypes to verify requirements. One of the simplest is to use mocked-up screen and report layouts to show the user what the inputs and outputs of the proposed system will look like. A more advanced prototype would simulate the online behavior of the system complete with dialogues (sequences of screens in the order in which the user would encounter them) and sensitivity to the use of function keys, menu selections, etc. These are particularly useful to gauge whether the user interface is appropriate to the target user community.

Prototyping can be incorporated in the lifecycle in a variety of ways:

- *As a replacement for the conventional specification process.* This is not recommended, except where very powerful tools are available and the project in question is creating new outputs from existing data (for example, an Executive Information System [EIS]). The danger is that we can end up with a product with poorly conceived and balanced functionality - a house without a sewerage system. This is illustrated in figure 8.8.

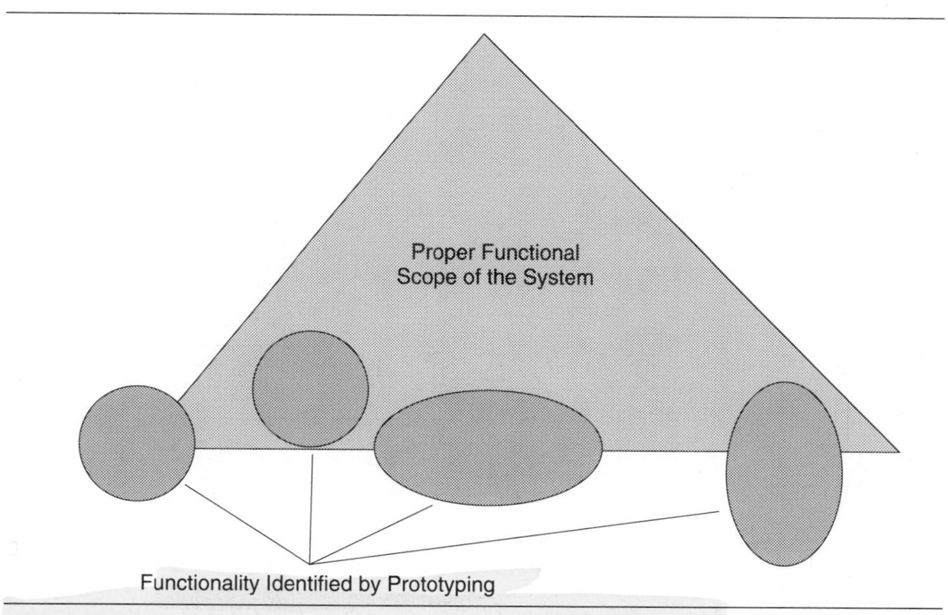

Prototyping as a Replacement for Functional Specifications *Figure 8.8*

- *To verify the requirements.* This is strongly recommended. Normal functional and data models of the proposed system are constructed. We then rationalize which functions are to be computerized in the current phase. These are then prototyped with reference to the data model and dictionary. In the process the detailed format of inputs, outputs and flow of dialogues is established and recorded for later incorporation in

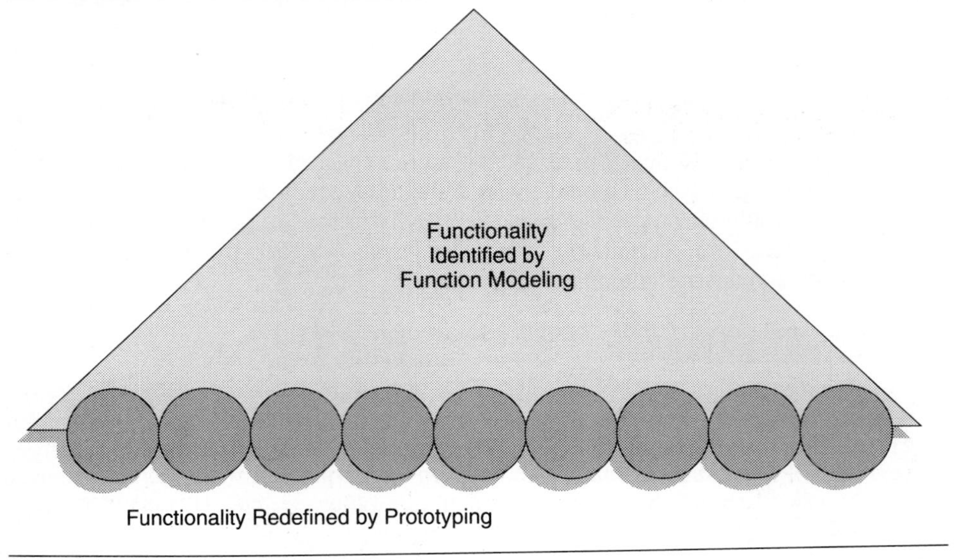

Prototyping Used to Refine Specifications *Figure 8.9*

the production system. This concept is illustrated in figure 8.9. A lifecycle to accommodate this is shown in figure 8.10.

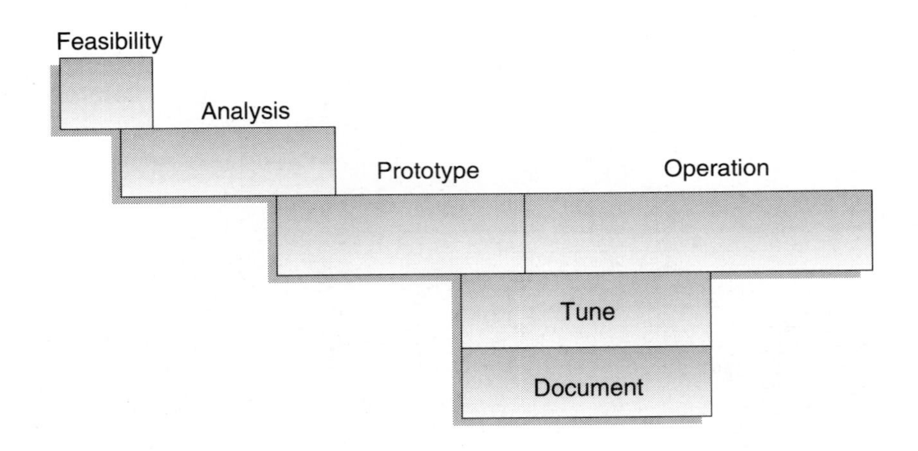

Prototyping Lifecycle *Figure 8.10*

- *To synthesize the user interface.* Again this is highly recommended. It allows the user community to assess how complete and practical the system is in operation. It should definitely be done where we are contemplating a change of user interface (e.g., from a character-based interface on a mainframe to a Graphical User Interface [GUI] on a workstation). Something as simple as getting the fields on the screen in the same order and relative position as they are on a familiar input form can greatly reduce training requirements, alleviate frustration, and enhance system acceptance.

- *Evolutionary prototyping* refers to the technique which builds a prototype and then adds functionality to evolve into the production system. This should not be attempted unless it is very carefully managed, the system is not mission-critical, and the technical environment supports the process very well. The latter would assume the availability of a comprehensive active repository for design information, such as would be found in an I-CASE environment. Some new object oriented development environments can successfully support this approach for small projects. Trying to do it with conventional technology is like trying to turn the architect's paper model into the real house - very difficult and usually unsuccessful.

Joint Application Development (JAD)

This is an approach first used in IBM Canada. It makes use of intensive, facilitated meetings with all players present to achieve rapid consensus. It aims to reduce the time required for a project by speeding up the data-gathering process, facilitating rapid decisions which carry commitment from participants, and improving the quality of communication between sponsor, users and other members of the team. The approach has considerable merit. It can be used in conjunction with various types of lifecycle. JAD activity is particularly high in the areas of strategy and planning (where it is referred to as Joint Requirements Planning [JRP]), gathering information, building models, and reaching concensus decisions (e.g., how best to proceed with implementation). Its typical usage profile in the lifecycle is shown in figure 8.11.

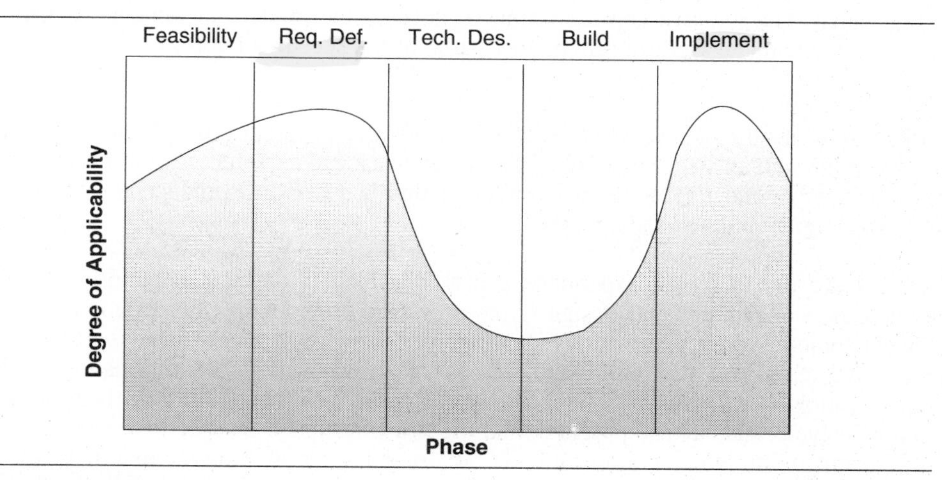

JAD in the SDLC *Figure 8.11*

The actual techniques employed will be discussed in more detail in the chapter on Human Communications which follows near the end of the book. For now, we will concentrate on the implications for project design. JAD should be used where possible to gather information rapidly, to shorten the project lifecycle, to enhance the quality of communication, models and specifications, and to obtain rapid decisions. To use it successfully, you will need senior management commitment, educated participants, and the necessary JAD facilitator and Scribe skills. These could be within the team but this is not optimal. The principle behind

JAD is that the facilitator and scribe have no vested interest in the outcome of the sessions, so that they can act impartially and extract maximum input from the participants. The facilitator and scribe should thus be from outside the team if possible. In some organizations, they may be drawn from a small pool of such specialists in a corporate planning area or development support group. Other organizations choose to use outside facilitators from consultancies. Although this can be expensive, the results obtained can often be superior, since the use of an outsider genuinely ensures impartiality, and focuses attention on the importance of the JAD sessions.

JAD used together with Rapid Prototyping and CASE has been described by Martin as Rapid Application Development (RAD). This is really just a synthesis of techniques, which we have discussed, by organizations with a sufficient level of skills, experience and technological infrastructure to use the combination successfully. Some spectacular results are quoted in terms of productivity achieved. While these are probably true, we suspect that there are just as many, if not more, spectacular failures of projects in organizations trying to employ the gamut of these techniques without the necessary management support, infrastructure, skills or experience.

Iterative Lifecycle

An interesting variation on the lifecycles presented thus far is the iterative approach proposed by Wong and others. This approach is based on the premise that software is built, not written. Thus it is unrealistic to expect to deliver a complete, complex working system in one "big bang". The iterative approach aims to develop the product incrementally, using feedback from the detail work to refine other parts, of the specification for example, as we proceed.

The lifecycle makes use of successive *builds*. In each build, a subset of the full target system is developed through all stages up to integration. Feedback and experience from this process is used to identify and tackle the next build. In this way the full product is eventually completed, integration tested and delivered.

Advantages are the increased experience and skill gained in successive builds, and the earlier detection of requirements/design errors or invalid assumptions. For example, if we have made some incorrect assumptions about the performance of a new database, this will be discovered during the first build, and we can take advantage of the information in all subsequent builds. If we followed a big bang approach, we would only discover the problem late in the lifecycle, and once all programs and file structures had been specified. Difficulties can be encountered in integration of different builds, and in partitioning the functionality in such a way that manageable chunks are obtained for each build. These have to have minimum interfaces to and reliance on other system components which will be delivered in later builds. If this is not so, excessive rework and bridging will be required to get each new build to work operationally. There can also be an undersirable increase in integration effort and testing if too many builds are specified. Approaches which we can adopt to partitioning the project will be discussed later in this chapter.

Phased Delivery

A variation of the iterative model is the evolutionary (or phased) delivery model described by Gilb, shown in figure 8.12. In this approach we again make use of successive builds, but

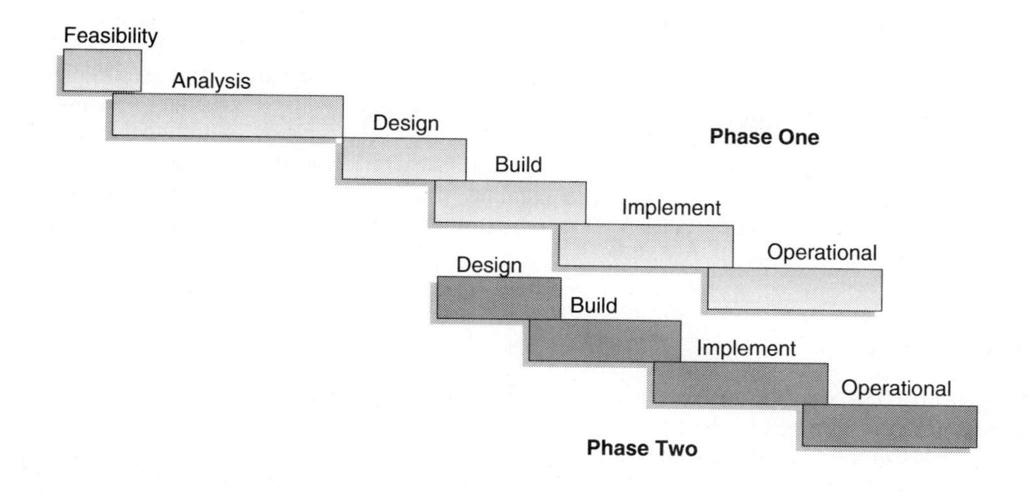

Phased Delivery *Figure 8.12*

each build is actually delivered as an operational subset of the final system. Essentially, a full functional specification for the complete system is prepared. This is then partitioned to provide useful discrete sets of functionality from a client perspective. These are then prioritized, with each subset becoming the focus of a build. The advantage is that useful functionality is made available to the user community much sooner. An example from a life assurance company will illustrate this. The target system is a new business processing system which will handle all application forms for 12 different types of policy. We might choose our builds as follows:

Build	Functions
1	Client details maintenance (add, change, delete, enquire, print)
2	All non medical applications (including endowments, annuities and provident funds)
3	Medical details processing and risk assessment Highest volume product with medical requirements (individual whole life)
4	Other medical required products
5	Combined plans and special packages

The choice of builds can be optimized in a business sense by making use of value to cost ratios. Basically, we determine the value of each subset of functionality, and estimate the cost to implement this. We then try, within the technical dependency constraints, to optimize the delivery sequence of builds so that maximum business benefit is derived as soon as possible for minimum expenditure. We may even decide that some components' yield is too low to justify their development and drop them. In practice, we normally find that the low priority builds are dropped in favor of more pressing needs at subsequent planning sessions. This may not please the purists who will argue that the initial system as architected will never be completed. We can counter this by saying: So what? It was an arbitrary package of functions which represented our best understanding of the requirements at that point. We now know what is more important! And so we can maximize the return on investment for the business, which is after all, our client.

Timebox

This approach is a novel and very useful one. The major criticism of most I.T. projects is that their delivery of results is unpredictable in terms of time and cost. In previous chapters we have highlighted the difficulties of achieving accurate estimates. The timebox approach guarantees a delivery date by fixing it and allowing the size of the product delivered to vary. It greatly simplifies planning by providing an essentially fixed timescale and lifecycle. Basically, we could decide to have a cycle lasting, say, four months. This could follow any of the lifecycles we deem suitable. The one difference is that the end date is absolutely fixed. If the requirements have to be reduced to meet the deadline, then so be it. At the final date we implement what is ready and quality assured. Anything which misses this date is carried over into the next cycle.

Advantages include predictability of the implementation date, control of costs, and reduced planning overhead.

The cycle can also be very easily adapted to include maintenance activity with no distinction from system development. Disadvantages include the risk of having something which requires integration only partially ready at the end date. This could delay implementation of a major set of functionality until the end of the next cycle. Careful partitioning is still required to try to get a manageable, but non trivial, chunk for each cycle. Project teams must be motivated or else the escape valve of dropping functionality will be used too liberally.

Spiral

Boehm has proposed a spiral lifecycle, which is really successive waterfall cycles (like the iterative approach), with a risk assessment between each build.

The Simulation Approach

We would now like to propose a new approach, which draws upon the good features of several of those previously presented, and which takes advantage of emerging software technology, particularly Object Orientation (OO). Object oriented techniques and tools are gaining widespread acceptance recently in the creation of complex software, e.g., Windows, the new Apple operating system (Pink) etc. Object Oriented Technologies have their roots in the early simulation language *Simula 67*. This was the first language to provide a specific syntax to represent real-world entities (or objects) which would have characteristics (e.g., size, color, location) and behaviors (e.g., ability to turn on or off; report their status; change color; and so on). It also introduced the important notion of *classes* which allowed the designer of a simulation to describe all similar objects just once. For example, once we describe the class automobile as having the attributes of make, model, year, number of seats and fuel consumption, and the behaviors accelerate, decelerate and turn, then any object identified as an automobile would *inherit* these characteristics and behaviors. Simula and LISP gave rise to Smalltalk, which in turn gave rise to many other object oriented languages. Like the structured programming concepts of the '70s, the OO approach has steadily expanded to include design, analysis, and now business modeling.

It is significant because it allows a much richer modeling of the real world in our architectures, analysis and design models, and ultimately in our systems. We are beginning to have the tools to create systems as *increasingly accurate simulations of the real world* - see figure 8.13. Another development is the emergence of commercially available *class libraries,* which embody the knowledge about a particular domain (for example, user interface model or application area [e.g. retail]). These facilitate massive re-use of pre-existing software components developed by experts in their own particular field. Re-use is the best way to obtain high quality and productivity simultaneously. These factors have far-reaching implications for the way that we build systems.

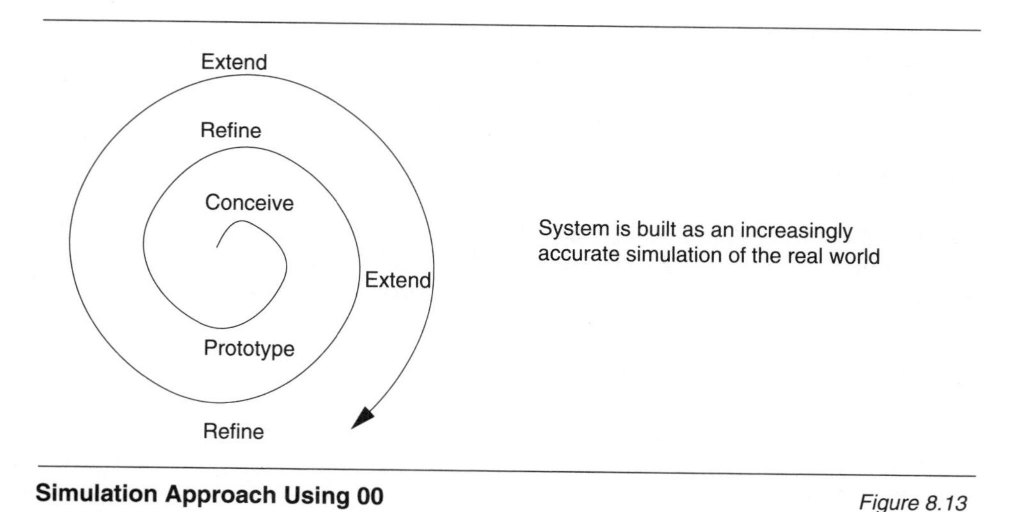

Simulation Approach Using 00 *Figure 8.13*

We can now build the core of an application by plugging together off-the-shelf components (from class libraries) and adding only a minimum amount of unique code. As we use the generated system, we will find that it does not exactly map to the way that we do things. Fortunately, the knowledge behind an object oriented system is explicitly spelled out in its classes and their methods (behaviors). We also have convenient ways to override and specialize attributes and behavior for certain cases and situations without disturbing the original knowledge. We can thus progressively refine the model of the real world which the system represents, until it is no longer cost-justifiable to do so.

The approach fits well with either the evolutionary delivery or timebox approaches mentioned previously.

Till now we have only discussed systems development projects, but, as we know, these are not the only kinds of I.T. projects that we will encounter. In the following paragraphs we will look at some of the others.

System Maintenance

Maintenance consumes between 60 and 80 percent of the software effort in most installations. Unfortunately, it is the unglamorous cousin and seldom sees the limelight. Any I.T. manager who reduces the effort in this area can realize major savings, and free resources for more productive work. Corrective maintenance can be virtually eliminated by high-quality system development, but that will still leave us with the other two categories: Adaptive maintenance which involves adapting existing software to new circumstances, requirements or technology, and perfective maintenance which refines and adds value to software which already meets the original specification.

The best way to view maintenance is as a series of rapid, high-powered, full lifecycle development projects. If you think about it, all the phases are there: Requirements, analysis, design, build/alter, test, integrate and make operational. They have to be done rapidly, without mistakes, and taking into account the already complex, functioning system. We should be using some of our best people for this. Don't hide them away - make it a vital area with a commensurate reward structure and visibility. Watch the turnaround. Incidentally, the timebox lifecycle is ideally suited to implementing a regular, rapid maintenance cycle. People in this area could experience the full SDLC in as little as a month, making it a great place through which to cycle trainees. Just make sure to place them under the control of a competent mentor - remember it is vitally important work.

It is gratifying to discover that virtually all the types of project that we can be called upon to manage can be handled by the generic lifecycle. Furthermore, they can be seen in terms of consistent phases, review points and baselines, as indicated in figure 8.14.

Technology Implementation

We will frequently be called upon to implement something other than a custom-written software system. This could be installing a new mainframe or communications network, or the implementation of a new software product, such as a database or development environment. These projects are different from development projects. Usually the requirements are much clearer. The hard stuff comes in the logistics, and the integration testing to make sure the new technology really works in the real situation. Here you will rely much more on your management skills than on your technical development skills. A sample lifecycle for technology implementation is shown in figure 8.14. You will notice that it still fits within our normal project lifecycle structure.

Package Implementation

It frequently makes sense to buy rather than build. Unfortunately, packages seldom fit "as is", particularly in operational areas of the business. This means that a package implementa-

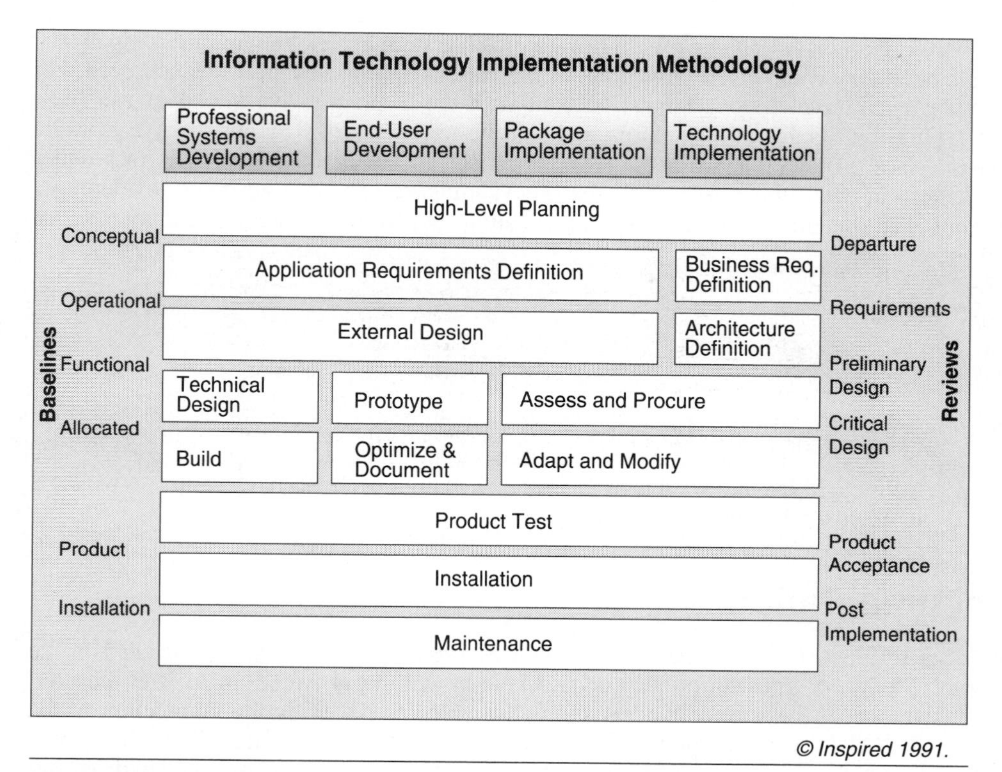

Alternative Lifecycles

Figure 8.14

tion project is a combination of a technology implementation project (putting in the thing you buy) and a system development project (specifying, designing, making and implementing the changes).

Consistent Management Approach

Referring to the Alternative Lifecycles diagram in figure 8.14, you will note that it is possible to manage a variety of project types using a standard management framework. This framework is consistent with the configuration management discipline mentioned in introductory chapters. We will encounter it again in chapter 14.

Methods, Techniques and Standards

It is vital before you proceed with the project to have clearly identified the methods, techniques and tools that you propose to use. These will include the methodology which you will follow for technical tasks (e.g., your SDLC, sequence of tasks, Quality Assurance approach and review process), the specific techniques that you will employ (e.g., entity modeling, functional modeling and prototyping) and the tools that you intend using (e.g., languages, compilers, utilities, and management software).

It is very important to achieve a high degree of integration across the various components. This is necessary if the benefits inherent in the various approaches are to be achieved. For example: our chosen lifecycle must fit well with the techniques we intend to use (e.g., prototyping); our development approach should be well supported by our tools (e.g., to do user interface prototyping, we need a screen painter and dialogue driver). The techniques should also be well integrated across the lifecycle phases. There is no point developing a rigorous and detailed requirements specification which the designers cannot use or understand. The effort involved in integrating all of these components is massive and should not, in general, be attempted by individual system development groups. Better to choose one of the well-developed, documented and supported proprietary or academic methodologies (e.g., Information Engineering, Method/1, Tetrarch, SSADM, Merise, Gane & Sarson, DeMarco, Booch) and adopt this with any local modifications necessary.

These issues should be standardized as far as possible within the organization. This will facilitate building skills, communication between projects, and collection of software engineering data. It also allows the progressive adaptation of the techniques in use which is vital to increased quality and productivity.

Documentation

We need to identify what documentation will be produced. This must be included in the Product Model. Suggested project documentation will be covered in a later chapter. Documentation of the various models and deliverables produced during completion of the various methodology tasks should be covered by your chosen methodology. In any case, you should ensure that you are clear how your work will be recorded and collated as you proceed.

Project Resource Requirements

When you have fully identified all the products which your project will produce (both technical and management) you can then identify the associated activities to produce them. These in turn will dictate what kind of skills and resources you will need to complete the project.

The product model and the work breakdown structure can be used to estimate the volume of work required. This should give an initial quantum of feasibility and resource requirements. The next step is to determine the skills required, and to ensure that these are available to you. They may be present in the team, elsewhere in the organization, or obtainable from outside contractors, vendors or consultants. Other resources required, for example a development machine to run your tools and utilities, must be identified and their availability established.

All resources required on the project will have associated time dependencies. We will not need the programmers at the outset, for example. Database design auditors will only be required at a few key points, and so on. For each resource (including people) we need to identify when they are required, to what extent (e.g., machine size, number of hours per day) and for how long they are required. A Gantt Chart (discussed in the next chapter) is a good medium for collecting this information. From this information we can construct a resource histogram as shown in figure 8.15. This will allow us to predict costs and cash flows.

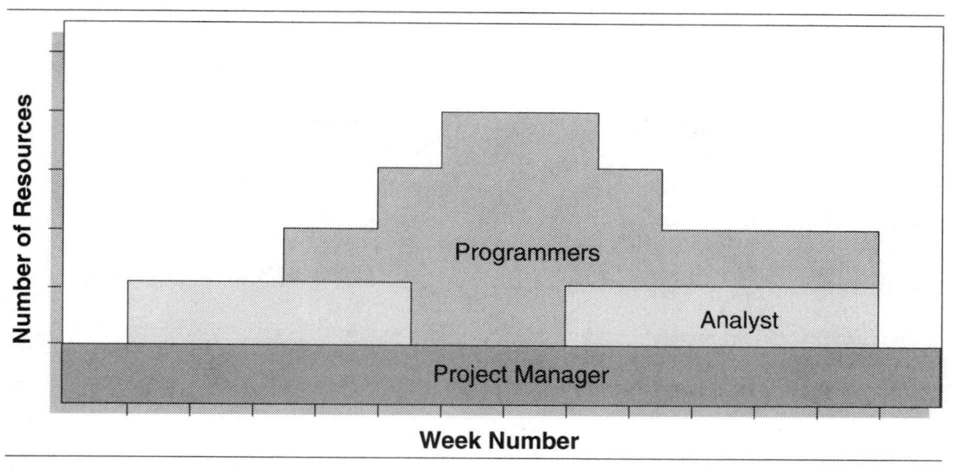

Resource Histogram *Figure 8.15*

A bit of bureaucracy here can save us a great deal of trouble later. We can use Resource Contracts to record the commitment of various sources to supply resources to us. A suggested format is shown in figure 8.16.

You may find that the promise "I will be available full time" is suddenly modified to "two hours per day excluding Mondays" when you ask for it in writing. Do not forget to include user personnel, management and your sponsor in your resource planning. In our experience as consultants, we have often been promised full participation from sponsors, only to find that they are always "too busy" to obtain decisions or resolve problems later on. In some cases we have even had to stop work on the project to get their attention and renewed commitment. Having the resource contracts up front lets people know that you are serious, and backs you up later if there are any difficulties.

Project Team Structure

Having obtained commitments for your resources, we can now structure the team. How we do this may be dictated by our organization. Where we have a choice, we should be aware that different structures will have a major effect on morale, productivity and the quality of the final product. Your job is to ensure that everyone works together to achieve the common goal - completion of a high-quality product on time and within budget. Since the '70s various project structures have been proposed. The pro's and con's of each will depend on the organizational culture, the leadership style and the individual's values and ability. We will concentrate mainly on the structures available.

Chief Programmer Team

The CPT concept is perhaps best known from the project that produced the IBM 360 operating system. As shown in figure 8.17, the chief programmer (a technical wizard) works with the project leader (the administrative wizard). The backup programmer is almost as

Resource Input Contract

Project_ID ______________ Project Name ______________

Resource Type ______________

Provider ______________ Date ______________

Name/ID	From	Till	%	Rate/Skill/Remarks

Resource Provider Signature ______________

Resource Input Contract *Figure 8.16*

technically competent as the chief programmer and can stand in at any time. The librarian function looks after program development and documentation. Thus the chief programmer is relieved of all administrative duties and can focus on the technical aspects of the project. This approach was found to be suitable where highly complex systems are being developed under extreme time pressure (sound familiar?). However, the approach is built around an individual, which makes the project vulnerable. This is reduced by the structure incorporating the backup programmer.

Egoless (Democratic) Programming Team

This is the other extreme, where the programmers share the decision-making and all report to the project manager. There is no management hierarchy. The approach can be pleasant, but difficult to manage. Difficulties can also be encountered in achieving a clean architecture for the full product - we may end up with a *camel.*

The next structures are similar, but require different management skills. The classical team structure is shown in figure 8.18.

In this structure, the project manager does the planning and design work and passes this to small programming teams, each led by a senior programmer. This structure complements modular design and has formal communication channels. The design work is focused on the project manager who may or may not have the necessary skills to do design work.

A modification to the classical structure can overcome these problems. As shown in figure

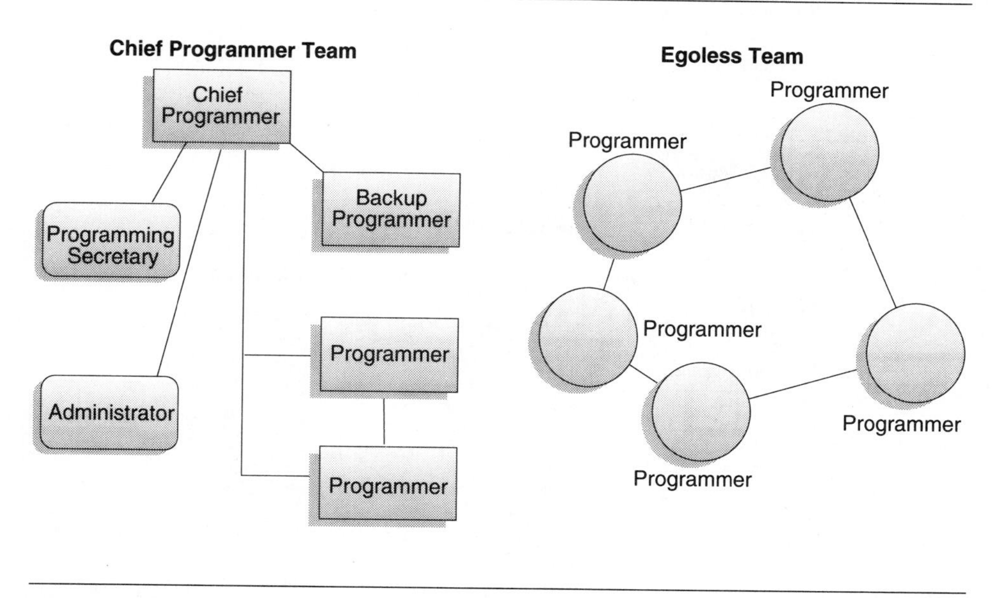

Structure of Team *Figure 8.17*

8.18, the senior programmers are replaced by analysts who manage small teams of programmers. The project manager looks after the managerial aspects while the analysts work with him as a team to produce the technical design. This structure works well, especially when formal development standards exist.

Participative Management

This is a concept whereby those who will carry out the work have a say in how it is assigned, estimated, and carried out. It requires careful management, but can lead to much higher commitment and motivation from project staff. If a manager consistently hands out work with unrealistic deadlines, sensible team members will soon ignore them. If team members discuss the specification with the manager and agree on a realistic timetable, they will have high personal commitment to meeting the deadline, since they have effectively promised this to their manager (and themselves). We need to balance this need for involvement with the need to use people with the right skills, and to build those skills through repetition (as we discussed in the preceding chapter on estimating). We may want a separate estimating group to remove subjectivity, but also want our people involved in the process for commitment. The solution is to put them together. Using the figures, techniques and experience of the estimating group, our people can participate in defining the parameters for the estimate, and can debate the reasonableness of the estimates produced. Often this discussion will add detail and knowledge not originally available to the estimating group, improving the estimating. Our staff will have had their say and been involved, and thus carry higher commitment. The estimates are likely to be very realistic and this further enhances their credibility and the team's commitment to meet them.

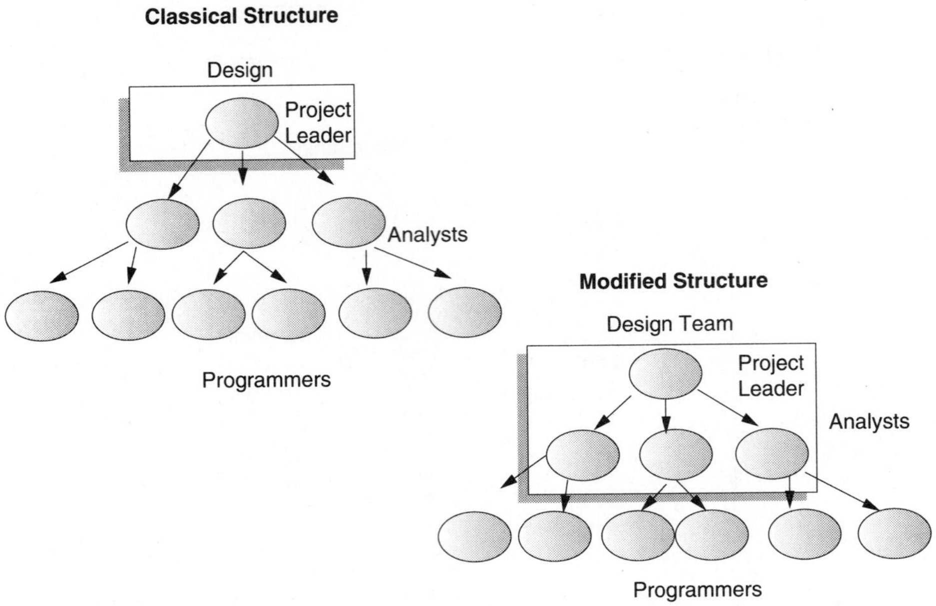

Classical Structure

Design

Project Leader

Analysts

Programmers

Modified Structure

Design Team

Project Leader

Analysts

Programmers

Classical Structures *Figure 8.18*

Multidisciplinary Teams

Increasingly, systems projects go to the heart of an organization's business. Often they are strategic in nature. This necessitates having highly skilled application area specialists (e.g., retailers, manufacturing specialists) on the team. We also have to deal with a plethora of information technologies (Comms, Database, Languages, File Systems, Operating Systems, CASE, etc.) and need to draw on experts in these areas. These factors mean that we will often be managing multidisciplinary teams. Research shows that multidisciplinary teams perform well on tasks which are nonroutine and provide significant challenges. Fortunately systems projects tend to be like this. A danger, however, is that we will not be able to manage all the individuals on the team in the same way. As we will see in a later chapter on managing people, the things that motivate I.S. personnel are not those that motivate the majority of the population. We will often need to adapt our management style to cope with this.

Selection of Team

As we saw in the discussion on estimating, the productivity of individuals varies dramatically. Whom you select to be on your team can make an enormous difference. Here are some guidelines:

- *Keep it small.* Try to find the skills you need in as few individuals as possible. This will drastically reduce communication overhead and difficulty

- *But not too small.* Do not put all your eggs in one basket. Try to make sure that your small team has backup of each required skill in another individual, and that you could survive the loss of key players. Effect of losses can be greatly reduced by having good standards and a strong commitment to producing necessary documentation. A team which communicates openly will also share their knowledge, thus reducing exposure. Of course the best thing to do is keep team members highly motivated so that they don't leave in the first place

- *Keep it high powered.* Galileo reportedly said "Give me a place to stand and I will move the earth." The project manager's refrain should be "Give me the right people and I will move mountains." Strong people will overcome the difficulties you may encounter. Weak ones, unassisted, will not

- *Think about the future.* Maybe you have to include some junior people who will not be as productive as your stars. But long term, you need to help them grow to become stars, so give them a chance. Assign a star to mentor them

- *Look for challenges.* Try to find a significant challenge for each person on the project, one that will stretch them beyond their current abilities, but one that they can master

- *Keep the user on board.* Make sure that there is strong and senior user involvement in the team

- *Look at personalities as well as skills.* How are the members going to fit together? Where will the likely points of friction be?

- *Have very clear responsibilities.* Goal and authority conflicts among ambitious, highly charged individuals can be a major source of friction

Relationship to Organization

This is another important element of project design. If we are to succeed in a useful goal from the organization's perspective, then we need to understand what that perspective is, both at the outset, and as the project proceeds.

It is also vital to stay in tune with the organization to ensure that our project gets the right resources and priorities, and that our staff do not end up in a political backwater.

Sponsor

The nominal sponsor of the project is of vital importance. This is the person who will champion (or kill) your cause at the highest levels in the organization. This person should be as senior as possible, commensurate with actually getting involved with the project. He should promote the value of the results you produce to the organization. He should have the necessary *clout* to get decisions made and resources committed. He should also be around for the duration. Having a sponsor change in mid-project can be very threatening, especially if the newcomer is only luke-warm to the idea of the project. This is not much of a danger with operational-type systems, but it is a major one with personality-driven projects like implementation of a DSS. Try also to stay in touch with your sponsor's likely understudy, in case he moves on.

I witnessed the anguish of a software development group who had spent some 18 months of intensive, committed, productive effort producing a relational database management system, and doing it well, only to have their project canceled summarily. They had their heads so buried in the technical challenges, that no one had noticed the acquisition of a subsidiary company which already had a relational database product on the market. The subsidiary's product was evaluated and adopted immediately to gain market share as quickly as possible. The organization did not want to have a fractured strategy supporting two competing products, so they axed their own inhouse project. This illustrates the danger of becoming too introverted when managing projects. The project manager should have seen the risk they were running and taken appropriate action to get his team involved in the evaluation and decision making.

GM

The Steering Group

This is normally a committee of I.T. and line managers responsible for realizing benefits from I.T. expenditure across the organization. You need to identify this body and its players, know who will support you, and who will not. You may be surprised at the divisional feuds that go on in corporations.

You should also look for senior line managers who will champion your cause.

Reporting Structures

We need to understand what the reporting structures for the project will be. Frequently we will need to report to I.T. management, the Sponsor, the Steering Group as well as possibly User Management, Quality Assurance, Internal Audit and other groups. Each of these may require different information at different intervals. Find out what these are. It's a bit of a chore, but it makes you a good corporate citizen. As a strategy, we should look to use one base of documentation and reports for the project, and merely collate them and alter the level of detail represented for various audiences. This will greatly reduce our effort in keeping the various groups happy.

Quality Assurance

As project manager, you are responsible for what your team produces. Ultimately the organization looks to you to produce the goods - that's why you were chosen. There are many aspects to achieving quality. We will expand on these in a later chapter. As part of our project design, we should ensure the following:

- That there is a proper Product Model

- That there is a proper Work Breakdown Model
- That all resources (and required skills), internal and external to the team, have been properly identified
- That estimates are realistic
- That a good methodology and tools have been chosen
- That we know who our sponsor is, and what the reporting structure for our project is
- That we know how to perform all tasks on the WBM (or will do by the time we reach them)
- That we know what standards we are expected to adhere to in producing the requisite deliverables
- That the format and structure of all deliverables is known and agreed
- That our project plan includes appropriate review points and recovery time to remedy deviations
- That we have arranged for all external resources (including reviewers) as necessary
- That resources have been committed in terms of contracts
- That all necessary training, leave and contingencies have been incorporated in the plan
- That we have done a risk assessment, that this is acceptable to management and that our project structure has been chosen to cope with the risk profile
- Clear responsibilities have been assigned for all work, at least for the next phase
- Activities and deliverables have been incorporated for any necessary data gathering

Project Partitioning

Most projects that we tackle nowadays are really too large to be single projects. There is a theory that says that beyond a certain point projects become "undoable". There is a serious danger that what should be a temporary structure turns into a department, with a life of its own. Things that should be provided by the environment become internalized (secretarial services, change control, librarian). When people start to see their career in the project, not the organization, then you are in trouble.

But some really large projects do succeed. Witness the creation of the space shuttle launch software or the creation of a major operating system. Why do these succeed? Because we know they are massive and difficult, we are very careful, and we apply top skills. Really small projects also succeed; it seems to be those in the middle that run into trouble (see figure 8.19). This is because we do not take them seriously enough.

We need to adopt some of the formality and skills from the big projects to the medium ones. We also need to break them down into smaller, more manageable chunks which are more commensurate with our skills and reasonable timescales. There are a number of ways to break a large job into several smaller ones:

- *Functional Partitioning* is where we develop a complete functional requirement, and then identify useful subsets to implement as phases, or possibly parallel projects. This is best done using a structure chart, rather than a dataflow model. We can enhance the approach by applying value/cost analysis as discussed earlier

- *Resource Partitioning* is an approach where we divide up the work according to the specialties of different groups of resources, or individuals. For example, we may contract the file access modules to the database programming team, the application modules to the retail team, and the communications interface to another team. The approach requires a strong methodology with good structured techniques and careful management

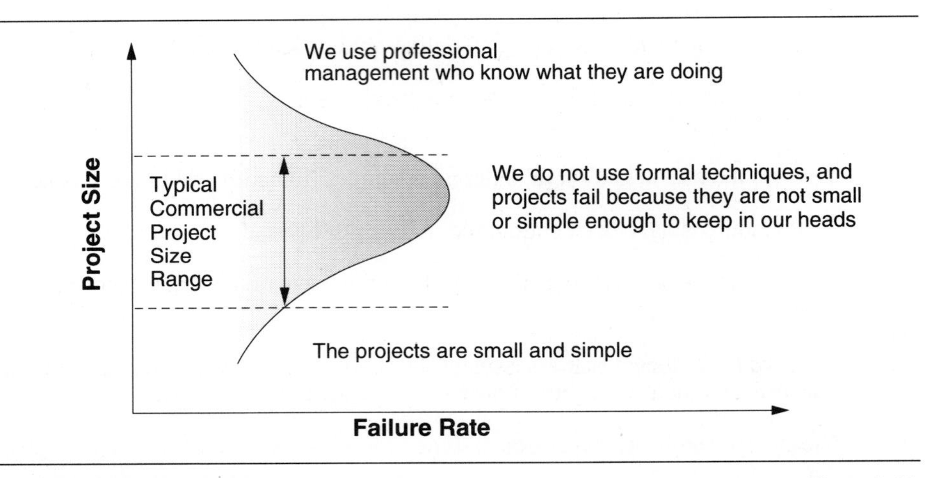

Project Failure versus Size *Figure 8.19*

- *Temporal Partitioning* is where we divide the work into time-related components. For example, we may do all the daily processing as a subset, followed by monthly processing, then year-end processing. This is one way of reducing the interfaces between subsets. It can, with forethought, also provide a nice way of handling installation in a phased way

- *Spiral Approach.* Boehm's spiral model allows us to partition by functionality or time, and then to deliver results in an incremental way

- *Data-driven Approach.* Here, instead of using the functional model, we use a data model (Entity Model or Normalized Data Model). On this we try to find cohesive subsets which have high internal relatedness (cohesion) and low external relatedness (coupling). These are referred to as *subject databases* in Information Engineering parlance. The idea is that the underlying data structure changes much more slowly than the required functionality and thus forms a far better basis for partitioning. From a technical perspective this makes sense and the approach will greatly simplify interfacing and bridging requirements. Unfortunately, it usually yields a development sequence which is the reverse of the business priorities. Some adjustment is usually necessary before you can sell it, unless you have particularly farsighted, patient and wealthy management

- *Simulation Approach.* Our simulation approach provides a unique scenario. What we are doing is a kind of functional partitioning, but in a subtle way. We are really implementing the core of the application first in an unsophisticated way to handle the normal case. Then we will gradually refine this over time, adding more detail and exception handling. The system will grow organically, specializing to handle the full complexity of the real-world situation (or as much as we can cost-justify). Careful thought should be given to reaching the right abstractions to describe the common attributes of the situation. This is the focus of Object Oriented Analysis and Design. Using Object Oriented Technologies, rework is minimal

  Unfortunately, it will be some time before most commercial organizations have the necessary skills and infrastructure to take advantage of this approach

We can obviously use combinations and permutations of the various approaches. Here are some guidelines to help you with the objectives we are trying to meet through partitioning:

- *Limit duration of each build or chunk to a maximum of nine months.* The longer the project, the more likely that the requirements will have changed before implementation. We have found in practice that a reasonable duration is a maximum of nine months. Beyond this horizon, environmental and resource changes will start to have a significant negative impact on the project

- *Limit the size to something a small team (< 10 people) can do*

- *Keep the interfaces clean.* This is the biggest source of problems. Try to partition so that minimal interaction is required between groups proceeding in parallel and between phases. This will greatly reduce integration effort and maintenance

- *Manage the boundaries.* Explicitly define the interfaces to your chunk, both to the system, and to the project. Keep these definitions stable. Subject any suggested change to careful scrutiny and formal change-control procedures. If any boundaries have to change, make sure you communicate

- *Communicate with related teams.* Make sure the communications lines are open and working. Talk to verify that things are still the same, rather than not tell each other that something has changed

- *Deliver something* before your sponsor and users get too nervous. *Make sure it works*

Risk Control

There is no doubt that developing software is a risky business, yet very few organizations perform any formal risk assessment to see if the risk can be controlled, reduced or managed. As project manager, it is your responsibility to ensure that your project, the organization sponsoring it, and the user community are exposed to minimum risk.

Risk is the chance of something going wrong. Since our definition of project success included delivering to specification, on time and within budget, we incur three major types

I have seen projects in a major financial institution where I was doing external quality reviews grow and overlap in scope until there was a 60 percent overlap in functionality (figure 8.20). This had gone undetected by either team or the organization. Only the users were somewhat confused as to why two lots of "I.S. people" kept asking them about the same things. My reviews of the two projects, about a month apart and at similar phases, highlighted the overlap. Fortunately, a consistent methodology was in place and the models could be easily compared. Also fortunately, they were taking quality reviews seriously. If this had not been the case, the organization could have spent some $300 000 on developing two different, noninterfaced versions of the same software.

GM

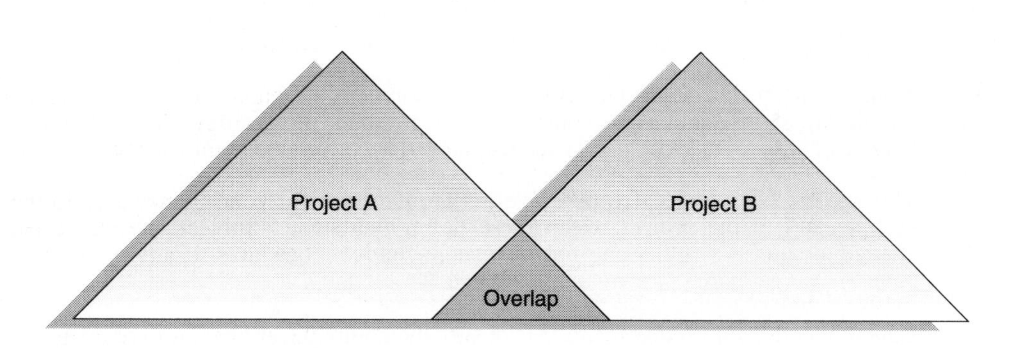

Overlapping Functionality *Figure 8.20*

of risk. These are that we will not meet the specification (quality), that we will be late, or that we will spend more than intended.

The sources of risk for our project also come in three major varieties. In a system of assessing risk developed by McFarlan and the Dayton Tire Co., they identified the factors as People, Structure and Technology. Let's look at each in turn.

People risk comes from inadequate skills (both technical and managerial) as well as inexperience. Inexperience could be of a general nature, or related to the specific application area or technology.

Structure risk comes from the degree of change which will be introduced into the user areas and business procedures, the number of discrete user communities the system must satisfy, and the number of other systems the new one must interact with. It is also affected by the experience of the organization with technology, and prior project history.

Technology risk comes from using new or untried technology. It is affected by the stability of products and suppliers used. Using one new technology on a project (e.g., a new compiler) is acceptable, but we are looking for trouble if we use a new operating system, database, compiler, code generator, CASE tool and communications monitor together.

McFarlan developed a questionnaire to assess the various risk factors. This is shown starting on page 143.

To determine the risk score proceed as follows:

Evaluate all questions on the questionnaire. Where there are multiple factors which affect risk (e.g., "What hardware is new?") you can check more than one answer. Take the score associated with your answer (or sum of multiple answers) and multiply this by the weight of the question. Add these up for each section (Size, Technology, Structure). This will give you three risk scores. These should be compared to the norms for the organization (which you will of course accumulate!) and any high scores should sound alarm bells. If the risk in a particular area is high, then the questions related to this section should be reviewed to determine the causes. Perhaps the project can be restructured to use less new technology, to reduce the size of the project, or the number of user areas involved? We should use this information for project design, not just a risk ranking. The questionnaire provided is illustrative only: you may want to amend it for your organization, or for different types of projects e.g., package implementation.

Risk should be assessed at the beginning of the project and then at the review of each phase as shown in figure 8.21. It should display a decreasing exposure as the uncertainties in the project are resolved and assumptions are confirmed (or adjustments made). Any increase in the risk score should be cause for alarm and remedial action.

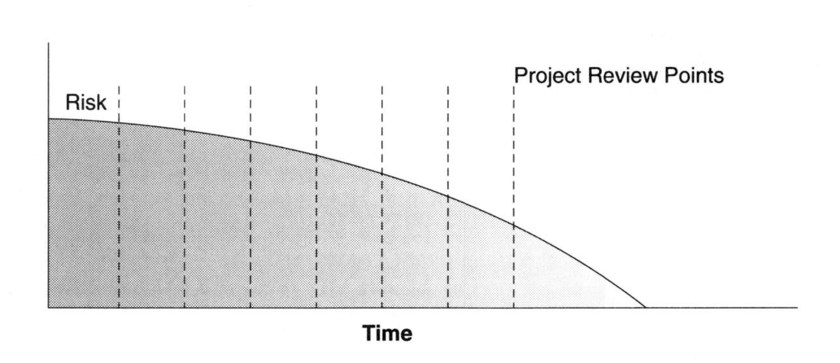

Risk Assessment *Figure 8.21*

Using the system portfolio approach to projects, it is acceptable to have some high-risk, high-return projects. These are the ones with which you attempt to gain competitive advantage, in the full knowledge that they are risky and may fail. They should represent an order of investment that the organization can afford to write off if unsuccessful. If you find that your whole portfolio is tending to be high risk, then serious re-examination is required, possibly leading to major strategy changes.

Where high-risk projects are tackled, we should be aware of this before we commence, so that we can apply the necessary rigor and formality to their management. In this way, the risk can be substantially reduced. Some ways to reduce risk include:

- Breaking larger projects up into smaller components
- Minimizing the number of interfaces and dependencies between projects
- Using high-level skills, and fewer project team members, without running the risk of relying exclusively on a single team member
- Not attempting too many new things on major projects
- Getting outside assistance the first time we attempt something, until we have built the internal skills
- Decreasing the Manageable Unit of Work which will result in identifying more tasks in the project plan, and closer supervision of individual activities
- Ensuring there is a strong, user-oriented steering body in place to supervise the overall project direction
- Ensuring high levels of user and management commitment and involvement in the project.

Summary

Project design, like other design activities, is a creative, experience-based activity. It is an outlet for the creative skills which you exercised earlier in your career in the application analysis and design areas. It is a crucial activity, and can have a major impact on the overall success of the project. Some organizations have set up central development support groups, with highly competent project managers seconded to them, to assist new project managers in devising good project plans and structures.

McFarlan Risk Questionnaire (Adapted)

Size | Weight

1 Total systems and programming mandays for system: (5)
- ( ) 12 to 375 — Low=1
- ( ) 375 to 1875 — Med=2
- ( ) 1875 to 3750 — Med=3
- ( ) Over 3750 — High=4

2 What is the project estimate in calendar (elapsed) time? (4)
- ( ) 12 Months or less — Low=1
- ( ) 13 Months to 24 Months — Med=2
- ( ) Over 24 Months — High=3

3 Number of projects within the system? (1)
- ( ) One — Low=1
- ( ) Two — Med=2
- ( ) Three or more — High=3

4 Most projects in the organization fall in the following range for systems and programming mandays: (1)
- ( ) Small - up to 250 Mandays — Low=1
- ( ) Medium - 251 to 750 Mandays — Med=2
- ( ) Large - over 750 Mandays — High=3

5 Average calendar (elapsed) time per project? (1)
- ( ) Less than 6 months — Low=1
- ( ) 6 to 12 months — Med=2
- ( ) 12 months or more — High=3

6 Length of economic payback: (2)
- ( ) Less than 12 months — Low=1
- ( ) 12 to 24 months — Med=2
- ( ) Over 24 months — High=3

7 Who will perform the work: (2)
- ( ) Mostly in-house personnel — Low=1
- ( ) Significant portions by in-house staff — Med=2
- ( ) Mostly contract/off-site personnel — High=3

8 Number of departments (excluding I.T.) involved (4)
- ( ) One — Low=1
- ( ) Two — Med=2
- ( ) Three or more — High=3

9 Approximate number of user department personnel required to run and operate the system? (1)

( )	Up to 20	Low=1
( )	20 to 50	Med=2
( )	Over 50	High=3

10 How many geographic locations will the system address? (2)

( )	One	Low=1
( )	Two or three	Med=2
( )	More than three	High=3

11 How many existing information systems must the new one interface with? (3)

( )	None or one	Low=1
( )	Two	Med=2
( )	More than two	High=3

Structure

1 The system is best described as: (1)

( )	Totally new	High=3
( )	Replacement of existing manual system	Med=2
( )	Replacement of automated system	Low=1

2 What percentage of existing functions are being replaced on a one-to-one basis? (5)

( )	0 - 25 %	High=3
( )	25-50 %	Med=2
( )	50-100 %	Low=1

3 What is the severity of procedural changes in the user area as a result of introduction of the system? (5)

( )	Low	Low=1
( )	Medium	Med=2
( )	High	High=3

4 Proposed methods and/or procedures: (2)

( )	First of a kind for I.T.	High=3
( )	First of a kind for User	High=3
( )	Breakthrough required for user acceptance	High=3
( )	Breakthrough required for I.T. implementation	High=3
( )	Routine/none of the above	Low=0

5 Does the user organization need to change to meet the requirements of the new system? (5)

( )	No	Low=0
( )	Minimal	Low=1
( )	Somewhat	Med=2
( )	Major	High=3

6 Is new/unfamiliar user-related hardware required? (1)

( )	None	Low=0
( )	Hardware user can easily adapt to	Low=1
( )	Hardware requiring extensive user education	Med=2

7 What degree of flexibility and judgment can be exercised by the systems designers in the area of systems outputs? (1)

( )	Very little	Low=1
( )	Average	Med=2
( )	High	High=3

8 What degree of flexibility and judgement can be exercised by the systems designers in the area of systems processing? (1)

( )	Very little	Low=1
( )	Average	Med=1
( )	High	High=3

9 What degree of flexibility and judgment can be exercised by the systems designers in the area of database content? (1)

( )	Very little	Low=1
( )	Average	Med=1
( )	High	High=3

10 What is the overall rating of predetermined structure for the new system? (2)

( )	Highly structured, requires little or no user procedure change	Low=1
( )	Medium structured, some user change	Med=2
( )	Low structure, high degree of user change required	High=3

11 Is this project highly or totally dependent on another project? (5)

( )	No	Low=0
( )	Yes, other project(s) low risk	Low=1
( )	Yes, other project(s) high risk	High=3

12 How many estimating questions were unanswered or had low confidence answers? (3)

- ( ) None — Low=0
- ( ) 1 - 10 — Low=1
- ( ) 11 - 20 — Med=2
- ( ) More than 20 — High=3

13 What is the general attitude of the user? (5)

- ( ) Poor - anti I.S. — High=3
- ( ) Fair - unsure or reluctant — Med=3
- ( ) Positive - good relationship — Low=0

14 How committed is senior management to the system? (5)

- ( ) Somewhat reluctant/unknown — High=3
- ( ) Adequate — Med=2
- ( ) Extremely enthusiastic — Low=1

15 Has a joint I.T./User team been established? (5)

- ( ) No — High=3
- ( ) Part-time user involvement — Med=2
- ( ) Yes, full-time user involvement — Low=0

Technology

1 Is additional hardware required? (1)

- ( ) None — Low=0
- ( ) Central processor type change — Low=1
- ( ) Peripheral/Storage device changes — Low=1
- ( ) Terminals — Med=2
- ( ) Change of platform, e.g., Mini/PC replacing Mainframe — High=3

2 Which hardware is new to your organization? (3)

- ( ) None — Low=0
- ( ) Central processor type — Low=1
- ( ) Peripherals or Storage devices — Med=2
- ( ) Terminals — Med=2
- ( ) Mini or microcomputers — High=3

3 Is special nonstandard hardware required? (5)

- ( ) None — Low=0
- ( ) Yes — High=3

4 Is system hardware the first of a kind for the vendor? (3)

- ( ) No — Low=0
- ( ) Yes — Med=2

5	How many vendors are involved in supplying the hardware?		(2)
	( ) One	Low=0	
	( ) Two	Low=1	
	( ) More than two	High=3	
6	Is system networked (online to multiple locations)?		(1)
	( ) No	Low=0	
	( ) Shared system in one location	Low=1	
	( ) Two or more locations	Med=2	
7	Is the system success dependent upon new hardware?		(3)
	( ) No	Low=0	
	( ) Somewhat	Low=1	
	( ) Very heavily	High=3	
8	The system is		(1)
	( ) Batch Database	Low=1	
	( ) Online data communications	Med=2	
9	Programming language used?		
	( ) COBOL	Low=1	(1)
	( ) Other 3GL	Med=2	
	( ) Well-established 4GL	Med=2	
	( ) Assembler/other 4GL/Logic Programming	High=3	
10	Is the system software (other than the OS) new to the I.S. team?		(5)
	( ) No	Low=0	
	( ) Programming language	High=3	
	( ) Database Management System	High=3	
	( ) Data Communications Software	High=3	
	( ) Other - specify	High=3	
11	Is the system software new to the vendor?		(3)
	( ) Yes	Med=2	
	( ) No	Low=0	
12	Are program packages being used?		(1)
	( ) No	High=3	
	( ) Yes, to a small degree	Med=2	
	( ) Yes, to a large degree	Low=1	
13	How good is vendor/supplier support of the program package?		(1)
	( ) Unknown	High=3	
	( ) Adequate	Low=1	
	( ) Good	Low=1	
	( ) Not applicable	Low=0	

14 What is the system complexity? (3)

( )	Straightforward	Low=1
( )	Average	Med=2
( )	Complex with many interactions	High=3

15 How knowledgeable is the user in I.T.? (5)

( )	First exposure	High=3
( )	Previous exposure, but limited knowledge	Med=2
( )	High degree of capability	Low=1

16 How knowledgeable is the user representative in the proposed application area? (5)

( )	Limited	High=3
( )	Understands concepts but no experience	Med=2
( )	Has been involved in prior implementations	Low=1

17 How knowledgeable is the I.S. team in the proposed application area? (5)

( )	Limited	High=3
( )	Understands concepts but no experience	Med=2
( )	Has implemented similar systems before	Low=1

Case Questions

MyWay Organizer

Q8.1

Consider what type of systems lifecycle would be appropriate for the MyWay Organizer project. Be prepared to justify your answer. (10 mins)

Q8.2

What kind of team structure would you like to use on the project? Why do you think this is the best way to organize the team? (10 mins)

Q8.3

Prepare a risk assessment for the organizer project. (20 mins)

Handover Trust

Q8.4

You have been asked to determine the risk profile for the New Business project. You have interviewed users and managers and determined the following:

- The system will be used by a variety of departments including: Quotations, New Business, Underwriting, Medical Assessments, Branch Sales Support. There will be about 2000 users of the system in total
- Most projects to date at Handover have been between 2000 and 4000 mandays
- Economic payback will be evident as soon as the system is operational
- The system will completely replace the old applications processing system
- The project will be dependent upon the successful implementation of the Quotations system now being re-written
- Users are eager to participate in prototyping, but nervous of potential changes in their work patterns
- Mr. Renfrew is busy forming a steering body which will include senior I.T. and line management
- The New Business Divisional Manager has been seconded half time to the project

- Vendor support has so far proven good
- Many members of the development team will be recruited externally because of the current lack of client server skills in the organization

Identify sources of unacceptable risk. How could you counter these? (1 hour)

Q8.5

Mr. Renfrew and the board are uncomfortable with long implementation times given the criticality of the New Business project to the business. Could you apply prototyping, phasing or timeboxing techniques to this project to address these concerns. Elaborate and indicate how you would proceed. (30 mins)

ThoughtWell Books

Q8.6

Using your (or provided sample) team selection from Q5.5, decide how you would structure the team. Justify your selection of this structure. (20 mins)

Q8.7

Determine the major source of risk on the ThoughtWell project. How could you reduce this risk area? (40 mins)

9 *Planning Techniques*

Introduction

A wide range of techniques is available to assist us in planning our project, portraying the plan, tracking progress and revising the plan as necessary. This chapter will introduce these techniques. They will be used whenever we plan or re-plan, as shown in figure 9.1.

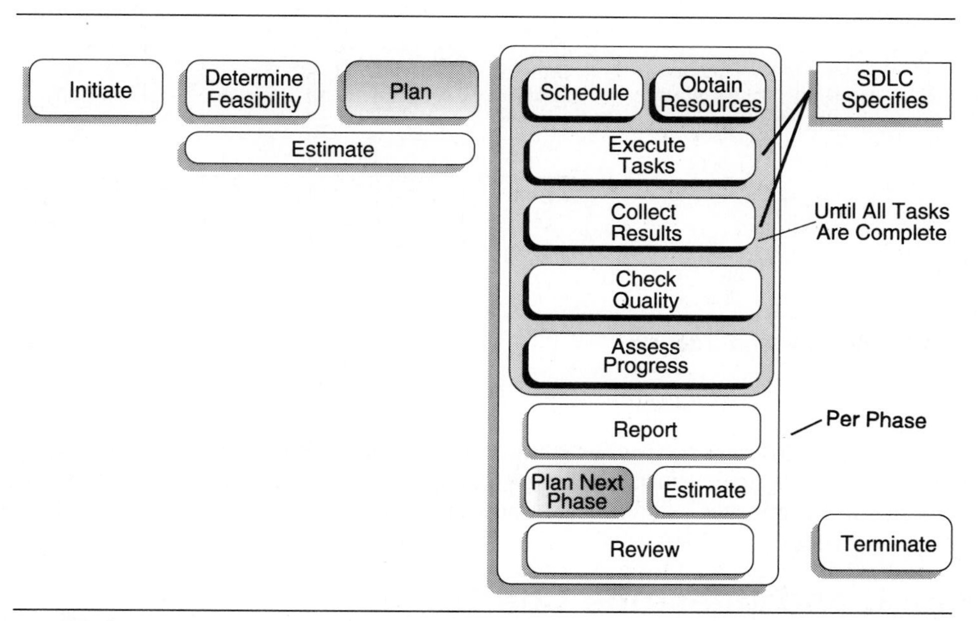

Project Lifecycle *Figure 9.1*

The simplest thing we can do when we begin to contemplate a project is to list the activities and the estimated durations. Let us assume that we are devising a project to produce a paper. We could list the activities like this:

Task	**Estimated Duration**
Select Topic	2 weeks
Locate Literature	8 weeks
Draft Outline	1.5 weeks
Write Draft	4 weeks
Revise Draft	4 weeks
Produce Final Copy	4 weeks

This will give us an idea of the duration and the activities involved, but may not give as much detail as we would like. We can decompose the tasks in the form of a Work Breakdown Structure to get a much better understanding of how the detail will be accomplished as shown in figure 9.2.

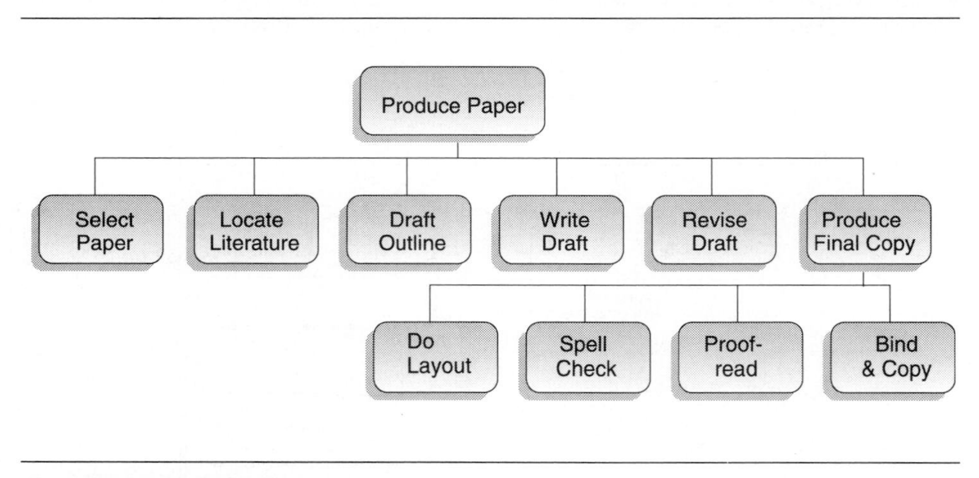

Work Breakdown Structure *Figure 9.2*

This gives us a different perspective on the composition of the project and the activities that will contribute to achievement of the overall goal. We still cannot see, however, when activities will commence and terminate, and which activities can occur in parallel. There are several varieties of bar charts which can assist us to depict these aspects.

Gantt and Milestone Charts

These are among the most popular and longest-standing project planning techniques. Gantt or bar charts were introduced around 1903 by the Frenchman Henri Gantt for planning and controlling military campaigns. They are simple and effective. Essentially we portray time on one axis, usually the horizontal. Bars are used to depict tasks or activities, and indicate their start and finish times. It is easy to see when tasks start and finish, and which tasks will occur in parallel.

Milestone charts are similar in depicting events against time, but do not show start times or durations, only the planned (and maybe actual) *completion dates* of events. They normally use a triangle as a symbol for the milestone. Only major events are shown e.g., completion of requirements specification, completion of technical design and installation. Milestone charts are very useful, providing management with an overall summary of the project in a visual and concise manner. They can obviously be combined with bars on the same chart, thus allowing us to see both the detail and a summary at the same time - see figure 9.3.

Both techniques are easily accomplished manually, or automated through either a magnetic board, or computer software.

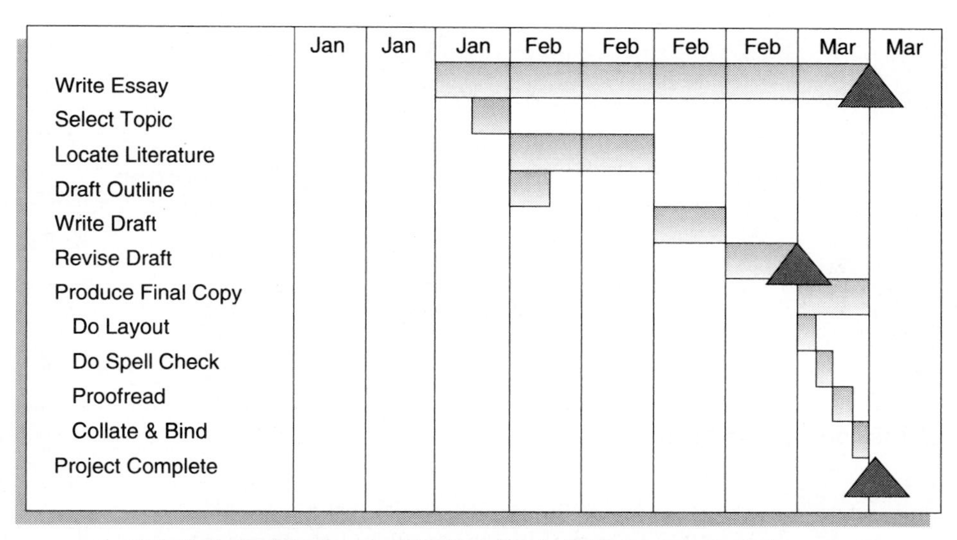

Note the indentation used to identify subtasks below a summary task

Gantt Chart with Milestones *Figure 9.3*

One limitation of the Gantt format is that it does not show dependencies easily (although some notations and software packages do have techniques for representing these). It is easy to see when tasks follow on from others, but it is not immediately apparent whether they are dependent upon the preceding task or not. This information becomes vital when we have to plan time-critical projects with many complex dependencies between a large number of tasks. In these situations, the network analysis techniques discussed in the next section become valuable.

Network Techniques

These allow us to portray the dependencies between tasks graphically, utilizing a network of nodes and lines. There are two major notations:

- *Task on line* notations put the activities on the lines, and use nodes to depict *events.* Dependencies are indicated by sequence of tasks in the network. In figure 9.4 *Select Topic* must occur before either *Collect Literature* or *Draft Outline* can commence; and both of these must be complete before *Write Draft* can begin. In this notation, task *Select Topic* would be referred to as activity 1-2; *Write Draft* as activity 3-4. Note that we need to distinguish between the two parallel activities, and these would be termed activities 2-3a and 2-3b respectively. Notice that the duration of tasks can be recorded on the lines representing the activities.

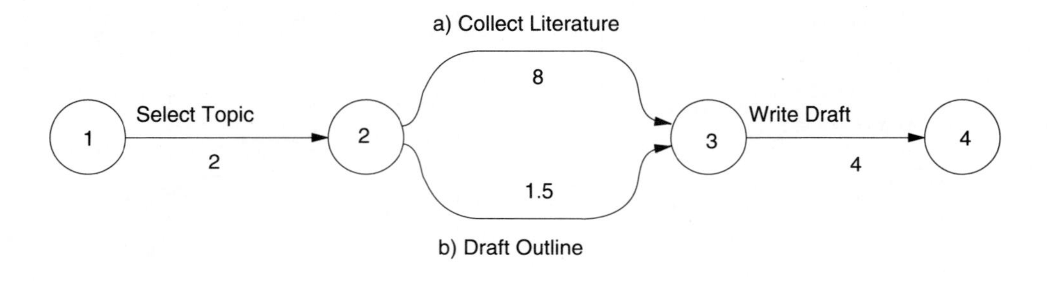

Task on Line Notation *Figure 9.4*

One advantage of this notation is that the length of lines between nodes can be scaled to represent the expected duration of the task as in figure 9.5. In this way, the diagram can give a very good visual indication of start dates, end dates and parallel activities, as we had with the bar chart notation. We, of course, also have the dependency information shown which is missing in the Gantt notation. Where a task is shorter than another performed in parallel, we may need to introduce "dummy" nodes to cope with this. Dependency lines with 0 duration are then used to link to the node from which dependent tasks will proceed. Two disadvantages are that the network can be very difficult to maintain and keep up to date manually, and that large networks can become confusing and difficult to lay out in a manner which is easy to assimilate.

The task 2-5, *Draft Outline*, is shorter than the other one in parallel between nodes 2 and 3. It is said to have *float time* of 6.5. This is the amount of time that its start could be delayed without affecting the date upon which the succeeding event, *3*, would occur. We will return to this concept when we discuss network analysis later in this chapter.

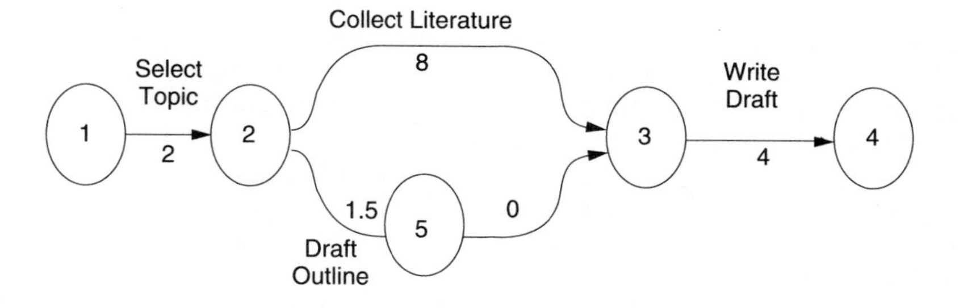

Scaled Network Diagram *Figure 9.5*

- *Task in node* notations put the activities in the nodes, and use lines solely to denote dependencies as in figure 9.6. This is also called a precedence diagram. Tasks are easily identified by the number or identity of the node, with no ambiguities. Unfortunately, it is not easy to scale the diagram based upon the duration of the task as we could with the previous notation. This notation is easy to automate and is particularly favored in project management software packages. It is easy to include a host of information for the task in the node box (e.g., Description, estimated effort, estimated duration, resources assigned, etc.). The same limitations as the task on line notation apply with respect to legibility of complex networks.

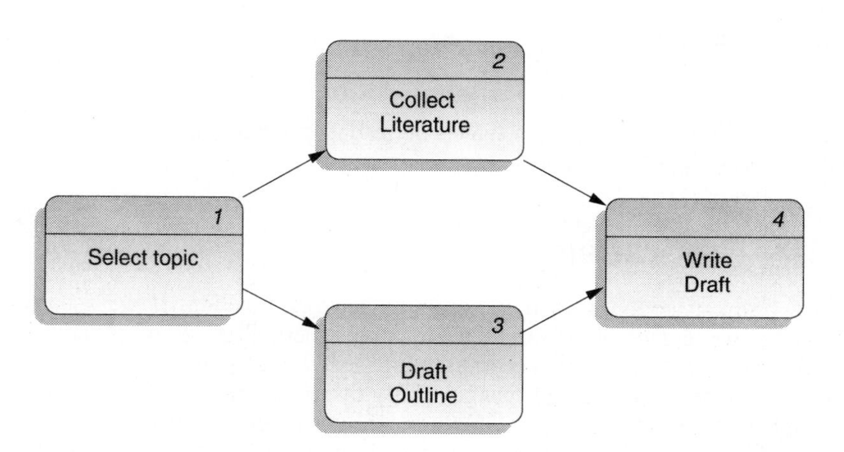

Precedence Diagram *Figure 9.6*

Ladders, Leads and Lags

These are terms which you will encounter with respect to network diagrams. A *lead* is a forced wait before a task can commence. An example would be scheduling a Post Implementation Review on a project for two weeks after installation date. There may be no intervening activity planned, but we have to wait anyway. A lead is thus a line on the network with a duration, but no description or resources. A *lag* is a forced wait after an activity before an event can occur. An example would be in building a floor - you might throw a concrete floor, then have to wait three days for it to "cure" before you cover it with vinyl tiles. In some cases, we can have both a lead and a lag, and this structure is referred to as a ladder. An example of this would be laying pipe in a trench. We may have 200 meters to lay, in 20-meter sections. Let us say that it takes us 2 days to dig 20 meters of trench, and half a day to lay a section of pipe. This could be represented as shown in figure 9.7.

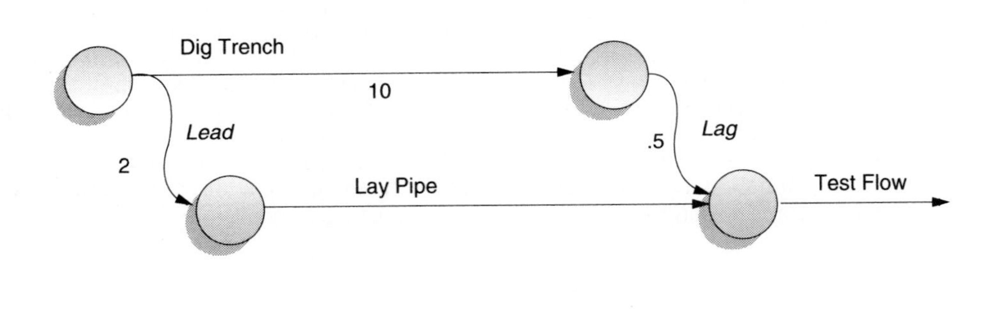

Ladders, Leads and Lags *Figure 9.7*

A systems example similar to this would be testing programs in parallel to writing more, once the initial program is complete.

Deadly Embraces

This is a problem common to both forms of network notation. If we specify a circular dependency, we create an impossible situation where *Task B* is dependent upon the completion of *Task A*, but *Task A* is waiting for *Task B* to complete. This seems obvious and easy to avoid, but can occur surprisingly easily once we start to define dependencies for a large number of interrelated tasks. Of course, if your tasks are truly described in this way, then you will need to examine the way in which the work has been partitioned and resolve this before proceeding. Some automated project management tools will warn you of circular dependencies gracefully, others will simply crash or go off into "never never" land.

The above presents a problem for iterative lifecycles, where you may wish to cycle through a set of tasks more than once. One way to handle this is to repeat the activities in the network, qualifying them each time as in figure 9.8. This also provides a practical way to limit the number of iterations. This is necessary to successfully use any of the iterative techniques in any case.

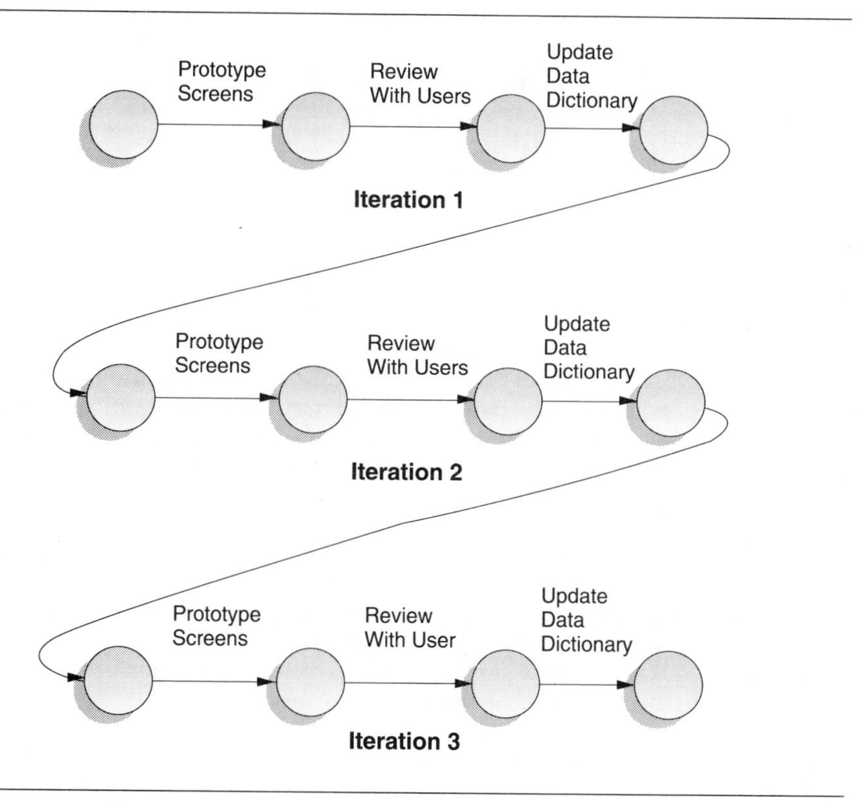

Controlled Iterations *Figure 9.8*

Optional Dependencies

Sometimes we will have the situation where we need to wait for one of two preceding tasks to finish, not both. That is, we can commence the dependent task when either of the other tasks completes, whichever occurs first. This can be shown in the network notation as in figure 9.9.

Showing Summary and Sub-tasks

Consistent with the principles of work breakdown and keeping presentation versions simple, we frequently want to hide some detail which we do not need to concern ourselves with at a

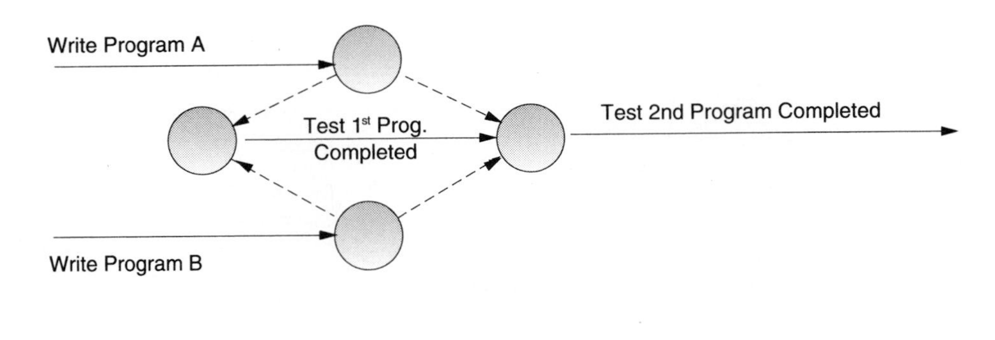

Optional Dependencies *Figure 9.9*

particular time, or for a particular purpose. An example would be a turnkey installation involving both hardware installation and software development. The team working on the hardware installation will want to see tasks related to this in detail, but only a summary for the software development, while the team working on the software development will only want to see their own tasks in detail, while showing the hardware installation at a summary level. We also saw in an earlier chapter that it is useful to collapse past and future phases of the project so that we can concentrate on the detail of the current phase. Sometimes we want to manage different parts of a complex project discretely, hiding the complexity of individual components from the overall management view. All of the above requirements mean that we need ways to show summary tasks which include a number of sub-tasks, and sub-projects which themselves decompose into more tasks.

The requirement is easily accomplished on the Gantt chart by using indentation for sub-tasks. Each level of indentation indicates that these tasks are sub-components of the parent task. The parent task is automatically a summary of the sub-tasks indented below it. This can be designed to accommodate sub-projects, or the hierarchy from the work breakdown structure. If the planning is automated, the summaries can be collapsed or expanded at will to suit the current purpose. This is much like the idea of an outliner used for developing the structure of a document. Some practitioners use a Dewey-style notation to indicate the levels, e.g., 1.2.2.4, etc. In our experience this should be avoided because of the maintenance overhead involved in keeping the numbers in step with the WBS structure, unless it is supported by the tool that you use.

Another approach is to show sub-projects as a single bar on a high-level Gantt chart, and to have a separate chart which breaks these down into their component tasks.

On the network notations, summary tasks are created as *hammocks* between the beginning and end nodes of the block of sub-tasks as in figure 9.10. The hammock represents all activities between the two points. In automated packages, it should summarize all durations, costs, etc. from the intervening tasks. We normally use these at least for the phases in an I.T. project.

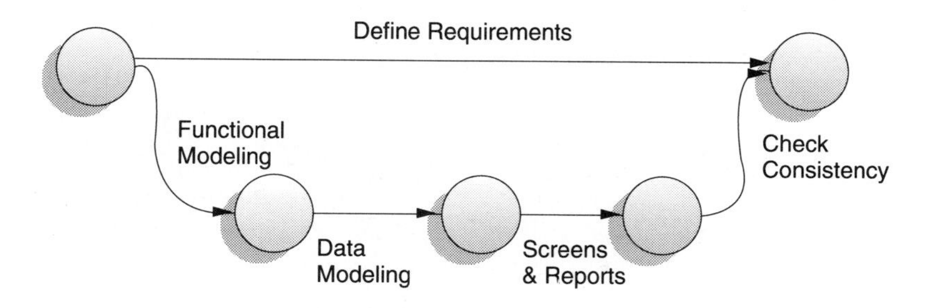

A Hammock (Summary) Task *Figure 9.10*

Some systems allow the creation of sub-projects by linking the start and end node of a separate network diagram to *interface nodes* in the higher level project plan - see figure 9.11. This technique can be useful for linking related projects, rather than those which are strictly sub-projects (where the sub-project would be viewed as a single task in the summary project).

Assigning Resources

Once we have determined our activities and their dependencies, we need to assign resources and responsibilities. On the Gantt chart, this can be achieved by showing a bar for each resource under each activity where that resource participates as in figure 9.12. This also allows for resources to participate in only part of a task's duration. It does not easily accommodate the situation where the resource is allocated for the whole duration, but only for some proportion of time, e.g., four hours a day.

Resources cannot easily be shown on the network *task on line* notation. They can be shown on the precedence network notation by coding the nodes or recording textual information in the node. One way to overcome the problem on a network diagram is a chart called a Gozinto chart (figure 9.13). This is not named after Mr Gozinto, but derives its name from *goes into*. It shows the necessary inputs, i.e., resources (human and otherwise), which go into each step of a process.

Analyzing the Network

Once we have set up our project plan with the dependencies and resources, there are other types of analysis that we can apply to see if it is a sound plan. We may wish to establish the risk inherent in the way that we have organized things, or to determine which activities are really crucial to completing the project on time, or what the likelihood of meeting the planned end date is. The following sections will deal with popular techniques addressing these issues.

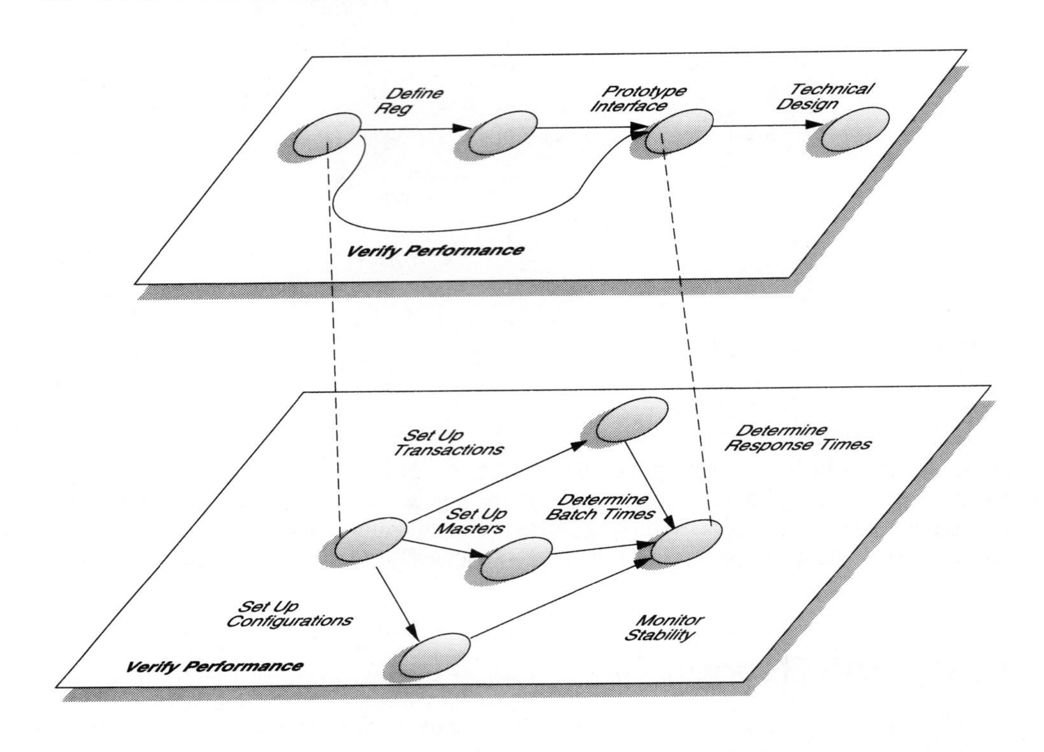

Subproject with Interface Events *Figure 9.11*

Total Project Duration (Calendar Time)

By adding the durations of the longest task between each pair of nodes, we can compute the total project duration. If a start date is assigned to the begin node, we can compute an earliest finish date. This will be the date on which the project will finish, assuming that no activities overrun their scheduled durations.

It is quite common to allocate a range for the estimate of each activity, as was advised in chapter 7. The low end of the range will be the shortest duration for the activity, and the upper end of the range will be the longest duration for the activity. If these are applied to the network, we can calculate the most optimistic end date for the project using the shortest durations, and the latest expected end date using the longest durations.

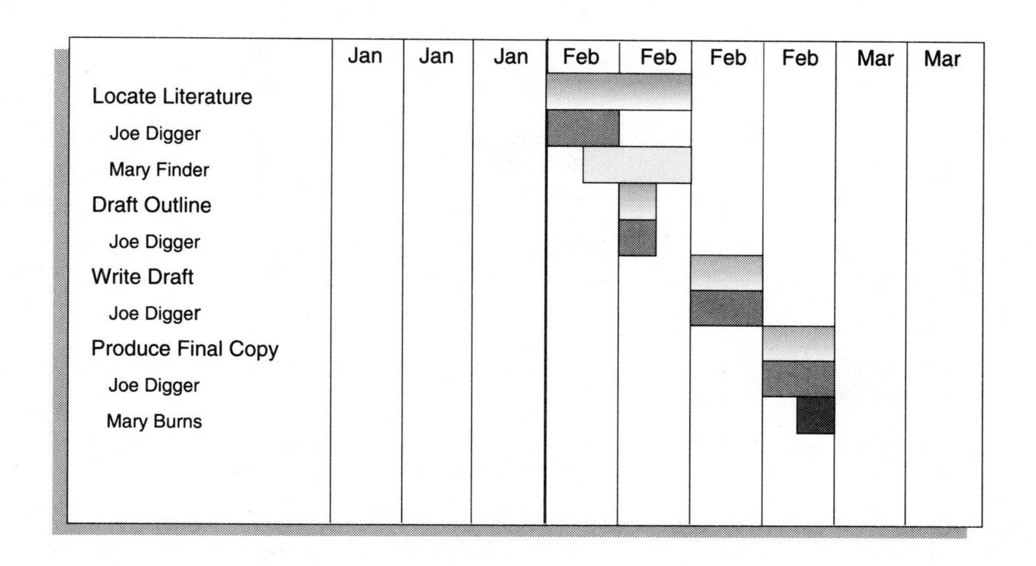

Gantt Chart with Resources *Figure 9.12*

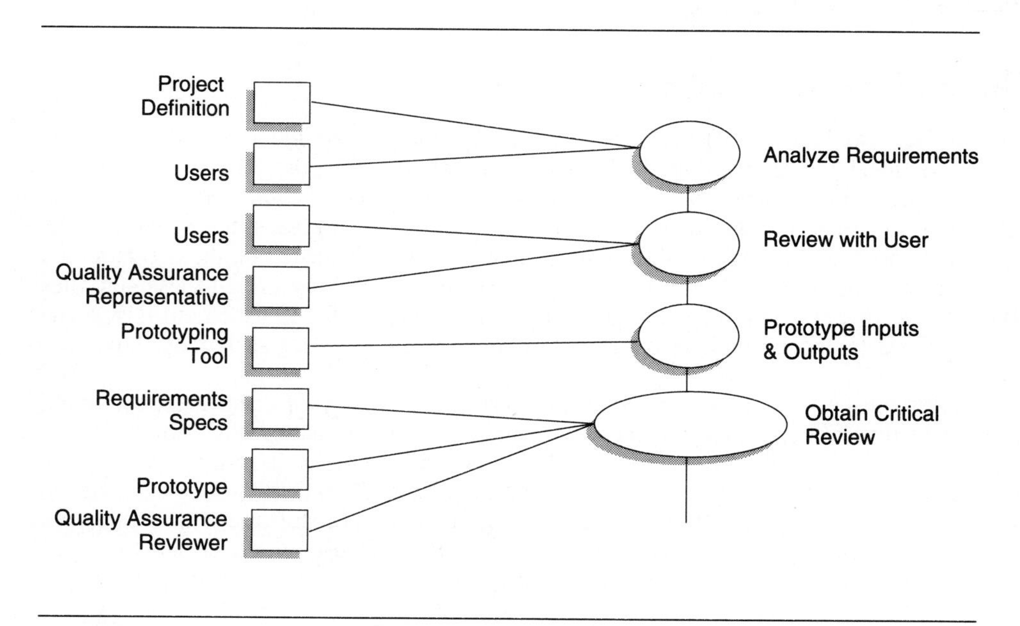

Gozinto Chart *Figure 9.13*

To compute the earliest time for an event:

- Assign an earliest time of the start date to the origin event
- For each event, add the earliest time for its predecessor event to the elapsed time for its predecessor activity. This sum is the earliest time for the event
- If an event has two or more predecessor events (i.e., if the event is a merge event) apply the rule above to each predecessor event-activity combination and use the *largest* sum as the earliest time for the event.

Another useful figure is the latest event time. This is the latest date by which the task preceding it should complete if the end date is not to be affected. To compute the latest event times:

- Set the latest time for the terminal event equal to the computed earliest time for the terminal event
- For each event, subtract the elapsed time for its successor activity from the latest time for its successor event. The result is the latest time for that event
- If an event has two or more successor events, apply the rule above to each successor activity-event combination and use the smaller number as the latest time for the event

The *slack time* for an event is given by subtracting its earliest time from its latest time. This is a measure of the delay which the event could suffer without causing the actual time for the terminal event to exceed its earliest time (in other words to allow the project still to be completed within the original estimate range).

Critical Path Analysis

Critical Path Method (CPM) was first used at Du Pont during the late 1940's. It uses network analysis to identify those activities which are critical to the early delivery of the final product of the project. These are the activities which have no *float time* - the longest activity between any pair of nodes. See figure 9.14. The *critical path* in the network is the path which proceeds through these activities. It is normally designated on the diagram by a bold line (or a colored line). The key thing to realize is that if any one of these activities takes longer than planned, the end date of the entire project will be pushed out (slip). The *critical events* are those which have the smallest amount of event slack as defined above.

By monitoring the critical path activities and events, the astute project manager can avoid delays which would cause the project to slip (not meet the end date). When the project is large and involves many interrelated activities, this can be a valuable tool to focus scarce management attention. It requires careful monitoring of actuals, and revision of estimates to succeed, however, since variations in the duration of events can shift the critical path to a new set of activities. It is perfectly feasible to have more than one critical path if events between the same pair of nodes have the same estimated durations, as shown in figure 9.15.

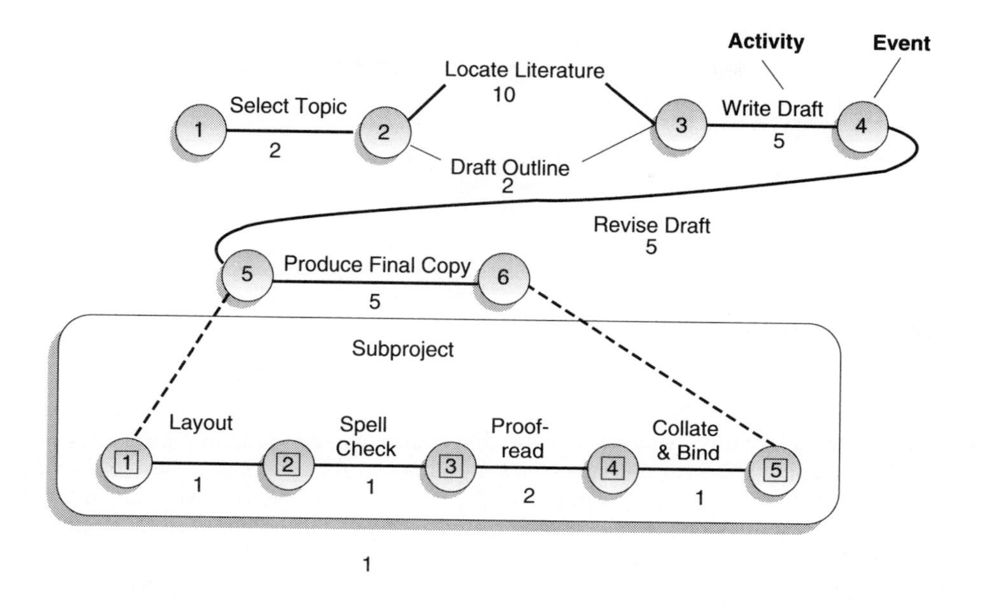

Critical Path Method

Figure 9.14

Program Evaluation and Review Technique (PERT)

PERT came to prominence during the Polaris submarine project, about 1958. It was jointly developed by the United States Navy Special Projects Office, consultants from Booz, Allen and Hamilton Inc., and Lockheed Missile Systems Division. The idea behind the techniques was to provide senior managers with better information on the status of projects, and particularly to control slippages. The team charged with coming up with recommendations for improving performance in these areas felt that the two major factors were detailed, well-quantified estimates for planned activities and precise knowledge of the required sequence in which activities are to be performed.

Because estimates, as we have seen, are frequently uncertain, the team strove to devise ways in which to quantify the degree of uncertainty. This led to the development of the statistical estimation technique which was the major contribution of the PERT approach. Dependencies were catered for by using network planning diagrams. It is interesting that all network diagrams are frequently referred to these days as "PERT charts". See figure 9.16. The unique feature of PERT is the technique used to estimate elapsed time for activities, which accommodates uncertainty.

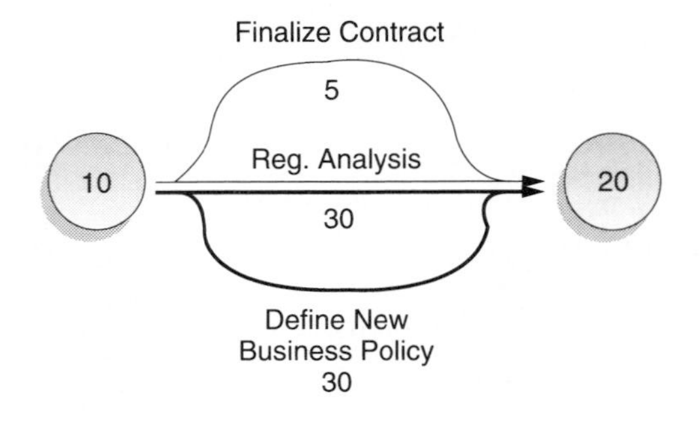

Multiple Critical Paths *Figure 9.15*

Essentially, it works like this: After the network has been prepared to represent the activities and dependencies, three elapsed time estimates are obtained for each task. These estimates are used to compute a single "best" estimate of the time that the activity will take, and the degree of variability that can be expected. The computed best times are used to calculate the critical path duration for the project. The variability associated with each task is accumulated along the network paths to arrive at a variability factor for each event. These figures are then used to make inferences about the likelihood of that event occurring at a particular time.

The PERT approach dictates obtaining the estimates from the people who will actually perform the task, or supervise it. This is done because the people involved should have the best feel for the degree of difficulty the task may present, and can give more accurate estimates. While there is certainly subjectivity in the estimates they will make, they should be able to predict the minimum and maximum times in a range with some accuracy. Between the pessimistic and optimistic estimates, there is a "most likely" point, which the estimators were also asked to identify. This can be done using the Delphi technique discussed in the chapter on estimating. The difference between the pessimistic and optimistic estimates (the range) is indicative of the degree of certainty or variability which can be expected in the task performance. Estimators will produce very narrow ranges for well-understood or routine tasks, and wider ranges for less well-understood, unique or unstructured tasks. The estimated durations should obviously take into account the various factors influencing durations previously discussed, for example, number and skill of resources available, complexity of task, etc. The three estimates should be made using consistent resource assumptions.

Most-likely estimates should be made first. This should be followed by the optimistic estimate using the same resource levels, but assuming everything goes particularly well. Finally, the pessimistic estimate is made assuming the same resources, but that difficulties are experienced (i.e., everything goes badly). This should exclude major catastrophes, such as acts of God, however. The estimate for a task should be done in isolation, and should not include the influence of a predecessor task being late.

Once the three estimates are obtained for each task in the network, the best estimate or *expected* time is computed as follows:

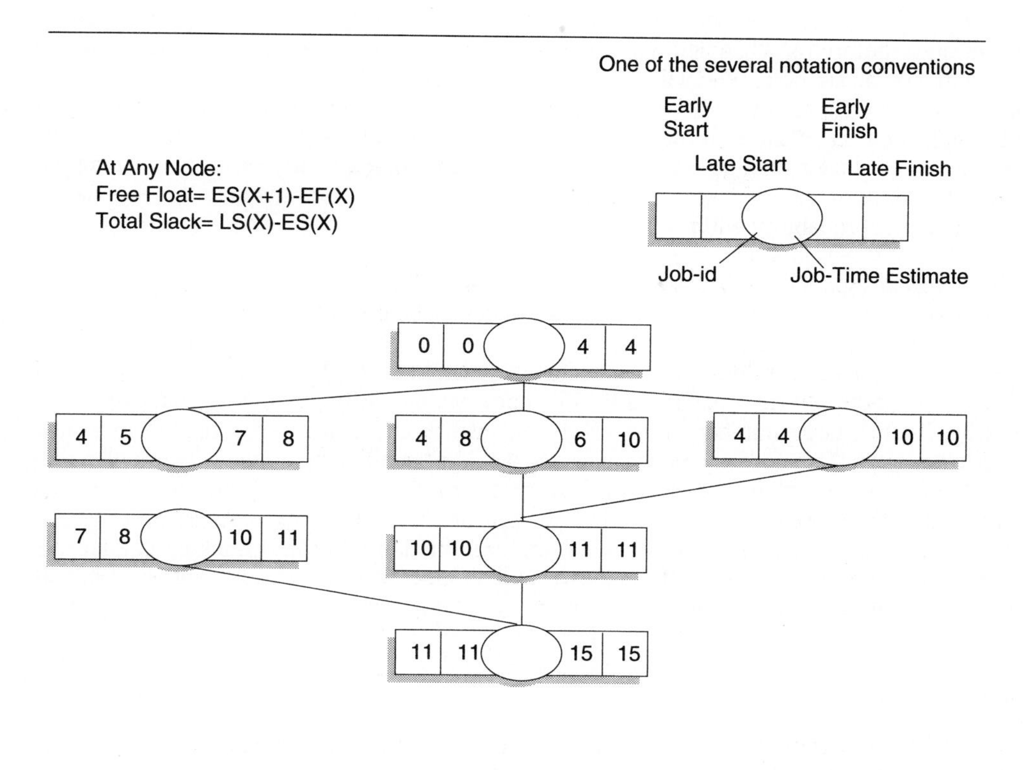

PERT Structure *Figure 9.16*

$$t_e = \frac{\text{optimistic time} + 4(\text{most likely time}) + \text{pessimistic time}}{6}$$

This formula is based upon the assumption that the probability of an activity taking a certain time is a beta distribution whose standard deviation is one-sixth of the range between the pessimistic and optimistic time estimates, and whose mode is equal to the most-likely time estimate.

The variability (degree of uncertainty) is called a standard deviation and is defined as:

$$SD = \frac{\text{Pessimistic time - optimistic time}}{6}$$

The standard deviation for the earliest time for an event is calculated from the standard deviations of the activities of the longest (in time) path leading to the event. This is done by squaring the standard deviation for each activity, summing the squared standard deviations, and then taking the square root of that sum (the "root mean square").

A further feature of PERT allows the calculation of the probability for accomplishing an event on or before its scheduled date (or other date). This uses statistical techniques which we will not detail here. We will touch on the behavior of the technique briefly. If the scheduled time for the event is before the earliest time for the event, then the probability of accomplishing the event by its scheduled date is less than 50 percent. If the scheduled time for the event is later than the earliest event time, then the probability of accomplishing the event before or on its scheduled time is greater than 50 percent, and the later it is scheduled, the higher the probability will become.

PERT was designed to provide continuous program evaluation. This involves the collection of actual completion dates, and revision of estimates as the project progresses.

PERT has received wide attention and been widely applied. In I.T. projects, we seldom need to implement the full rigor of formal PERT, unless we are dealing with very complex, large projects with a large number of activities and interdependencies (e.g., creation of a new operating system). In commercial systems, we would normally employ the network planning techniques, and possibly the approach of obtaining three estimates and calculating an expected time, but seldom go on to do all the probabilities and revise these as we progress. We contend that we can get just as much useful information for project tracking in a far less arduous manner, by using Work Breakdowns and Product Models to track progress. We will discuss this in more detail in the chapter on Project Reporting.

Risk

The network diagram can also be analyzed for undue risk in the construction of the project plan. If we examine the network, we should regard the following as danger signals:

- *An excessive number of activities scheduled in parallel.* Typically, the manager can only handle about five parallel activities. If more activities are scheduled in parallel, risk is increased. The chances that "something will go through the cracks" become much higher.
- *Too little float time in later stages* of the project. If we find that virtually all the activities toward the end of the project have little or no float, this may mean that we have scheduled all the noncritical activities early in the project, leaving all the really important things for the end. Move more important activities to earlier in the project, or build in some padding toward the end to minimize risk.
- *Activities which are too long.* No activity, with the assigned resources, should exceed the Manageable Unit of Work (MUW) discussed previously in the book. If a month-long activity overruns by 100 percent, we are a month late. If a one week activity overruns by 100 percent, we are a week late. Try to break longer activities down into meaningful, manageable chunks, preferably with a well-defined deliverable.

Tracking Progress

The Gantt chart can also be usefully employed to track progress graphically by introducing an "actual" bar for each task, shading the bars, or color-coding the bars (see figure 9.17). Progress can also be shown on network charts by coding events which have been achieved in some way, or by recording actual completion information on the nodes when using a precedence-type chart.

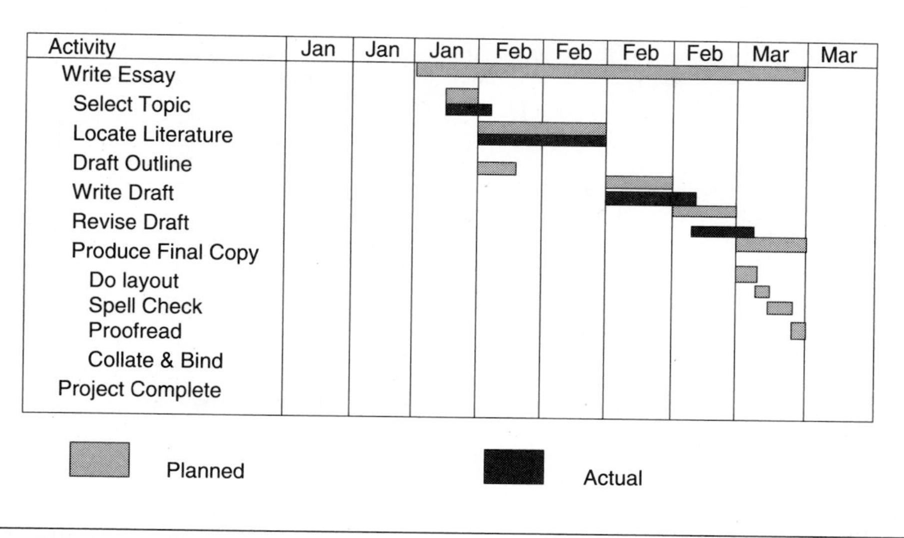

Gantt Chart with Actuals

Figure 9.17

A variation on the network chart, known as a slip chart (figure 9.18), allows us to depict slippage visually. On this chart, the time originally scheduled for each event is shown, and then as time passes, the updated estimates, and finally the actual completion date. It is easy to pick out the deviation from plan visually. In the figure, events 1, 2 and 3 have been achieved. Events 4 and 5 are running behind schedule, but have not slipped further since the last estimate. The diagonal line indicates that for completed tasks, scheduled time equals actual time.

Summary

To summarize, Gantt charts are most useful when we wish to depict information in a simple visual manner for presentation or summary purposes. They do lack the more detailed information about task dependencies which is easily portrayed on network-type charts. These are thus more suited to detailed planning of dependencies and critical paths. Some notations do allow us to show the critical path activities on a Gantt once they have been determined. Milestone charts are useful as a high-level summary and for communication with steering bodies. Resources are easily shown with activities on Gantt charts, but are difficult to show in task-on-line network notations. They can be included in task-in-node network notations. Gozinto charts can be useful to show the activities and the necessary inputs or resources required to fulfill the activities.

No matter which notation we use, we should realize that this is merely a portrayal of the same collection of data related to tasks, milestones, and resources. It is thus perfectly possible to initially use one technique, and then to portray the same data in another form. If

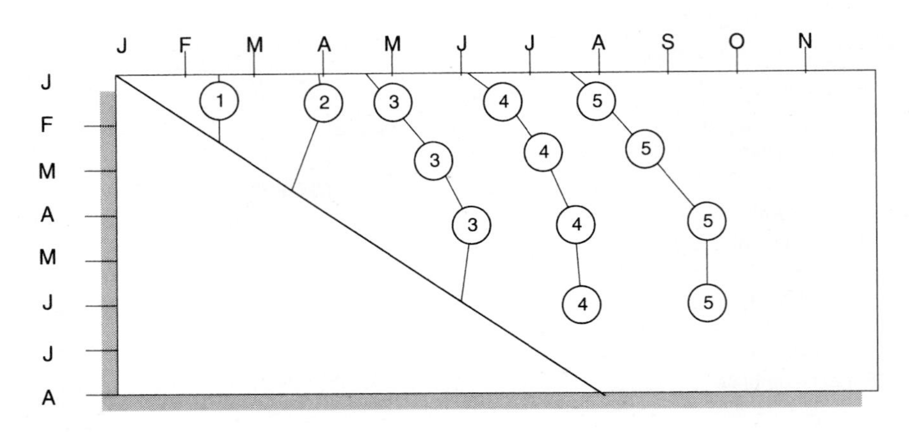

Slip Chart *Figure 9.18*

you have used these techniques manually, you will have realized that doing them on a large scale can involve a significant amount of work, and quite a lot of calculation. This work needs to be repeated whenever actual figures or new estimates change the position. The effort involved causes many project managers to give up on using them. We do this at our peril, however, since we can then easily lose track of critical issues in our project. A much better solution is to automate a lot of the drudgery by using a software package to do it for us. This is discussed in the next chapter, together with some more sophisticated planning options which are really only practical with automated support.

Case Questions

MyWay Organizer

Q9.1

A colleague has produced the following calendar-time estimates for activities within the design phase of the organizer project:

ID	Description	Duration	Dependent on
a	Prototype user interface	10 days	-
b	Design physical file structures	2 days	a
c	Map functionality to module structure	3 days	a, b
d	Define interfaces	2 days	c
e	Define global in-memory structures	2 days	c
f	Define and test compression algorithm	5 days	c
g	Design common modules	5 days	e, f, d
h	Design normal modules	8 days	g
i	Predict performance	3 days	g, h
j	Define standards for coding	5 days	-
k	Devise test plan	5 days	e
l	Prepare test cases/data	5 days	k
m	Review phase deliverables	2 days	all

Prepare a Gantt chart for the design phase, respecting the dependencies and using the calendar-time estimates. Try to complete the phase in the minimum time. (15 mins)

Q9.2

Using the data from Q9.1, prepare a PERT style (task on line) chart of the phase showing dependencies graphically. Show calculated durations only - do not worry about ranges or most likely times. (15 mins)

Q9.3

Using the data from Q9.1, prepare a precedence chart of the phase showing dependencies graphically. (15 mins)

Q9.4

Using the answer for Q9.2, or Q9.3, determine the critical path for the phase. Which activities does it include? What is the minimum time taken to complete the phase? What effect will it have if task c takes 5 days? What effect would there be on the project duration if task k takes 8 days? What effect if task e takes 7 days? (15 mins)

Q9.5

Use the data from Q9.1. Assume that the durations are for working days. We have decided to plan work for 5 days per week. This will give us a contingency, since the team actually work 5.5 days per week. Plan the phase using a Gantt chart, respecting task dependencies and the above criteria. What is the minimum time for completion? (20 mins)

Gleam Stores

Q9.6

The Gleam roll-out planning is well under way. The macro-plan estimates (in calendar months) and resource allocations look like this:

ID	Task	Duration	Responsible	Resources
a	Document & Trng Pkg	5	Anthony	Peter, June, Ray
b	Local Region Stores	4	Michela	Peter, Ray, Mark, Mary Jacob, Milly
c	Northern Region Stores	6	Peter	Ray, Mark, Jacob, Fred
d	Southern Region Stores	5	Mary	Milly, Vincent, Hilary, Martin
e	Western Region Stores	12	Peter	Ray, Mark, Milly, Vincent, Hilary
f	New Eastern Stores	5	Martin	Jacob, Fred

Add a contingency of 1 month to all estimates. Create a Gantt Chart with milestones to show the completion of implementation in each region. Include resource assignments on the Gantt. Assume the project starts in June of this year. Arrange sub-projects to achieve the shortest overall implementation time. Ensure that there are no conflicts in terms of resource assignments. Work out the approximate cost if a person-month costs $6 000. (40 mins)

Q9.7

Using the data from Q9.6, prepare a macro-plan of the whole implementation effort in the form of a network chart (task on line). Show the various region implementations as sub-projects. Expand one of the sub-projects into its own chart, linking the nodes to the main project. In the sub-project, show about ten tasks with their dependencies. (30 mins)

Q9.8

Using the data from Q9.6, prepare a macro-plan of the whole implementation effort in the form of a precedence chart (task in node). Show the various region implementations as sub-projects. Expand one of the sub-projects into its own chart, linking to the main project. In the sub-project, show about ten tasks with their dependencies. (30 mins)

Handover Trust

Q9.9

Using your knowledge of the system development process, plan the New Business project for Handover Trust as follows: Assume that the Feasibility Phase is complete. Show this as a single activity. Plan the Requirements Definition Phase in detail (about 20 to 30 tasks). Show a summary task for each subsequent phase. You may use previously derived estimates (chapter 7) or reasonable assumptions for durations. Show all dependencies. Present your answer as a summary Gantt chart for management, and a detailed precedence chart for the Requirements Phase. Determine the critical path through the Phase, as well as the project. (45 mins)

Q9.10

Given the following data (obtained by Delphi estimates) for the Quotations Project, calculate the earliest time, the latest time and the most likely time for completion of the project using PERT techniques:

ID	Description	Minimum	Most Likely	Maximum
a	Feasibility	4	4	4 (compl)
b	Requirements	10	12	20
c	External Design	14	20	26
d	Technical Design	8	10	14
e	Build	24	36	48
f	Integration Test	6	12	20
g	User Documentation	10	12	12
h	Installation	4	6	8

All times are in calendar weeks. The project start date is the beginning of August this year.

Task dependencies are as follows:

b on a
c on b
d on c
e on d
f on e
g on d
h on f and g

(45 mins)

ThoughtWell Books

Q9.11

Using your previous work on the ThoughtWell Books project from Q8.6 (team structure) and the list of tasks provided below, do the following:

- Add any necessary QA or Management tasks you think necessary
- Change any tasks, descriptions or dependencies you think necessary
- Develop a network diagram showing the dependencies
- Allocate resources to tasks in an optimal fashion

Prepare a plan for each team member showing which tasks they will perform and the likely timing of these. You can assume the project starts September 1.

ID	Task	Estimated Effort	Dependent upon
a	Object Relationship Model	4 person weeks	-
b	Business Process Model	2 person weeks	-
c	High Level Event Models	6 person weeks	a, b
d	Prototype User Interface - Generic	1 person week	a, b
e	Prototype Event Dialogues	4 person weeks	a, b, d
f	Persistent Data Design	1 person week	a, e
g	Design Help System	2 person weeks	d
h	System Architecture	1 person week	e, k
i	Design Reusable Components	2 person weeks	e
j	Design Unique Components	4 person weeks	e
k	Communications Prototype	4 person weeks	c
l	Populate Help System	4 person weeks	g
m	Code Reusable Modules	4 person weeks	i
n	Code Unique Components	6 person weeks	j
o	Write User Manual	4 person weeks	i, j
p	Unit Test Reusables	2 person weeks	i
q	Unit Test Uniques	2 person weeks	j
r	Integration Test	3 person weeks	p, q
s	System Test	2 person weeks	r
t	Client Operations Training	2 person weeks	r

(If unfamiliar with tasks a and c, treat these like "entity model" and "high-level logical data flows".)

(1 hour)

10 Project Management Tools

Facilities

Modern project management tools are very affordable and user friendly. Time was when packages were available only on mainframes, took hours to run, cost tens of thousands of U.S. dollars, and produced output that looked like a hexadecimal dump. All that has changed. Very good packages are available for around $500, run on high-end personal computers and produce stunning output. Really high-end packages supporting huge projects with thousands of activities, scores of sub-projects and hundreds of resources still require a minicomputer and a well-padded wallet, but these are seldom needed in I.T. projects. Typically we need a good PC package such as Timeline™, SuperProject™ or Microsoft Project™. All of these will run under Microsoft Windows™on an Intel™386 processor machine. For large projects (>250 activities and > 10 people), an Intel™486 level machine or better is recommended. There are also good packages available on the Macintosh™ and Unix.

Typical facilities include the following:

- Support for the creation and maintenance of calendars
- Capture and maintenance of information related to activities
- Capture and maintenance of dependencies between activities
- Capture and maintenance of resource categories and individual resources
- Assigning resources to activities
- Structuring activities into hierarchies (Work Breakdown Structures)
- Viewing activities as Gantt charts, or network charts
- Scheduling of activities, including checking resource availability, and calculation of expected dates
- Recording actual delivery dates

- Resource costing and calculation of actual expenditure based upon resources assigned and actual durations

- Comprehensive reporting, including production of graphic Gantt and network charts

In addition, the more sophisticated packages offer facilities like:

- User customization of display screens, reports, and menu structures

- Resource leveling (discussed below)

- Partial assignment of resources

- Support for sub-projects, and consolidation of these into super-projects

- Open interfaces via standard file structures to other products such as Lotus 1-2-3™ and DBase™

- Realtime interfaces to other Windows™ products via Dynamic Data Exchange (DDE) or Object Linking and Embedding (OLE)

Repository

All the good packages have an underlying repository (see figure 10.1), or database, containing all the project information. This makes it possible to capture data on a new project using a Gantt or Work Breakdown outline, then view it as a network when desired or vice versa. This flexibility is a major advantage over manual methods. In general, any information should only need to be captured once. For example, when I define a resource, I may do this on a special resource-editing screen, or I may decide I need a new one while working on the project task outline. Once the resource is defined, I can add it to other tasks by simply using its assigned name, or even by selecting it from a scrolling window with a mouse click. The repository should contain the following information:

- Calendar information which specifies work and holidays, and the hours worked on various days (weekdays versus Saturdays)

- Resource information which will detail the name, type, quantity and cost of each resource. In I.T. projects, our major resources are usually people, and since they are so individual, we normally identify them by name as discrete resource types. One exception to this could be where we are using a pool of contract programmers, when we might specify programmer as a resource category, and the number available to us. Even there, we will normally want to refine this once actual resources are assigned, to take into account individual skills and characteristics.

- Task information, which would include: Name of the task, description of the task, estimated duration, planned start date, calculated end date, dependencies on preceding tasks, assigned resources, costs unrelated to resources, actual start date, actual end date and various other information such as float time, earliest start, latest start, calculated costs, whether the task is critical path or not, and so on.

- Structural information such as the collation of tasks in a Work Breakdown Structure. Frequently we can also access tasks by various other means, such as an assigned category (e.g., Technical, Management, Quality Assurance).

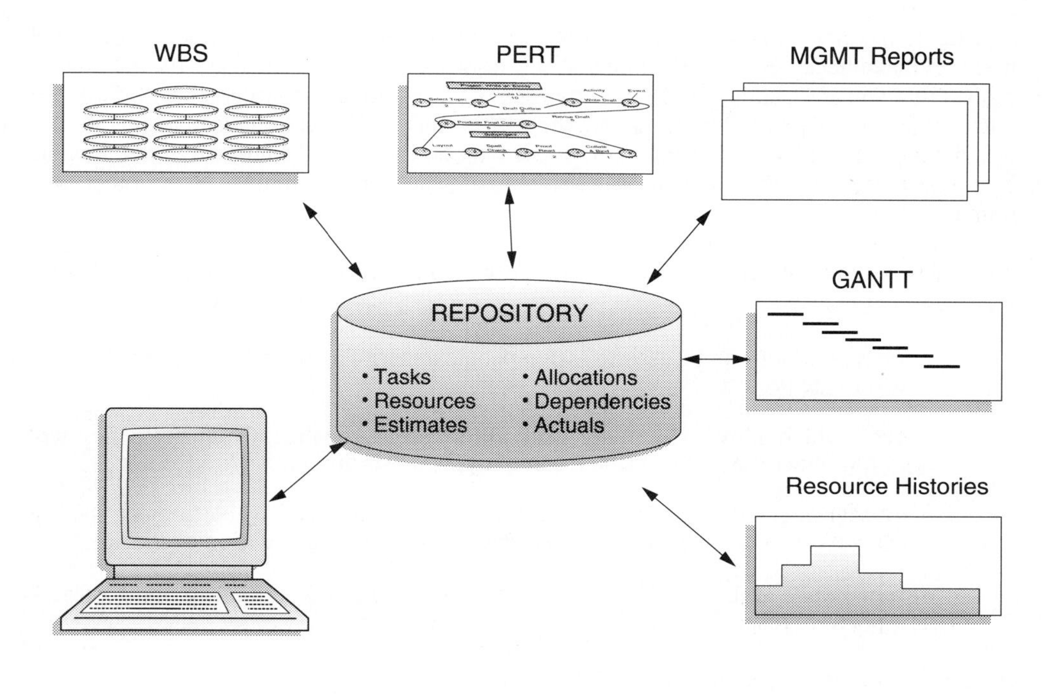

Project Management Tool *Figure 10.1*

Calendars

Two types of calendars are useful. The first is a project calendar. This will record which days of the week are normally worked, and the usual hours on those days. It will also record which days are nonworking days for the team (e.g., public holidays). This information allows nonworking time to be skipped automatically when scheduling tasks and working out durations. Having the duration of the working days allows automatic scaling of durations based upon effort estimations.

The second type of calendar is a resource calendar. This allows customization for a particular resource type or individual resource. An example would be a person who works half days. We could define a calendar for this person, which stipulates four hours of work time per day. The package will then use this information in scheduling and calculating duration of tasks assigned to this person.

We will normally define the project calendars as the first step in setting up a project plan on an automated tool. We will now discuss the following steps in the order they would normally be performed.

Task Definition

Having defined the calendar(s), the next step is defining the tasks in the project. These will normally have been determined using a WBS, drawing on the methodology for the type of project that you are tackling, and using the generic project lifecycle discussed earlier. At the outset, it is normally sufficient to capture the task-id (this may be automatically generated), a task description, and the estimated effort or duration for the task. It is normally more flexible to define the effort and allow the software to calculate the duration, than the other way round.

Tasks which may not occur to you at first, but which are very useful to have in your plan include:

- *Training* - which can be used to assign people undergoing training and have this time costed to the project
- *Leave* - which allows you to see who will be on leave when. Company policy will decide whether the project bears the employees' costs during this time or not
- *A management task*, which stretches for the duration of the whole project, against which we can allocate the project manager's time and administration overheads.
- *Review points* at the termination of each phase, and a task at the beginning of each new phase to respond to the findings of the review

A suitable structure is shown in figure 10.2.

Dependencies

These are normally captured after the tasks have been captured. Many packages allow us to do this in a simple fashion by clicking with the mouse on the precedent task and the dependent task. This can be done in an outline (Gantt style) view, or on a network representation as in figure 10.2. Dependencies should be indicated by your methodology. Where you have to determine these yourself, the best way is usually to examine the deliverables associated with tasks, as discussed previously under the Product Breakdown concept. Think about which tasks produce which deliverables, and which are required as input to other tasks. If this is not apparent on the surface, then examine the data content of the deliverables. An example would be that to prototype effectively, we need a data dictionary in place, as well as a functional model of what is required. Figure 10.3 shows this relationship as a flow diagram, from which it is easy to derive task dependencies.

Care should be taken to avoid definition of circular dependencies. This occurs when a task is dependent upon another, which in turn is directly or indirectly dependent upon the first task. These are all too easy to create in a complex project plan. If they occur, most modern packages will warn you and ask you to resolve them before network analysis can commence.

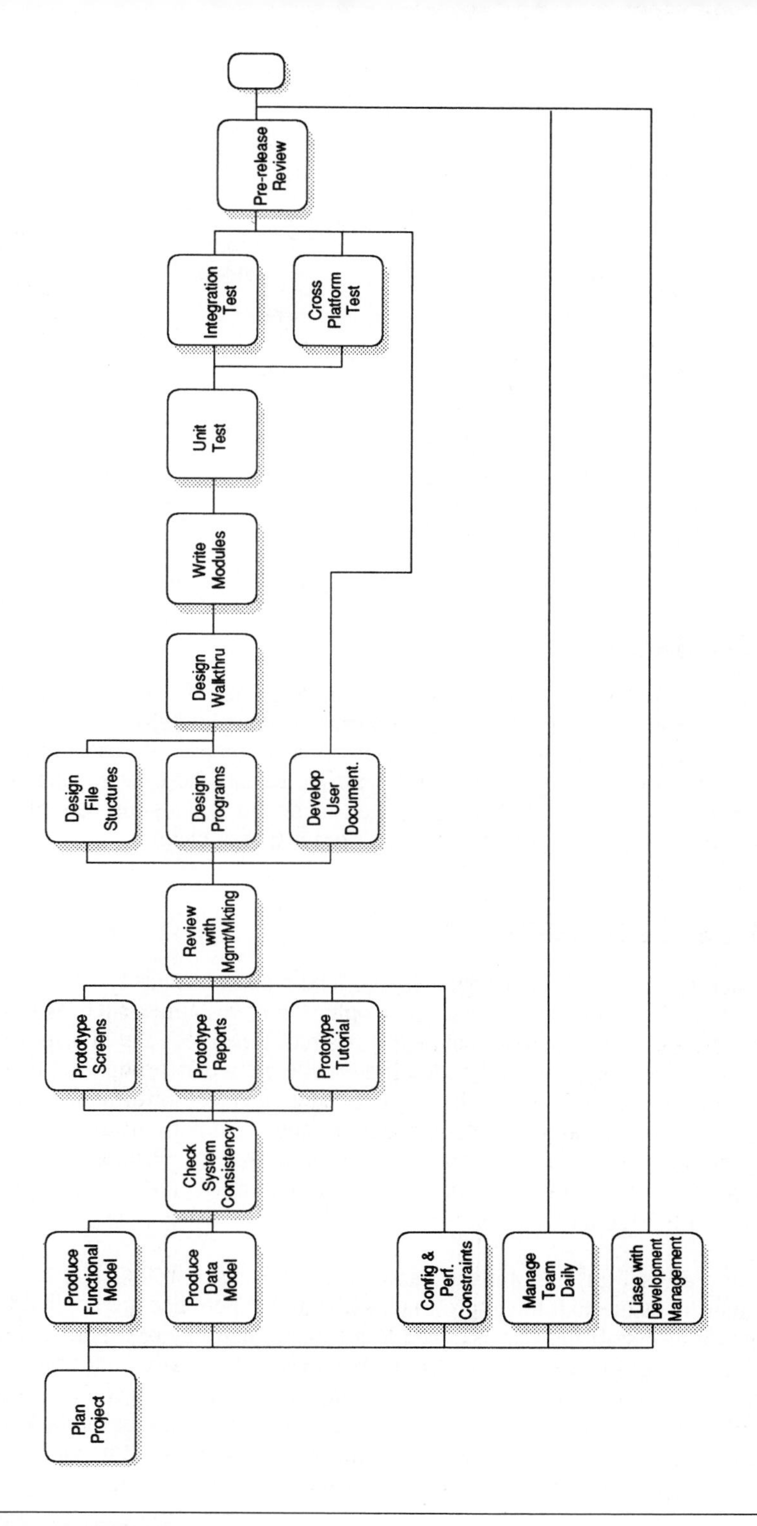

Network Diagram Example *Figure 10.2*

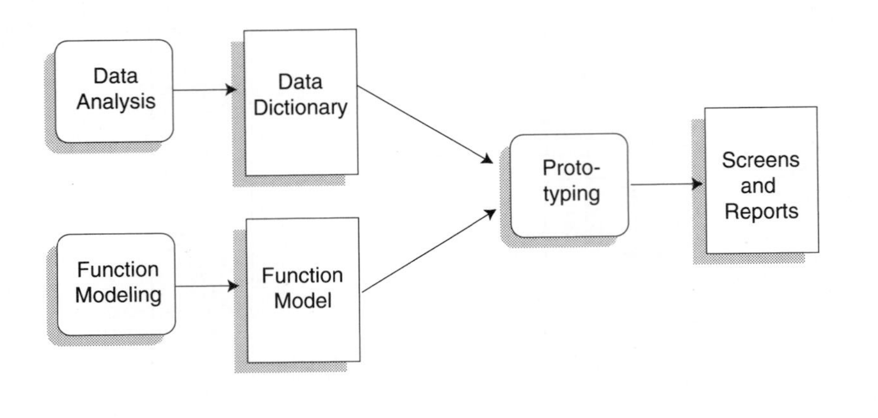

Determining Dependencies *Figure 10.3*

Work Breakdown

It is extremely useful to group our detail tasks under summary tasks within a WBS. This allows us to collate estimates and actuals upward in the hierarchy automatically. It also allows us to collapse past and future phases to a summary representation, while we concentrate on the detail tasks of the current phase. Recall the *creeping window* concept from chapter 7. Having a good WBS on which to record actuals also simplifies gauging project progress and reporting. This will be explored in a later chapter.

Resource Definition

Having defined the tasks and dependencies, the next step is usually to define resources. In I.T. projects our major resource is usually people. Other resources such as machine time, outside services, and capital can be included if desired. Resources can be captured first, and then allocated to tasks, or they can be identified while we work through the tasks, and then we can provide full details for each resource later. Typical information held for each resource would include a unique id, description or name, cost per unit time, and availability. Availability could be expressed as the number of a type of resource (e.g., we have 3 programmers), or it can be taken to the level of an individual (e.g., Susanne is available 5 hours per day, from March till June).

Once resources are set up, they must be associated with tasks in the project. This requires careful thought to ensure that the appropriate skills and experience are applied to each task. In some cases junior staff may be assigned to a task so that they can learn necessary skills. This will represent an investment cost, since they may not actually contribute to producing the deliverables more quickly. If you find that you must allocate too many resources to a task to get it done in an acceptable time, this could be a warning that the task has not been decomposed successfully. You should re-examine your WBS to see if it can be broken down into independent activities which can be performed by very small teams.

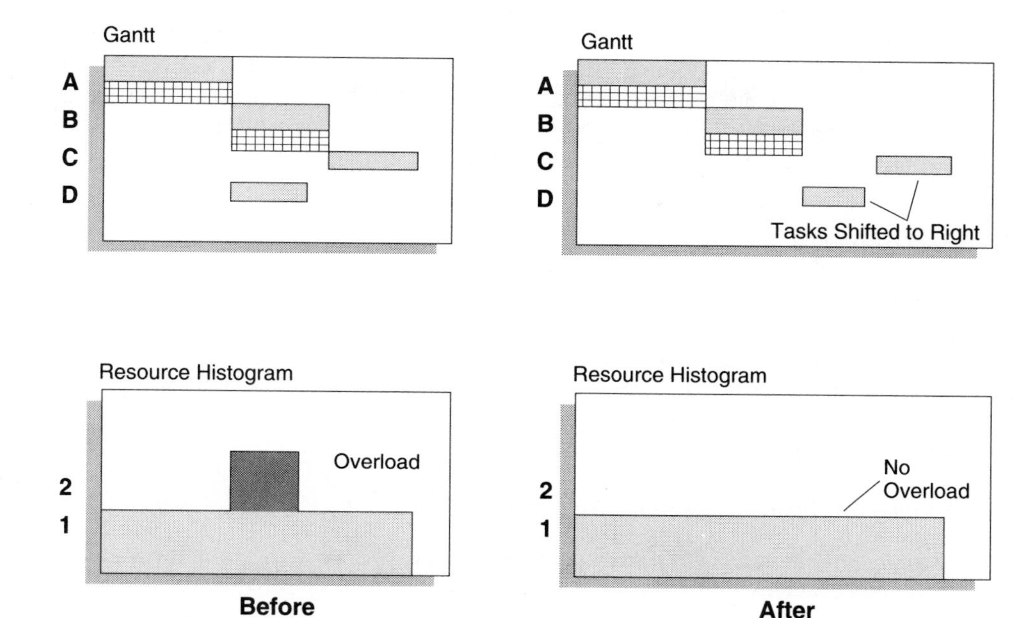

Resource Leveling *Figure 10.4*

Resource Loading

When resources are allocated to activities which occur in parallel, the load on the resource will be cumulative. If a person is allocated fully to two tasks occurring at the same time, the person will be overloaded - we are asking them to work 16 hours a day! This can be resolved in two ways: First, by scheduling the tasks so that, even though there is no technical dependency, the one will occur after the other to resolve the resource overload; or second, by allocating the resource only partially to each of the parallel activities. For example, we could have an analyst who is splitting her time equally between writing user documentation, and integration testing of modules. Different packages allow different options, so you should decide which options you need and try to find these in the short listed packages before choosing a tool.

Many packages provide resource histograms which make overloads easy to detect. An example is shown in figure 10.4.

Some packages offer automatic resource leveling. This will shift tasks, which compete for resources, so that some start later, preventing resources overloading. If this were applied to the situation shown on the left of the diagram, we would get the result shown on the right.

Notice that no resources are now overloaded, and that activities have been shifted to resolve the problem. This feature allows scheduling optimally, taking into account both dependencies and resource constraints. Basically, we set up our project plan with only the necessary technical dependencies, allowing maximum parallelism. This will give us the shortest possible project plan, given unlimited resources. Next we add the resources available and apply auto-scheduling as well as smoothing. This will then give us the shortest duration possible, respecting dependencies and resource limitations. This does a very complex job (considering the calendars, dependencies and different resource availabilities as well as task effort estimates) for the project manager in a painless manner.

Scheduling

As we proceed through the project, we want to ensure that the current or imminent phase is planned and scheduled in detail. As we complete each task, the actual date and effort should be recorded into the system. This allows us to track progress against plan effectively. We should avoid overloading resources as far as possible. It may be necessary for certain short periods of time, but should never be routine. People under pressure do not work more productively.

Different packages provide a variety of scheduling options. These include:

- *Minimum time with no resource constraints.* This will allow maximum parallelism, with no consideration for risk or availability of resources. Not frequently practical.
- *Minimum time within resource constraints.* We have discussed this previously under resource smoothing.
- *Least cost for completion by a given date.* Here we specify an end date that we want to meet, which is later than the finish date for the preceding option, and let the software work out the optimum resources to apply (minimum to get the job done in time), thus minimizing cost.
- *Least cost.* This will allow the end date to be totally flexible, and optimize solely to minimize cost. This tends to give very long project durations, and is seldom practical for I.T. projects.

What If?

Part of the power of automated project management tools lies in their ability to work through scenarios rapidly. They are useful if we want to capture a project plan and calculate an end date. As our estimates change during the project, recalculation is needed. To examine a variety of different ways of handling the project and compare the outcomes, to assist us in project design and make crucial decisions of a structural nature during the project, such tools are vital. They allow us to play what if? and see what the implications of various choices are in a short period of time - something that would never be possible manually.

Reporting

As indicated in the summary of chapter 9, various types of representation are useful for different purposes. The beauty of automated tools is that they let us capture the information in one format, work with it in another, and then print it out in a third suitable for review with management. This flexibility allows us to tailor the output for the audience, thus enhancing overall communication.

Some packages also allow the tailoring of the specific information on the reports and graphs to match the user's requirements exactly. We can, for instance, choose which of the fields related to tasks we want to see on the report, and in which order, as well as how we might like the report sorted.

Standards within the Organization

As with CASE tools and programming environments, it is important to have a corporate standard for project management tools. This is necessary to build a productive level of skill, to allow interchange of project information in a consistent form, and to encourage re-use. The latter is particularly important. The idea is that a project plan from a previous similar project in the organization is a very rich source of pertinent information for a new project in the same organization. It will already contain the required tasks (technical, management and quality assurance) as well as amendments which the previous team found it necessary to make along the way. It is normally only necessary to alter the start date, estimates and resources to arrive at a comprehensive new plan. This can save an enormous amount of effort, as well as enhancing the overall quality of the plans.

Some organizations with which we have worked, have built standard skeleton plans tailored to their particular methodologies, types of projects, and organizational circumstances. These are made available to project teams from the Development Support area. Once they have been modified by the team for a new project, a copy is filed with the Development Support group. They consolidate these into super-project plans. Since all underlying project plans use similar concepts and terminology, they are able to identify, on company wide basis, what the demand for scarce resources will be at a particular point in time. This allows them to avoid resource overloading of central service areas like Data Base Administration, or Network Support. Teams will be warned of clashes for these resources, allowing them to work around and thus ensure that external resources they need are there when they need them.

If actuals are fed back to a central area in a similar manner, I.T. management can track overall performance against deliverables and cost budgets very effectively with little overhead on the part of individual project managers.

Costs and Overheads

People costs are usually the major component of I.T. project costs. Costs are reflected in the tool by associating a cost per unit time against the resource. A task cost will then be calculated by multiplying the task duration by the unit cost for each resource, and summing these. Other costs, such as purchase of a component, will have to be reflected in different ways. In some packages these can be captured directly as nonresource costs linked to a particular activity or milestone. Others do not allow costs which are not linked to resources, forcing you to add dummy resources, or maintain these types of costs outside the package.

Product Definition

As yet, few project management packages support defining a product breakdown, or relating a configuration to the tasks in the project plan. There are separate configuration management packages available which do this, but integration is still lacking. We would like to see project management tool vendors incorporate facilities to easily include product definitions linked to tasks in the project plan.

Version Control

High-end packages provide facilities to keep several versions of the project plan concurrently on the system. This is particularly useful to support the configuration management concept. We can keep a copy of the project plan at each review point, which includes actuals to that review, before we modify estimates and expand tasks for the next phase. This allows us to accurately recall how the estimates, plan and actuals have shifted over the life of the project. Measures like the Estimating Quality Factor discussed in chapter 7 rely upon this sort of information.

Selecting a Package

In many cases you will not have this luxury (or onerous task!) as the organization will already have a standard package in use. Should you be required to select a tool, some of the criteria which you may consider are listed below:

Feature considered as essential

- Support for WBS, dependencies and Gantt Charts
- Reliability
- Friendly, intuitive graphical user interface (e.g., Microsoft Windows™, Macintosh™ or Motif™, Presentation Manager™). This should be taken advantage of and used to good effect by the package. It should not merely be a "Windows" version of an earlier text-oriented package. We should be able to visually insert dependencies, drag time dependent tasks around, etc.
- Support for capture of resources and allocation to tasks
- Good reporting and printing facilities (check formats, speed, limitations, etc.)
- Repository, single source of entry, i.e., if we enter information in one format (e.g. Gantt view) it should be stored semantically and available for use in other formats (e.g., PERT or WBS)
- Support for calendars which allow handling working and nonworking days, holidays, etc.
- Support for capturing of actual effort and durations and comparison with planned figures
- Ability to play "what if?" scenarios without disturbing the base plan

- Support for the concept of milestones
- Critical Path Analysis and identification of slack time

Desirable features

- Network (LAN) and multi-user support
- Support for sub- and super-projects and consolidation
- Full PERT support including calculation of early, likely, late dates and probabilities
- Ability to handle effort-driven and duration-driven scheduling. Effort-driven is where we specify the effort required and the resources allocated and the package will scale the duration accordingly. Duration-driven is where we specify the duration and this time will be assumed to be used for each resource assigned to the task.
- Resource smoothing and leveling
- Resource histograms and calculation of costs versus time
- Performance to handle large numbers of activities (>300) and human resources (>10)
- Support for fixed-date activities which may not be adjusted by automatic scheduling, e.g., a board meeting
- Plotter support
- Online, hypertext style help facilities
- Online, progressive tutorials
- Allowing user to customize screen views and reports in terms of position, scaling, content, sequence and filtering
- Ability to collapse and expand tasks easily based upon WBS collation
- Openness, that is, easy file exchange or real-time exchange of data between the package and other applications, including spreadsheets (e.g., Lotus1-2-3™, Excel™); graphical presentation packages (e.g., Harvard Graphics™, Corel Draw™, PowerPoint™); and databases (e.g., Microsoft Access™, Paradox™, dBase™)
- Ability to bring in standard plans from a library
- Individual resource calendars which can be used to handle training time, leave and differences in work-time availability
- Handling of nonresource related costs, such as rental of equipment, etc.
- Calculation (and perhaps graphing) of value of work complete versus budget
- Derivation of productivity statistics
- Support for versions and baselines

- Multiple views (e.g., Gantt and Resources) simultaneously visible on screen
- Ability to drive package in batch mode for data entry and/or capabilities to drive package in real-time from other applications, e.g., as an OLE server

Future of PM Packages

In the future we are likely to see much greater integration of project management software with other types of packages and facilities. Things we may see:

- Integration with e-mail, so that resources are alerted to impending tasks, or to serious deviation of actuals from planned delivery dates by automatic generation of e-mail messages
- Tie in to groupware which will allow various collaborators to work on aspects of the plan in parallel simultaneously, and to be aware of the other participants' actions, if these affect them
- Integration with group calendaring software to allow for automated scheduling of meetings and activities when necessary participants are available, but still respecting planned dependencies
- Ability to analyze the history base of completed projects to determine estimating performance, calculate and plot productivity trends and other useful data
- Support for Product Structure Models linked to Work Breakdown Models within the packages
- Tied to the above, the ability to define and manage a configuration. Control of versions and changes (see chapter 14)
- Links to methodology management tools and to CASE products. This may also include automatic recording of actuals as deliverables, in electronic form, are completed and verified in other tools

Case Questions

The questions for this Chapter assume you have a representative software package such as Timeline™, SuperProject™ or MsProject™ available to you. Before attempting these questions, you should work through the tutorial provided with the package.

MyWay Organizer

Q10.1

Capture into your package the activities defined in Q9.1. Specify durations as given and add dependencies to your plan. Print a Gantt chart of your plan. Compare the overall plan and timing with the one you generated manually for that question. (20 mins)

Q10.2

Print a PERT or Precedence chart of your plan. Locate the critical path and determine its length. Adjust the time for activity c to five days and observe the change in the end date. Change activity k to 8 days. What is the effect? Change activity e to 7 days. What is the effect? (20 mins)

Q10.3

Using the calendar facilities (if available) or dummy activities (if no calendar available), plan for a work week of Monday to Friday, with 8 hours available per day. Record two public (or bank) holidays. What effect does this have on the plan captured in Q10.2?
(20 mins)

Q10.4

Add resources to the plan as follows:

Name	**Roles**	**Activities where involved** (* indicates responsibility for task)
Yourself	PM	You chose
Mary Lloyd	Sen Analyst	a*, c* ,d* ,k*, l, m
John Fowler	Analyst/Programmer	b*, c, e*, f, g*, h*, j*, l*, m
Bill Goates	Technical Specialist	f*, i*, m

Ask the package to recalculate the schedule. Do you get any warnings? Has the end date changed? If the package allows, add costs to the resources ($275 per day) and cost the design phase. (30 mins)

Gleam Stores

Q10.5

Using the data provided in Q9.6, construct the macro-plan for the Gleam implementation projects. Do not capture any details for the sub-projects. Using the file provided by the instructor, import or link the sub-project for the Western Region. Calculate the duration and end time for the overall implementation process with this data included. You may find that some resources are overloaded. Resolve this manually and/or automatically, depending upon the facilities your package offers. No resource should be scheduled to work on more than one macro-activity in parallel (they are in different geographic locations!) (30 mins)

Q10.6

Using the plan captured in Q10.5, and the list of tasks from the imported sub-project, expand one of the other implementation tasks. Choose your own durations. Recalculate the overall project. (30 mins)

Handover Trust

Q10.7

Using your (or a provided sample) answer for Q9.9, capture the tasks and activities for the New Business Project. The Requirements Definition phase should be captured in detail. Capture all activities in the form of a Work Breakdown Structure. Define necessary dependencies between phases, and within the Requirements phase. Assuming the Requirements Phase took 7 months and 42 manmonths of effort, capture this actual completion information into the plan. Now collapse the Requirements Phase for further reporting purposes. Expand the External Design Phase into detailed activities (about 20). Revise the estimate for this phase based upon the variance in the Requirements Phase. (1 hour)

Q10.8 (Requires PERT facilities in package)

Using the data from Q9.10, capture the activities, including minimum, most likely and maximum times. Determine the critical path. Determine the earliest, most likely and latest completion dates. (45 mins)

ThoughtWell Books

Q10.9

Using the data from Q9.11 (provided by instructor) capture resources and allocate your team members as selected in Q8.6. Print an overall summary Gantt for presentation to management. Next produce individual resource calendars/schedules for each team member.
(40 mins)

11 *Project Execution*

Micro-scheduling

During project execution, the planning does not stop. There are always more details to be worked out and adjustments to be made. These occur at each project revision point and as we move through the tasks in the phases as highlighted in figure 11.1. It is particularly crucial for the project manager to allocate work optimally to team members and outside resources. This should be done with respect to skills, as well as availability of resources, workloads, and not ignoring the need for training and development. Matching the skills to the job is vital to ensure high motivation and productivity. We have seen in the chapter on

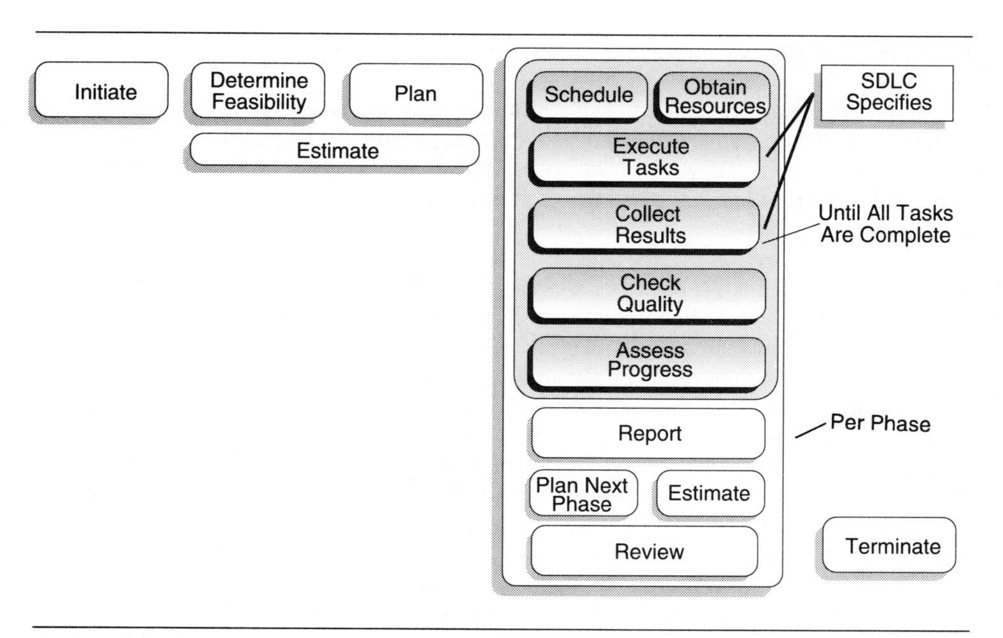

Project Lifecycle *Figure 11.1*

estimating the vast effect that individual skills can have on the effort required for a particular task. The level of detail required on a daily basis should probably not be committed to the formal project plan - it simply becomes too much effort and would divert attention from the real issues at hand. It is usually sufficient to adjust the plan weekly if necessary or at longer intervals.

The project manager should not be hidebound by the plan because he is related to it by sweat, but should monitor the actual performance of activities closely, so that any necessary adjustments can be made. We may in our planning have made some incorrect assumptions. Say we assumed that a particular programmer was proficient with a certain utility when in fact he was not. These things will crop up frequently. It is not imperative that the initial plan should be perfect, but it is vital that we adjust as necessary while the activities occur. If we find that things are going well for one person while another is struggling, we should not hesitate to ask the person who is ahead of schedule to help the other out. This builds team spirit and boosts morale, as well as providing learning opportunities for weaker project members.

Estimates at the summary and macro-levels may have been made with generic factors and average resource productivity in mind. These will need to be adjusted when you have an actual person assigned to a task. Another important psychological issue is commitment of people to the scheduled times. At the summary level, estimates are frequently made without the concurrence of the people who will eventually perform the work - often we do not know yet who this will be. Before tasks are actually assigned and begun, we need to re-visit these estimates and discuss them with the people assigned. This process can provide valuable adjustments to make the estimates more realistic, as well as increasing the team members' commitment to them. People are far more committed if they have had a say in the scheduling and have agreed to complete the work by a certain date.

Obtaining Resources

Care should be taken to confirm availability of outside resources well ahead of time. Providers may have promised them to you six months ago, but are they still available and have they remembered? They may have had staff changes and be able only to offer someone with lower skills. This will have an impact on the duration of tasks and requires adjustment of your plans.

If you find that the team assigned is not coping, and the work must still be completed by a given date, you may have to negotiate for more resources with management. Be very careful, however, of adding resources to a project in midflight. To quote Fred Brooks, "adding resources to a late software project only makes it later." New resources can be very disruptive to the team and can divert the attention of team members, who have just invested a huge amount of effort in getting themselves through a learning curve, from the task at hand. Bear in mind the extra communication overhead which you will incur, thus decreasing productivity for everyone and increasing costs. Generally, if a project is in trouble and needs more resources, the best thing to do is to bring in a single very highly skilled person - a consultant to address the specific problem. The brief given to the consultant should be to work with the existing team members, assisting them to solve the problems and transfer to them the necessary skill so that they become self sufficient when faced with a similar problem in the future.

It is also vital to keep management, the steering bodies and the user and sponsor areas fully on board as the project proceeds. It is all too easy for them to forget and focus on other things once the initial enthusiasm and fanfare of a new project has died down. If you are not visible to them and constantly informing them of progress, a request out of the blue for additional resources, time or funding will not go down at all well.

Monitoring Parallel Projects

During project definition, we identified parallel projects upon which we might depend for implementation. For example, we may need a menu system created by another team in order to run our system. We need to keep abreast of the progress and developments in these projects to ensure that the expected deliverables are in place when we need them. One company completed a very expensive sales analysis system on time and within budget, only to find that it was useless because the functionality to capture the input data in the point of sale system was delayed by some 12 months.

In a similar vein, we need to inform any other projects which depend upon our deliverables of any changes in our schedule which will affect them.

Executing Tasks

While we should keep a close eye on progress, we should not be acting as a policeman and breathing down the necks of our staff. Tasks should be allocated to individuals to complete. They should understand that they assume responsibility for their completion and the quality of the deliverables. They should also know that the project manager is available to help them resolve any issue outside their control which could impede the work or negatively impact quality. Managers with a technical background should avoid getting stuck in and doing the job. Sometimes you have to take a back seat and let the team members work through it themselves. It is a little like being a parent, even though you know the child may fall, you have to let him climb the tree, otherwise he will never develop the skills to do it unassisted. Of course, you should provide instruction, guidance and advice and support when requested. Successful managers are those who achieve results through their people.

Staff should be encouraged to develop a results-oriented work pattern. This is achieved by measuring results, not busywork. There is a management principle that says you get what you measure. If you measure people's performance by the cut of their tie and whether or not they are at work at one minute past eight, you will get employees who are at work on time and wear the kind of tie you like. This may not make them productive though. If we want productivity, we should measure it. This is a fairly involved task for systems work, and we will tackle it in some depth in the chapter dealing with measurement. The principle is clear: We encourage people to produce good results by defining what good performance is, providing examples, supplying a supportive environment, and being demanding. We should be flexible on things which allow individuality and do not negatively impact performance. If your star designer likes to work barefooted, and he is not working with clients who find this a problem, so be it.

Checking Quality

Quality can never be checked into anything - it must be built in. That said, it is vital to ensure that we do check quality to ensure that anything which is substandard is caught as early as possible, and corrected before it is visible outside the project. This involves various activities, such as walkthroughs by peers, inspections and quality reviews. These will be explored in detail in the chapter on Quality Management. For now, let us accept that each and every deliverable produced by the team should be quality assured before it is accepted and credited for progress, and before it is passed on as input to a subsequent task. We cannot build a quality house from inferior bricks and timbers.

Assessing Progress

As we progress, we need to collect actual work results (deliverables) as well as actual completion dates and actual effort expended. The latter should be captured into our project management tool. This will allow us to accurately establish how much of the work is complete, as well as how accurate our initial estimates were. If we have spent twice as much effort on requirements definition as we expected, chances are that design will also take about twice as long as we initially planned. Progress assessment is vital to catch any deviations from plan as early as possible. The earlier we catch them, the easier, cheaper and faster they are to correct.

Project Meetings

These can be a bore and a pain, or alternatively, an opportunity for team-building. Meetings should normally be held weekly. Friday afternoon is a good time. This is a time when people are generally less productive than normal anyway (the weekend beckons!). People also have an incentive to get finished and this can make progress through the agenda swift. The meeting should involve all members of the team, as well as any outside resources who are currently directly involved. On certain occasions, such as major review points, there may be management, sponsor or user representatives present as well. An agenda should be prepared to ensure that the meeting has focus and does not waste the time of participants. Progress on all activities should be checked with the parties concerned. Work assignments and workloads for the next week should be reviewed, and consensus and commitment sought. Problems being encountered should be raised for discussion. If they can be solved with the input of those present at the meeting, this should occur and the resolution recorded. If they cannot be resolved in the meeting, an individual should take responsibility to pursue a solution and advise the group. This responsibility (action item) should also be recorded. Minutes should only list responsibilities for actions and decisions taken, not all of the discussion. Minutes for a normal project meeting should not exceed one page. It is essential that the minutes are prepared and distributed rapidly. If we hold the meeting on a Friday afternoon, everyone should have them on the next Monday morning. Neatness, spelling and pretty fonts are less important. Get the information to the people who need to act upon it. Electronic mail facilities can be a boon here.

Walking About

In the 1980's Management by Objectives (MBO) became very popular. This is still a good approach for project management since we have definite objectives which we wish to achieve. We should couple it with a far more basic approach which is regaining favor at the moment, namely Managing by Walking About (MBWA). Don't get so involved in talking to the project management tool, I.T. management, and the steering groups that you fail to pay attention to the issues facing your staff at the coal face. Get out of your office, chat with the staff and see how things are going. Lend advice where you can. Pick up problems in an informal way and act to resolve them just as if they had been raised in the formal weekly meeting. Often staff, particularly juniors, will have problems raising issues in formal meetings with seniors present. They may be much more relaxed and open when discussing one-on-one. Keep an eye out too for the morale of the team and the quality of work being produced.

Crises of Realization

It is a fact that humans don't relate easily to long-term goals. If the next deadline is six months away, we tend to go into "manyana" mode. We feel that there is lots of time, and so we can spend time now thinking, relaxing and generally getting ready for when the real work starts. Usually we wake up to the fact that we have not really begun the task when the deadline is imminent (no more than a few weeks away). This is called the "crisis of realization" illustrated in figure 11.2. Suddenly we begin to work hard, put in extra effort, work overtime and try to meet the deadline. Unfortunately, we frequently leave it too late, and cannot catch up. The solution to this phenomenon is to schedule in detail so that we always have an imminent deadline, no more than one month away. In this fashion, we will have many mini crises of realization, but we will catch up each time, because we can sustain the extra effort for a short burst. The moral is that we need to plan in detail and have frequent deadlines with associated deliverables. This will keep the team under constant slight pressure, which is positive.

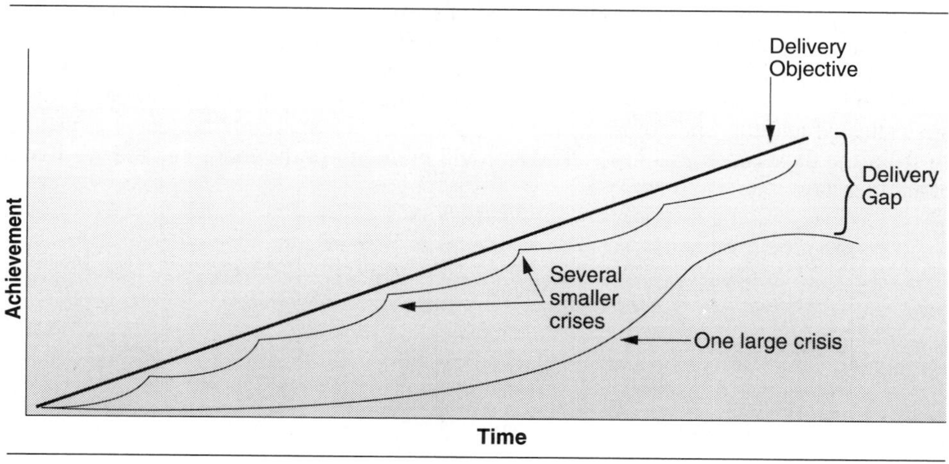

Crisis of Realization *Figure 11.2*

Abandonment

It is an unfortunate fact that projects can be abandoned during their execution without meeting their objectives. This can occur for many reasons, including the following:

- The original scope of the project was too ambitious and it proves too expensive or technically infeasible. Solution: Do a proper scoping and feasibility analysis at the beginning. Consider scaling back instead of canceling

- The skills to complete the tasks are not available. Solution: Consider lower-tech ways of achieving the same results. Draft consulting skills in from outside. Consider skill feasibility at the project outset

- Budgets dry up (This can sometimes be beyond your control, but is avoidable in many cases.). Solution - try to ensure that your project is seen as a critical one by senior management. Sell the benefits that it will deliver and keep them informed of real progress. Keep within the assigned budgets to assure them that the goals can be achieved for the agreed amount

- Loss of key staff members. Solution: Watch staff motivation carefully. I.T. staff are highly motivated by their work (more on this in a later chapter) and are extremely unlikely to leave in the middle of an exciting project unless they have become very frustrated. They like to see a job complete and in production, so something must be wrong if they want to leave in the middle. Of course, sometimes there are circumstances beyond our control, but we can certainly minimize the problems by being sensitive

- Change of corporate policy, direction or priorities. Solution: Stay close to the politics, policy-making and direction-setting in the organization so that the project can track these if necessary

When the project still has to be terminated early, this must be handled with great care. It can be extremely demoralizing for the staff to see their hard work getting shelved. You need to pay particular attention to reassigning individuals to try to ensure their continued career growth. Also, emphasize the positives - for example, the skills which they have gained through the project. Try to see if there are things that the project created which can be salvaged and used in other projects which will continue. Watch your own morale - projects do get canceled and this should not be taken as a sign of personal failure. See what you can learn from the experience and grow. The sun will come up in the morning, and we live to fight another day.

Case Questions

MyWay Organizer

Q11.1

Using the plan you built in Q9.1 through Q9.4, record actual effort and durations as follows:

ID	Description	Planned Duration	Actual Duration	Planned Effort	Actual Effort
a	Prototype User Interface	10 days	11 days	10 days	11 days
b	Design Phys. File Structures	2 days	2 days	2 days	3 days
c	Map Funct to Module Structures	3 days	1 day	6 days	3 days
d	Define Interfaces	2 days	2 days	2 days	3 days
e	Define global memory structures	2 days	1 day	2 days	1 day
f	Compression algorithm	5 days	10 days	10 days	9 days
g	Design common modules	5 days	7 days	5 days	9 days
h	Design normal modules	8 days	6 days	8 days	8 days
i	Predict performance	3 days	4 days	3 days	6 days
j	Define standards for coding	5 days	12 days	5 days	4 days
k	Devise test plan	5 days	4 days	5 days	5 days
l	Prepare test cases/data	5 days	7 days	10 days	10 days
m	Review phase deliverables	2 days	1 day	6 days	5 days

The project manager has expended 40 mandays which are not included in the above.

How is the project doing in relation to planned completion dates? How is it doing in relation to budget? (15 mins)

Q11.2 (requires the use of a project management package)

Record the data from Q11.1 against the plan created in Q10.1 through Q10.4. How is the project doing in relation to planned completion dates? How is it doing in relation to budget? How would you adjust estimates for the next phase (Programming and Unit Testing)? (15 mins)

Q11.3

Use the plan you built in Q9.1 through Q9.4 as a starting point. Assume the project is about to commence. You have heard that Mary Lloyd will be leaving the company. You have no choice but to assign her work to John Fowler until a replacement is available, which will be in two months time. What will the effect be on the overall plan in terms of delivery dates and cost? (20 mins)

Q11.4 (requires the use of a project management package)

Use the plan you built in Q10.1 through Q10.4 as a starting point. Assume the project is about to commence. You have heard that Mary Lloyd will be leaving the company. You have no choice but to assign her work to John Fowler until a replacement is available, which will be in two months time. Make these changes to resource assignments within the package. Recalculate the schedule. What will the effect be on the overall plan in terms of delivery dates and cost? (20 mins)

12 Measurement

Why Measure?

Figure 12.1 indicates that measurement should be performed continually throughout the project, as we execute and complete tasks.

Few organizations would embark on major capital expenditure without a thorough feasibility study, a budget specifying how and when the money will be spent, and then carefully tracking delivery of the products they were purchasing, or having built. It seems strange then that organizations routinely do this for software projects. But are software projects really capital investments? Absolutely! In a recent study conducted by one of the authors involving some

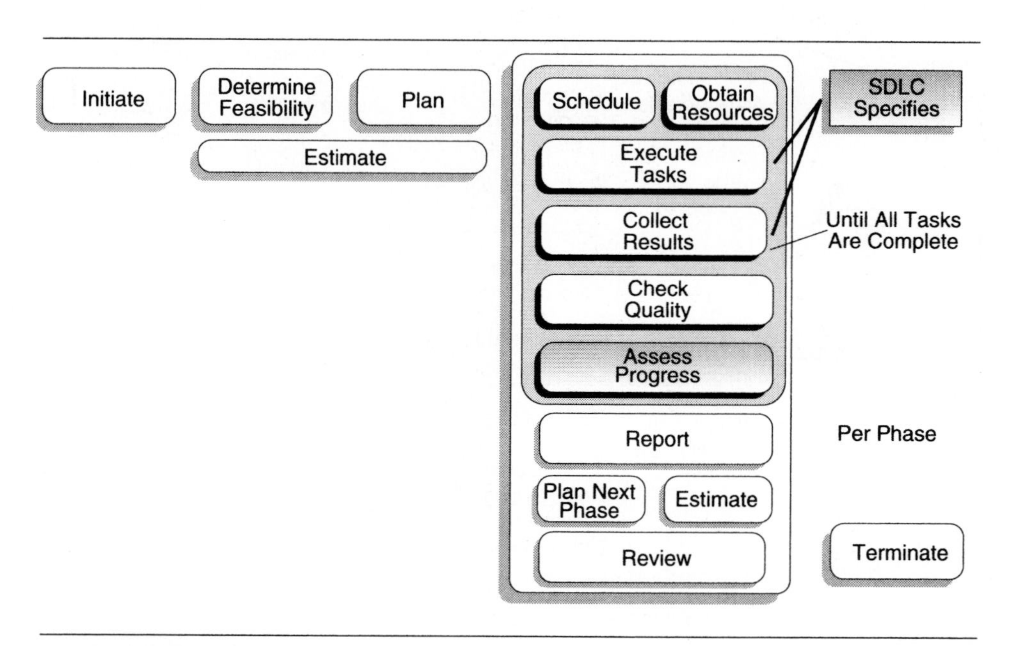

Project Lifecycle *Figure 12.1*

45 projects across 20 organizations, the average project size exceeded 400 manmonths, giving a cost of over $1 million per project. This money will be spent over a period of time to create an asset from which the organization hopes to gain future benefits. Sounds like capital investment to us.

Another very good reason for measuring, is to gain control of the software development, maintenance and implementation processes. Costs tend to migrate from any area which is placed under scrutiny. If we closely monitor bug removal during testing, we will find that the testing process becomes more efficient and effective. If we concentrate on rewarding people for creating reusable code and thus introduce ways to measure reuse, we will achieve higher reuse. Thus measurement of appropriate things will encourage development in desirable directions.

We may also want to measure to establish a baseline. If we are going to introduce a new technology (e.g., CASE) or methodology, we cannot verify the benefits claimed for it, or understand the impact on our projects, unless we have a basis for comparison. What we need to do is measure the situation before introduction of the new approach, and again afterward. Only then can we actually see whether the change is yielding the desired results, and to what extent. Watts Humphrey points out in his model of software process maturity (figure 12.2), that to successfully use technologies such as I-CASE, an organization has to have a stable,

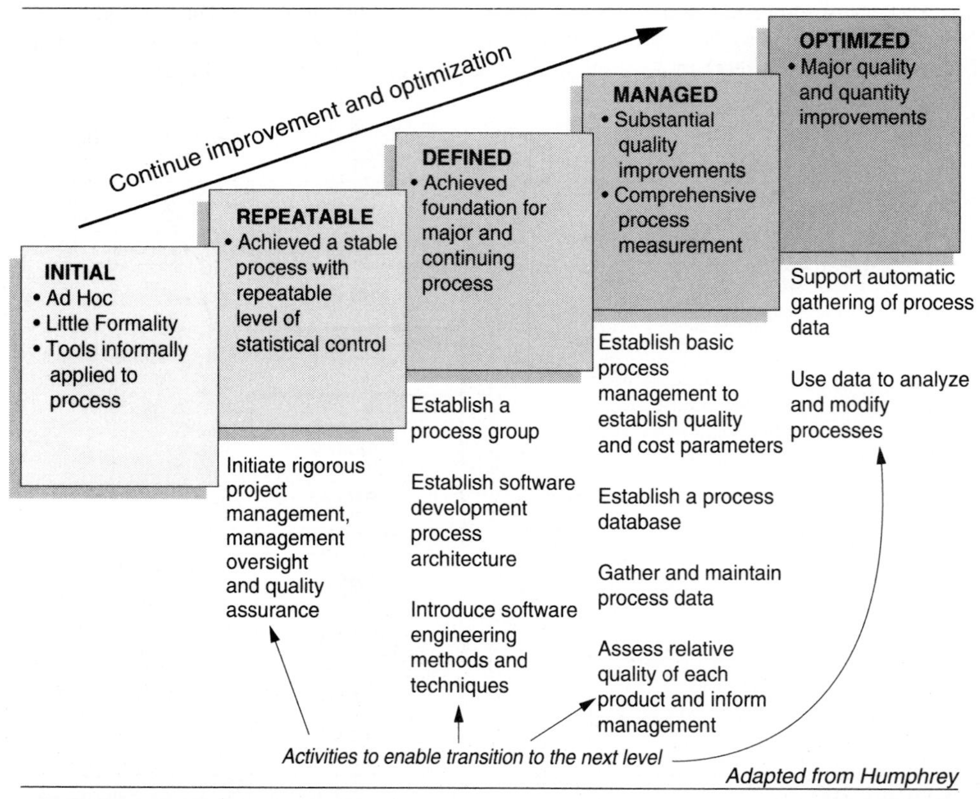

Humphrey's Five Levels of Maturity *Figure 12.2*

managed, repeatable development process which is under statistical control. This means that we have to understand how we do things now, and have the baselines against which to compare when we introduce any changes. To move to the higher levels of maturity, where significant productivity and quality benefits are achieved, organizations must implement a measurement process, collect data, analyze it, and act upon the results. Measurement of both quality and productivity is necessary. We will return to this theme later. Peter Drucker sums it up well:

If you can't measure it, you can't manage it.

It seems surprising then, that in numerous surveys conducted among software development organizations, less than 10 percent have a formal metrics program, or any real quantitative measurement of their software production process in place. Is it really so difficult? Or perhaps we are just not too keen on what we might find.

Returning to our recurrent theme - if we want to become more professional as managers and software developers, we must institute proper disciplines of planning, estimating and measurement. We acknowledge that measurement in the software field can sometimes be difficult, but to quote Tom Gilb:

Anything you need to quantify
can be measured in some way
that is superior to not measuring it at all

What Should We Measure?

Viewed from a management perspective, a system development organization can be thought of as a systems factory, as in figure 12.3. We feed resources (money, people, skills, materials, machine time, etc.) in one side, together with requirements, and we receive products (software, documents, services) out the other side. Some fundamentals to measure are the quantum and value of resources we are putting in, the quantum and the quality of

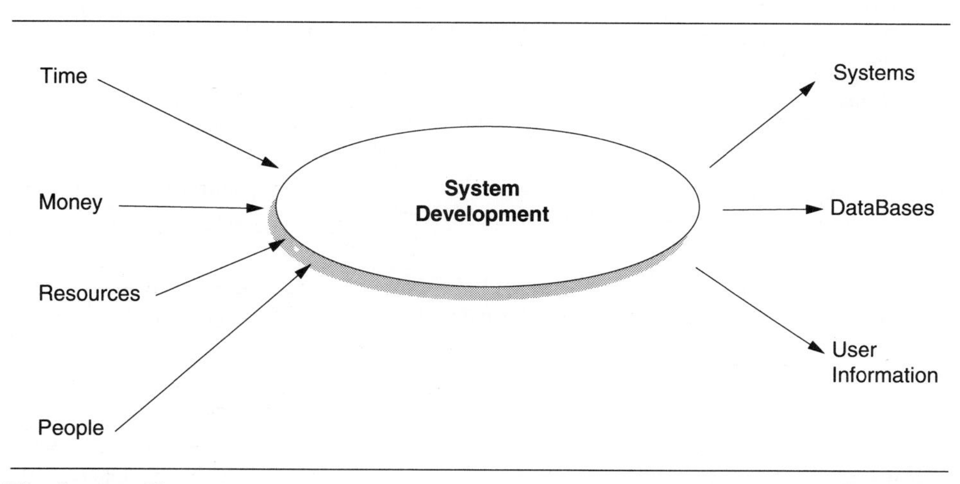

The System Factory *Figure 12.3*

what we are getting out, how long it takes to receive a certain amount of output, and what the ratio is between what we feed in and what we get out. This will give us an indication of how effective our "factory" is.

Other measures are appropriate for other types of projects. For example, a maintenance project might be measured by how responsive the team can be to the changing business requirements. A hardware implementation project might be measured by the number of hardware devices installed per unit time, and so forth. Some factors remain constant. We need to know:

- The cost of the resources going in, so that we can compare this with our initial budget on which the project was justified
- The amount of output produced, and when it was delivered. This allows us to gauge progress against the planned delivery
- The quality level of the output. There is no point in receiving masses of deliverables if they are garbage
- The ratio of input costs to value of output will provide us with a measure of team productivity and the development process and tools
- *When* the deliverables have been produced, as this can affect their value in terms of anticipated benefits. It also allows us to track progress versus the original plan.

Figure 12.4, "Measuring Productivity", provides a framework showing how different measures can be related. From this it is apparent that, to measure productivity, we need a number of more basic measures in place first. It is somewhat like accounting for a business.

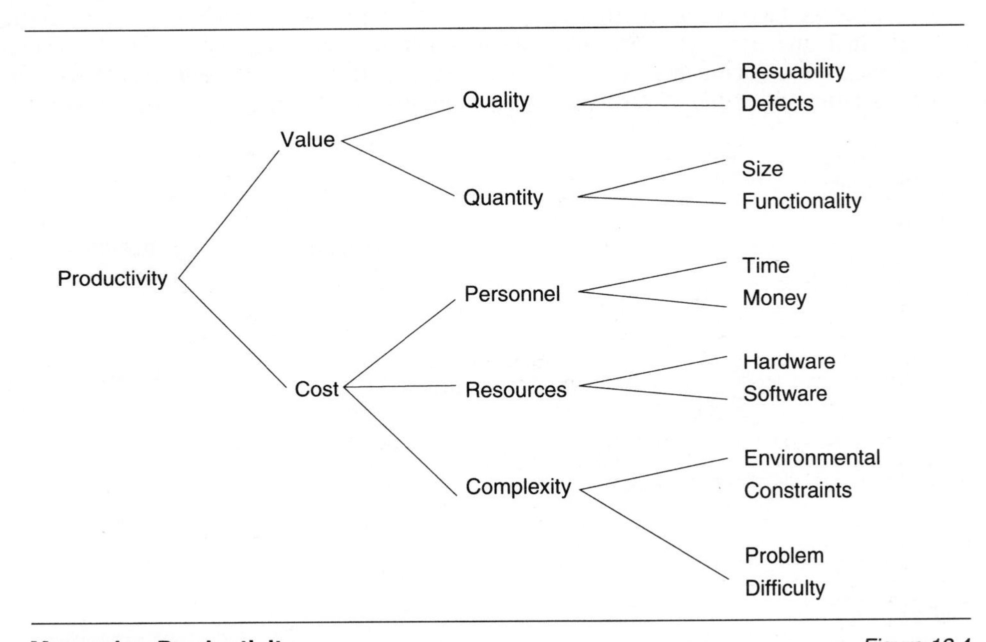

Measuring Productivity *Figure 12.4*

We might really want the profit figure, but we need the expenditure and revenue figures before we can calculate it.

Limitations

In measuring systems projects, the inputs are relatively easy to quantify. They are normally related to people costs, equipment costs, and direct expenditure. People costs are normally measured by time spent on the task multiplied by a cost per unit time. The cost is usually a composite of salary and benefit costs, together with overhead costs. Overhead costs include the provision of office space, typing services, telephones and other infrastructural items.

Equipment costs can normally also be translated to an hourly or daily rate and calculated fairly easily. Direct costs include such items as purchase of manuals, copying charges, project team lunches and so on and are simply tabulated.

Measuring output is much harder. We have already discussed (in chapter 7) the limitations of Lines of Code for measuring the size of a product. To recap:

- They are language dependent
- They do not measure functionality
- You get what you measure - measuring lines of code encourages verbosity
- Nobody can really agree on how to count them

We came to the conclusion that, although it too has its limitations, the best available measure at the moment is Function Points. We will use this technique to try to gauge such things as relative productivity of a project team. Other suitable measures will have to be introduced for nondevelopment projects.

Measuring Progress

One of the most difficult things to get a handle on is just how far along, i.e., how complete, a project is. We are all familiar with the 80 percent syndrome. We ask the project manager how things are going and we are told, "We are 80 percent complete." Unfortunately, this remains the status for 80 percent of the project lifecycle. How can we *really* find out what is going on? The answer lies in assigning a value to each deliverable or task. When we set up the project plan, we devised a Work Breakdown Structure (WBS) and a Product Structure Model (PSM). Either of these can be used to track progress in a very simple, quantitative way as shown in figure 12.5.

The model we choose to use must be completed at the planning stage (although of course it can be expanded at each review point). It must also record an estimated effort and duration for each task. To record progress, as we work through the project, we record on the model, for each deliverable *received and quality checked,* the actual effort and delivery date.

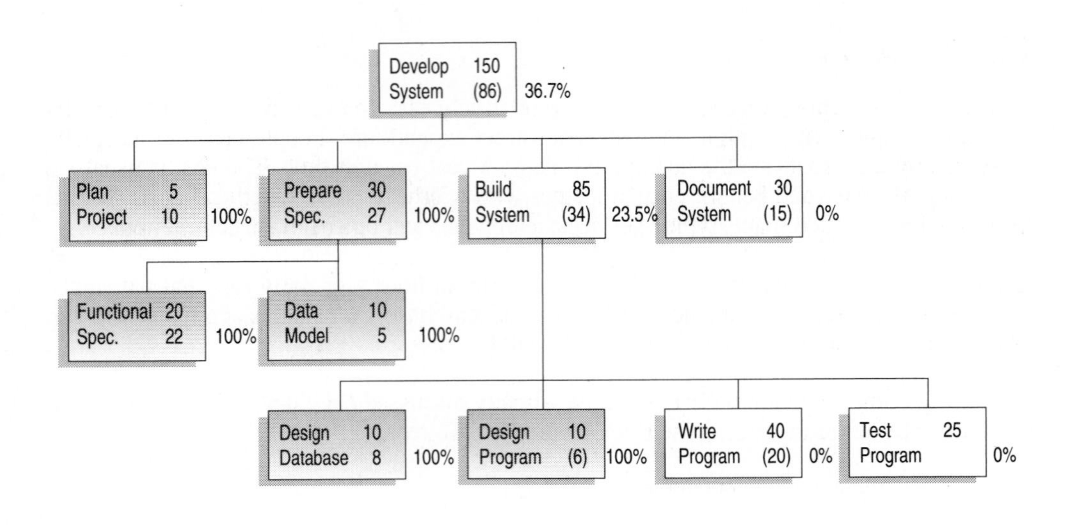

Recording Actuals on WBS *Figure 12.5*

Value of Work Complete

At any time, we can then compute our Value of Work Complete (VWC). This is done by summing the *budgeted* effort for each completed task/deliverable as shown in figure 12.6. This is the value of the work delivered to date according to the original plan which represented what we agreed to deliver at what price.

To determine the actual expenditure to date, we simply sum the *actual* effort for each completed deliverable *as well as those which have begun but are not complete*. The actual effort expended against these activities should be updated on the model at each project meeting.

The Binary Deliverable

Please note that no deliverable is ever counted as partially complete. It is a binary situation: either it is finished (including quality checking) or it is not. Less than 100 percent is 0 percent. We *can* derive percent complete figures for summary tasks or products higher up in the model by calculating the proportion of the child-box effort which is complete. This is shown in the accompanying diagram, "Recording Actuals on WBS". The ratio of the parent attributable to a child-box should be derived from the budget figure of the child-box divided by the budget figure for the parent-box. Thus if the parent was budgeted to require 85 days, and the children 10, 10, 40 and 25 days respectively, and the first two child-tasks are complete, then the parent can be said to be 20/85 = 23.5 percent complete. It will still be 23.5 percent complete if the third child has begun and has used 20 days, but is not yet complete.

Task	Budgeted	Complete	Actual
Plan Projects	5	y	10
Functional Specifications	20	y	22
Data Model	10	y	5
Design Database	10	y	8
Design Programs	10	y	6
Write Programs	40	n	20
Test Programs	25	n	0
Document System	30	n	15
	150		**86**

Value of work completed = 55
% Complete = 55/150x100 = 37%
% Budget Expended = 86/150x100 = 57%

Calculating Value of Work Complete *Figure 12.6*

This philosophy is extremely important. It will encourage the team to plan in detail, since they will not get credit for deliverables which are not totally complete. If they plan for a deliverable which is 40 days long, they will appear to be behind schedule for 40 days while they spend time on it without receiving credit for work delivered. The smaller the deliverables become, the sooner they will receive credit for work complete, and the less apparent pressure there will be. This is very positive since short horizons eliminate the "manyana mode" which we spoke of earlier. A slippage on a short task is also far less damaging to the overall schedule, since it allows us to catch problems early and to catch up with a sustainable burst of energy.

We will show in the next chapter how the same figures can be used very effectively for reporting progress.

Timesheets

Unfortunately, collecting actual effort involves that great bugbear, the timesheet. Project staff hate them, and we can usually sympathize wholeheartedly, having had to endure them ourselves. They are, unfortunately, a necessary evil. We can minimize the problems associated with their use by:

- *Not recording in too much detail.* It is not realistic to try to track every 15 minutes. One organization we saw had timesheets that broke the day down into tenths of hours. That means each division was 6 minutes! When the time spent by staff was analyzed in this excruciating detail, what did they find? You guessed it: About a quarter of everyone's time was spent completing the timesheets. That is clearly ridiculous. People also resent it if they have to fill in that they went to the rest room, made a private telephone call, or filled in the crossword over lunch. We all need some personal private space

- *Making the purpose clear.* We should make it clear to staff that the timesheets are not being used as a policing mechanism to see if they are working hard enough. They should understand that they are being used to gather information which will help us to find out where the time goes, to pick up problems early when people are having to put in inordinate amounts of extra effort to remain on schedule, and where our initial estimates were very inaccurate. Stress that the information is to be used to help all of us as a team to work smarter rather than harder. Of course, having promised this, you have to stick to it. The first time you berate an employee for not putting in the required amount of effort based upon the timesheets, you immediately lose all good data since others will be too scared to put in real numbers that you may not like

- *Don't insist that the numbers add up* to some predetermined "required" number of hours per week. If you do this, people will pad activities to make it true. The half hour spent on testing will become 1 hour to cover the time spent discussing a new design idea with a colleague over the partition. Allow unspecified time; we only want to know about time that was expended against the project, and possibly time that couldn't be spent on the project because of problems needing resolution

- *Make it as painless as possible*. Design a simple, easy-to-complete timesheet which is not too detailed. Consider figure 12.7. We find that the most effective manual format is a sheet for the week which looks like a week-to-view diary. Time divisions should represent hours. Put space at the top to record projects being worked on and a code for each. Then, during the day, as we change activities, all we have to do is bracket the time we spent and put a project code and activity next to it. At the end of the week all we have to do is add up the bracketed times for each project or activity

Name *Hilary Rose* **Week Starting** *12 May 1995*

Project	**Activity**	**Code**
Debtors Rewrite	*Impact Assessment*	*DRSimp*
do	*Spec changes*	*DRSspec*
Maintain Slea	*Test Change 100-103*	*MSLStest*

	Monday	Tuesday	Wednesday	Thursday	Friday	Saturday	Sunday
07h00							
08h00							
09h00	*MSLStest*	*DRSspec*					
10h00			*DRSspec*				
11h00							
12h00							
13h00							
14h00	*MSLStest*						
15h00							
16h00		*DRSspec*	*MSLStest*				
17h00							
18h00	*DRSimp*						
19h00							
Other		*2 hrs DRSspec*					

A Sample Timesheet *Figure 12.7*

combination. It helps if this can be done by a secretary or administration person, taking the drudgery away from the project staff

- *Set an example.* You have to do one too, on time, and to the required quality level

Use of automated project management tools can assist with gathering actual effort expended. Some packages have optional or built-in modules which facilitate capture of individual time expended (a kind of online timesheet). An example is the Project Management Workbench which has a well-integrated actual collections module. This summarizes times on a weekly basis and loads them automatically into the actual values of the relevant project plans.

Budget

To measure our performance against budget, we must of course have a budget set up. This is normally done at the outset based upon the expected expenditure per month. Expected expenditure should include:

- People costs
- External resource costs
- Direct costs
- Charges incurred for use of equipment, license fees, etc.

As actual expenditure is incurred, this should be summarized on a monthly basis and graphed against budget. Where inflation of people costs is a problem, we can choose to budget and record these in resource time units, e.g., manhours/person-days. This cancels the effect of inflation. For other expenses, a similar effect can be achieved by using an adjusted currency value for all amounts. For example, amount in 1995 Escudos.

The simplest form of automation here is a spreadsheet, although some project management packages have facilities for recording budgets and actual expenditures integrated with the project plan. Where these facilities are available, they should be used in preference to a stand-alone facility.

Staff Turnover

On large projects, or if you are managing several projects, it is very important to have a handle on staff turnover. Often people leaving are expressing their final frustration with the organization or project by voting with their feet. This should never happen if you are keeping tabs on morale on an ongoing basis, but it can happen if somebody is not paying attention. Any resignations should involve an exit interview where any information which requires action can be picked up. The person leaving may finally feel free to express his views in a totally honest and open way (He should, of course, have felt this way all along if we created the right climate). He may be able to give you clues as to how to correct problems, or alert you to the unhappiness of other team members. We probably do not want to achieve zero turnover, since this would probably lead to stagnation of the team, but we should try to aim for a figure below the industry norms. Remember that recruiting, training and getting a new member up to full speed is a very expensive business - it can easily cost half a year's salary and benefits for the position you are filling. There can also be negative effects on the productivity of the entire team. Members who are under pressure on their own assignments, and who have spent a long time on the project, may resent having to take time out to show a newcomer how everything works.

Productivity

Productivity is the ratio of useful output achieved for a level of input. In systems terms, we normally measure the input in resource units (e.g., manmonths) because this is independent of inflation, and people represent our largest cost. Output can be measured in terms of delivered functionality (Function Points). Productivity would then be expressed in Function Points per Manmonth.

This is an external measure of productivity, and does not take into account the difficulty of the problem (except as provided for in the Function Point calculations), the ease of use of the development environment, the time pressure under which the team worked, the team size, or any other special conditions or circumstances that affected productivity. We saw in the chapter on estimating the large influence that deadline pressure can have on the effort required to complete a project. Function points per person-month should thus not be used as a measure of productivity of the team members, unless we can adjust it for the factors mentioned above.

A commercial metrics approach and methodology, based upon the Putnam-Norden approach, which does compensate for the effects mentioned, is the Productivity Enhancement Programme (P>E>P) developed by the U.K.-based Butler Cox consulting organization. Their approach uses the concept of a Productivity Index (P.I.) calculated as a composite productivity score for a team. The calculation compensates for the effects of deadline pressure, technological environment and project size.

Requirements Change

Monitoring the level of requirements change is a useful activity from several perspectives:

- It allows us to make the team and the user community aware of the severe impact that changes have on productivity, costs and schedules. This encourages people to do it right the first time

- It allows us to monitor the effectiveness of our specification gathering and design processes. If we are frequently picking up major changes late in the lifecycle, then we should look to the approaches that we use in the early phases

- It encourages a formal change management process which is good for quality and reliability

Degree of change can be derived from the function point count of the changes expressed as a percentage of the total function point count for the project.

Quality

As we have seen, it is not sufficient to measure output only in terms of quantity: the quality of work produced must also be measured. Quality is a composite of many things, but essentially it can be summarized as shown in figure 12.8.

Quality is conformance to (client) requirements - *Philip Cosby*

We will cover various aspects of quality in detail in chapter 15. For now consider the relationships between various aspects of quality shown in the figure. We should also realize that quality is not intangible, it can be measured in a very quantitative unit, namely money. Of course, achieving quality is more complex, and involves many cultural and social issues.

Reusability

One of the best ways to increase productivity is to create output of high quality with no or minimal input. This is obvious given our definition of productivity. What is less obvious is how to achieve this in systems work. The simple answer is to reuse components which were developed previously, and which have already been tested. We know that they are good, since they are in production and have a measured, stable performance. We can obtain this output at minimal cost: i.e., the cost of cataloguing what we have, and identifying the correct component to reuse.

Reuse is not restricted to code modules, as many would believe. We can, with careful analysis and design, create many reusable components of various types. An engineer designing a building does not usually invent a new type of roof structure for each building. Instead he will draw upon a library of plans of suitable roof structures and choose one which meets the requirements for the particular building, project and circumstances. All he will then have to do is to scale the design, and make any necessary unique adjustments. In a

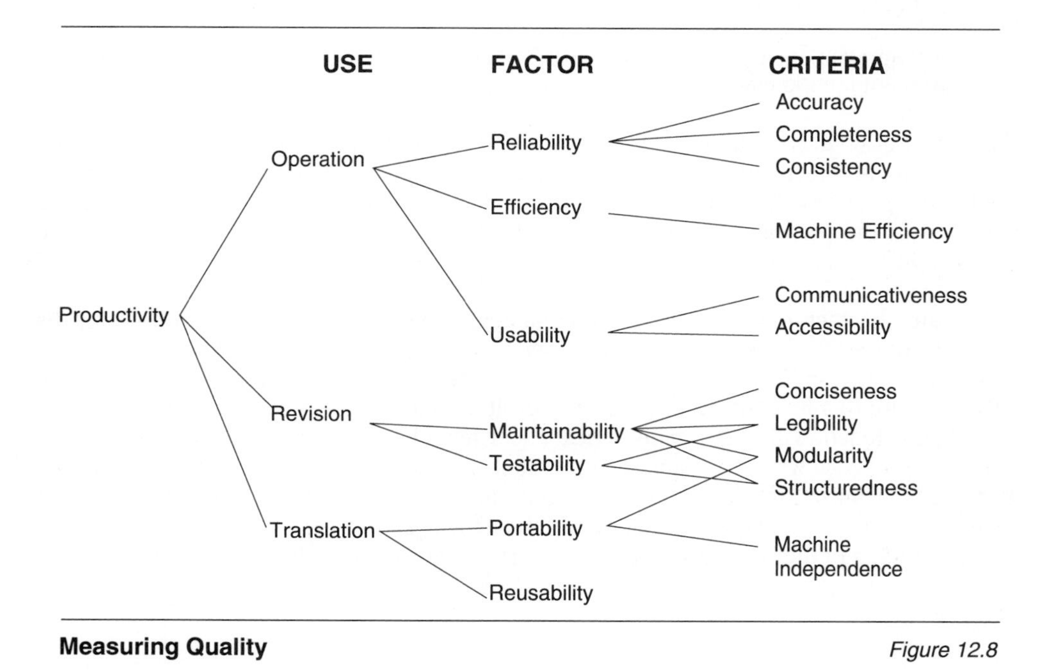

Measuring Quality *Figure 12.8*

similar way, we can reuse specifications and designs, as well as file structures, program code, test data and many other components - even parts of user documentation.

Reuse can occur in several forms:

- *Cloning* - taking an existing component and modifying it slightly to fit a new purpose
- *Scavenging* - working through existing code not originally written with reuse in mind to identify components which could potentially be useful in other contexts and projects
- *True reuse* - using components in a completely unaltered state

Cloning and scavenging do not realize the full benefits, since modified or adapted components will have to undergo the same testing as a new component. This can consume 50 percent or more of the development effort. True reuse is certainly the first prize. It does require some cultural changes in the organization, including:

- Promulgation of reuse as a goal and a strategy by management
- Establishing a library of reusable components into which projects will make contributions
- Pro-active creation of modules of code, and other components, with the express purpose of making them general and reusable
- Active use of reusable components during projects to reduce the amount of effort required to deliver the required product
- Monitoring of the level of reuse being achieved, and of the reusage count of components in the library
- Encouragement of reuse by rewards for creators of reusable components, and for users of components

The above, of course, requires careful selection of the manner in which reuse is calculated and the units in which it is quantified. Our recommendation is that components should be sized in terms of function points when collated into the library. Reuse levels should then be computed as the percentage of function points in the completed system derived from the library and used in an unmodified form.

Organizations which have adopted reuse in an aggressive way with very impressive results include Celite Sales Corporation with the Application Software Factory concept, reported by Swanson et al. in *MIS Quarterly*. Over a period of 3 years, they were able to increase software reuse to a level of over 90 percent resulting in huge cost savings and the ability to deliver new functionality extremely quickly and reliably. These benefits can be obtained with very little new technology, and without increasing staff levels. Object-oriented approaches promise high levels of productivity and quality as well as maintainability and flexibility, largely through a philosophy which encourages the identification of the generic, and the building of highly cohesive, slightly coupled components which are easy to reuse.

Integrating the Measures

To obtain a composite measure of effectiveness of a development group, we suggest that the aspects of quantity and quality should be integrated into a single measure. This is rather difficult, since they are normally measured in different ways. Productivity could be measured in function points per person-month; and quality in the Cost of Non-Conformance (CONC). We will detail the computation of this fully in the chapter on quality. Reuse should be actively pursued to decrease cost and increase quality, but it is not necessary to include this as a component of the final measure, since it will be reflected in higher productivity and quality where it is successfully achieved.

To obtain a single measure, we recommend turning both measures into a single unit, namely money per quarter. We can do this by calculating the value of a function point in the following way:

$$\text{Value of Function Point } (V\mathit{fp}) = \frac{\text{Cost of systems department per annum}}{\text{New function points delivered per annum}}$$

We can then determine the value added to the organization for a given quarter as follows:

Value added (V*a*) = (Function points delivered this quarter x V*fp*) - CONC this quarter

This measure will allow us to track on a quarterly basis how the value added to the organization (including aspects of productivity and quality) has changed. A bar chart showing the two components (value of FP delivered and CONC) and the resulting total, provides a good graphical representation allowing assessment of progress at a glance (figure 12.9). It also allows us to see the relative situation with regard to productivity and quality.

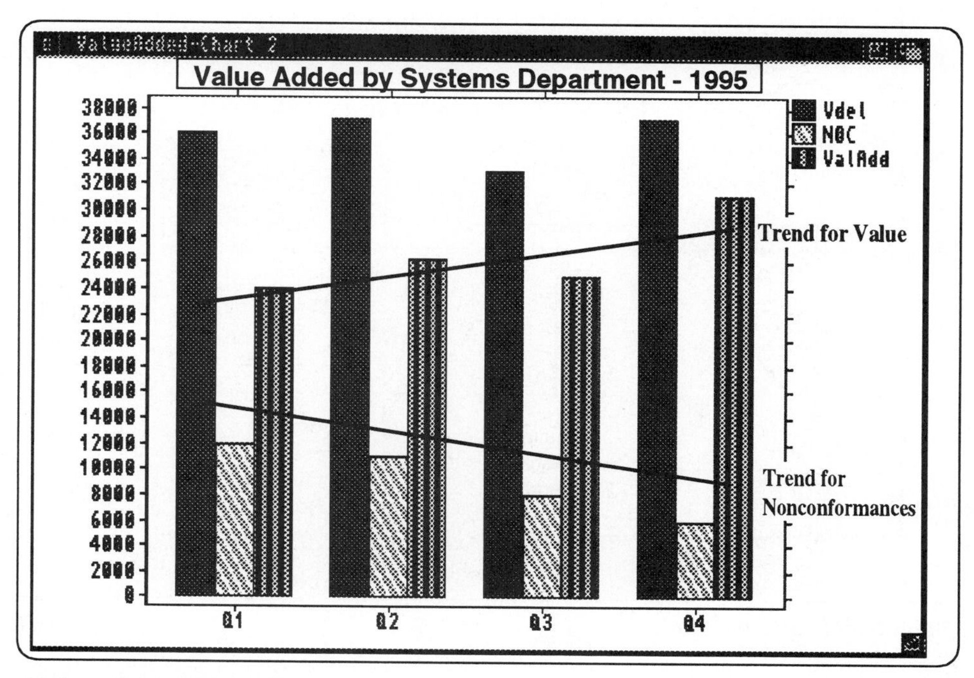

Value-Added Performance

Figure 12.9

Purists will have noted that our baseline figure for the value of a function point is calculated on a yearly figure. This is arbitrary, and you could use the total data for as many years as accurate figures are available. This should be calculated only once at the outset, and should be used as a basis for comparison thereafter. We can if we like, of course, compute new values for the V*fp* each year and graph these. The overall trend, if we are succeeding, should show that the cost to deliver value to the organization is declining.

Statistical Process Control

Statistical process control refers to a state in which a process is repeatable with known outputs being produced within established limits. As applied to software development (or installation, implementation of packages, etc.) it requires that we have the following:

- *A defined process* which is followed. This is normally defined by the methodology in use and the organizational standards
- *Appropriate measurements* in place to determine the level of output, the quality of output, and the efficiency of the process
- *A change management process*, whereby changes to the defined process are assessed for likely impact, implemented in a controlled way, and monitored through measurement for effectiveness

These concepts are illustrated in figure 12.10.

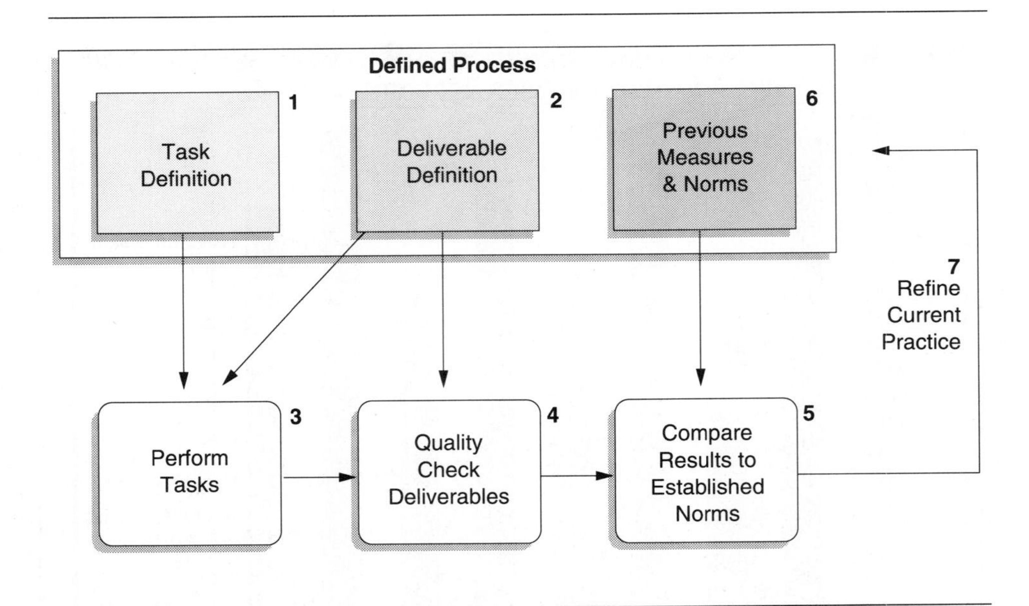

Statistical Process Control *Figure 12.10*

We begin with a definition of how the task should be performed, and what a high quality deliverable resulting from completion of the task should look like. We perform the task according to this standard, producing the required deliverables. We then measure the actual resource consumption, product quality and other relevant factors, e.g., performance of a piece of code. Next we compare these measurements to established norms derived from previous measurement. If this is the first time through the cycle, we will have to use measurements derived from other similar installations, or best estimates. On subsequent iterations, we will be comparing to our own history. This analysis will tell us how we are performing. If performance is falling behind previously established expectations, we should look to the process to try to identify the sources of problems. Alternatively, where a new approach to a task is suggested, it may be performed with careful monitoring, and the effectiveness of this new technique compared with previous benchmarks. If this shows that the new technique is more effective, then it may be wise to change the standard definition for the task, or deliverable(s).

Once a process is under statistical process control, we are then in a position to improve it continually. Measurements will tell us which parts of the process are not working as effectively as they should. We can then examine the activities which are carried out there, the techniques and tools employed, the skills of the people performing the task and various other factors to determine how the problem can be solved. Once a solution is proposed, it must be put into operation with proper measures to determine whether it is effective. If it is proved effective in a pilot project, then we may want to scale this up for general use.

The approach has been successfully employed by companies like Hitachi Software and Computer Sciences Corporation. We will discuss the details in chapter 15, which deals with quality.

Organizational Issues

Work performed by De Marco and Lister has produced some extremely interesting results related to organizations and relative productivity. They conducted "Coding War Games" over several years, starting in 1984. These involved the development of a consistent program from a given specification by participants from a wide variety of companies using their normal technical environment during normal work time. This obviously required the knowledge and support of their organizations. Over 600 developers from 92 organizations have participated over the years. The objective was to examine the issues of programming quality and productivity over a wide range of organizations, technical environments and programming languages. The surprising result was that productivity was affected far more by organizational issues than the technical environment or language. For example, they found that productivity varied by a factor of about 1 to 10 across all participants, but only by an average of 21 percent between members of pairs evaluated from the same organization. Significantly, there was no correlation between productivity and programming language, years of experience or salary.

This holds a very important message for managers: We should maybe look somewhere other than technology when we seek to increase productivity - maybe we need to look at organizational factors. De Marco and Lister began to do just that. They found that some work-place factors had extremely high correlations with the productivity and quality results obtained. These issues are summarized in table 12.1, and show a significant correlation between quality and quiet, as well as a direct relationship between density and noise.

Environment Factor	1st Quartile Performers	4th Quartile Performers
Dedicated work space	78 sq ft	46 sq ft
Quiet	57%	29%
Private	62%	19%
Silence phone	52%	10%
Divert calls	76%	19%
Needless interruptions	38%	76%

- Significant correlation between *quality* and *quiet*
- Direct relationship between *density* and *noise*

Coding War Environment *Table 12.1*

Tom DeMarco and Timothy Lister, *Peopleware: Productive Projects and Teams.* © 1987 by Tom DeMarco and Timothy Lister. Adapted from Table 8.3, p.49, by permission of Dorset House Publishing, 353 W. 12 St., New York, NY 10014.

They point out that the density of seating within a given area is directly related to the level of noise. Noise, in turn, has a detrimental effect on both productivity and quality. Research suggests that this may be due to the interruption of the mental condition of "flow" required for sustained creative work or work on complex models or problems. We are all familiar with the scenario where we feel that "nothing gets done between 9 and 5" because we are continually interrupted by the telephone, people looming over the partitions, noise from other offices, etc. It takes about 15 to 20 minutes to fully immerse oneself in a complex task - to "get your head around it". Only then can you really perform at a nonsuperficial level. A single interruption, no matter how short, destroys this condition. Several small interruptions during the day can thus rob us of an enormous amount of productive time. This is illustrated in figure 12.11. We will frequently find developers coming in early, or working late into the evening to escape these problems. "I get more done between 6 and 8 in the morning than the rest of the day."

Considering that the cost of providing office space for professional workers is only about 10 percent of their total cost, it seems silly to scrimp and save here when we could be costing ourselves orders of magnitude in productivity. These figures should be used to convince management to provide systems staff with productive accommodation. Key attributes include:

- *Sufficient office space, and work surface.* Most computer professionals need space to accommodate a terminal and/or personal computer, as well as room to spread out several design documents
- *Privacy.* Interruptions by people walking past, sticking their heads over the partitions, or having to answer someone else's telephone in an open plan area are extremely disruptive. We once saw a consulting firm where about 12 consultants not working at client premises were crowded into an open-plan seating area with one shared telephone per 3 consultants. This place was a nightmare when most of the people were present - interruptions were constant. When most of the people were away, the few left there spent all day answering phones and taking messages. These

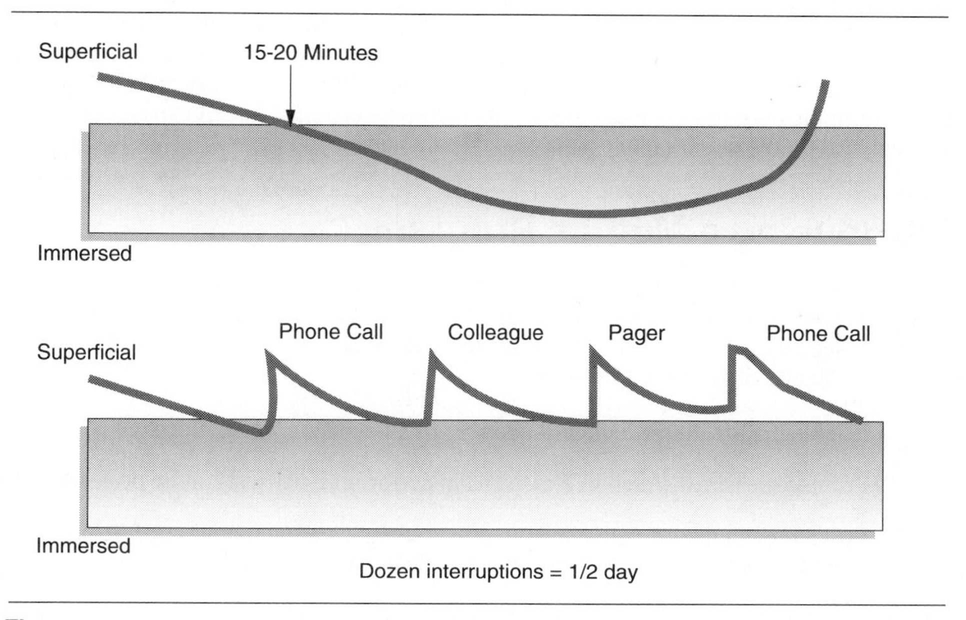

Flow *Figure 12.11*

were highly skilled and very expensive people. Management's excuse was that they wanted to encourage consultants to get back out to clients. This had some merit, but it also meant that anyone who needed to do some productive work in a quiet place (which was not available at the client site) couldn't do it at the office

- *Ability to redirect one's own telephone calls.* This allows us to concentrate on the task at hand, and to batch telephone work to a couple of half-hour slots per day. Calls should be fielded by a secretary who can judge urgency. Urgent ones could be put through, others have messages taken. A good strategy, if you have the facilities, is for the secretary to capture messages into an electronic mail facility. This makes them available to the team member immediately when they want to look at them, without interrupting them. It also provides a permanent record of important messages

- *Freedom to choose work hours* within reasonable limits. Generally people will put in many more productive hours if they can choose their work times to suit themselves. A working mother may want time off to fetch children from school at 2 pm, while others may choose to come in early and leave early to pursue sporting activities. Normally, there will be an agreed "core time" during which most people will be available. This is when meetings should be scheduled. Again, e-mail makes this kind of environment more practical, by providing an easy communication medium to people who may not be in their offices when you want to talk to them.

De Marco and Lister have even invented an "Environmental Factor" or *E-factor* based upon the interruption characteristics of the environment. This is calculated as follows:

$$\text{E-factor} = \frac{\text{Total Uninterrupted Hours}}{\text{Total Hours Present on Job}}$$

Good environments show E-factors around 0.38 and poor ones can be as low as 0.1. This means that to do the same quantity of meaningful work in the one environment, you have to be present 3.8 times as long as in the other environment. More typically, if you are present the same number of hours in each environment, you will produce only about 26 percent of the work in the poor environment that would be produced in the good environment. Clearly, we should fight to get our people a decent environment.

Introducing a Metrics Program

Introducing measurement in an organization is always an emotive issue. There is the danger that personnel will think that they are being checked up on, and respond negatively. We therefore need to sell the idea and the purpose of measurement very carefully. The objectives we set out should include the following:

- Establish a baseline which will allow us to know when we are improving
- Provide measures which will assist us in performing our work more professionally
- Provide evidence of value delivered to the organization to support I.T. organization motivations for increases, better funding, better facilities and office accommodation
- Allow us to gain statistical process control over our activities, thus providing us with the means to continually improve quality and productivity, without the need for individuals to work extra hours
- Illustrate to management and the user community the effect of changes in requirements and unrealistic deadlines, thus paving the way for a more amicable relationship

On no account should measures be used punitively at an individual or team level. The very first time you do this you will lose the trust and cooperation of the people you need to carry out the work and to provide you with accurate data. Your team members are not stupid. If you use the numbers to measure them, and respond in a negative way, they will fudge them. If we want meaningful figures, our use of the data gained must always be in a positive way, to assist our people to do the best job that they can, and to remove obstacles which prevent them from performing to their potential.

Self-Monitoring

One way to ensure trust is to allow people to monitor themselves. We have used this very successfully, even down to junior-programmer level. Essentially, we discuss with the individual concerned the project plan and the estimates for tasks assigned. We agree on reasonable times for completion of the assigned work, and also establish the estimated effort for the tasks as a measure of the value of the deliverables to be produced. The team member then monitors his own progress toward completion, only requiring the intervention of the project manager (or other senior staff) to quality assure components before they are formally credited in the overall project plan. Having this information available to individuals visually has proved to be a major motivator.

Allowing Mistakes

While the objective of measurement is surely to improve performance, we should be careful not to expect perfection in every case. If we create an environment where there is no room for error, we will stifle creativity and risk-taking. Both of these are necessary for a vital and high-achieving team. What we need to do is manage the overall trends over a longish period of time, not to expect that no one will ever make an error. We should, on the contrary, create an environment where the occasional mistake is acceptable, provided that this is when we are trying to stretch our capabilities. Mistakes in routine tasks should not be tolerated, but we do need to allow people to experiment in new areas which show promise. When these mistakes do occur, the individuals concerned should know that they can rely on the support of their team. We should, however, take every opportunity to learn as much as we can from the error, and not repeat the mistakes again. Having an environment where mistakes are accepted will allow those involved to share the experience fully and openly with the rest of the team, thus maximizing the learning opportunity.

Commercial Products

We have already mentioned one commercial measurement program, namely the CSC Index (previously Butler Cox) P>E>P program. There are several others which you should be aware of. These include:

- An integrated toolset from Quantitative Software Management (Putnam's organization) which includes: software to manage a database of project data, PADS (Productivity Analysis Database System); software to size systems; a resource estimating and planning tool, SLIM™ (Software Lifecycle Management); and SLIM Control, a project tracking tool allowing monitoring of actuals versus plan
- A product called METKIT™, which is a byproduct of the Alvey research in the United Kingdom and the European ESPRIT projects dealing with software construction techniques
- LOGICSCOPE™ is a commercial software measurement tool marketed by the French firm Verilogic. It offers static and dynamic testing tools
- QUALIGRAPH™ is a tool marketed by the Computor Kontor company in Germany. It helps to document the structure, complexity and other variables of software. These can aid in determining productivity, maintainability and quality

Summary

We feel that measurement is vitally important. If we do not know where we are, it is difficult to plot a course to go where we want to be, and to know whether our actions are taking us in the right direction. We are shooting in the dark if we introduce new technologies and methods without first establishing a benchmark, defining the objectives for the new approach, and measuring the results. Measurement does not constitute a large overhead relative to the kinds of inefficiencies we typically encounter in systems shops. Yes, it is difficult, and the techniques and measures available have their limitations, but even if the measures are not that accurate, the very act of measuring conveys a subtle message that this is important, and that alone can yield significant benefits. As time goes by, we can improve the overall process and project management approach that we use, and the metrics we employ as an integral part of this.

Case Questions

MyWay Organizer

Q12.1

You are nearing the end of the programming phase of the MyWay Organizer project. The Project started on August 1. Today's date is November 25. Your analyst has collected the following data:

ID	Phase	Planned End	Actual End	Effort Estimate	Effort Actual
a	Requirements	30 Aug	28 Aug	36 Mandays	46 Mandays
b	Design	1 Oct	30 Sep	74 Mandays	77 Mandays
c	Programming	5 Jan		165 Mandays	119 Mandays
d	Integration Test	1 Feb		44 Mandays	
e	Cross Platform Test	1 Mar		31 Mandays	
f	Marketing Material, License Agreements	1 Mar		15 Mandays	

Determine the Value of Work Complete, the % Work Complete and the % Budget expended. When do you think the project will be completed? (This may not give you enough detail to satisfy management.) (15 mins)

Q12.2

To obtain more detail, you have spoken with the programmers and obtained further data:

Of 12 modules, 6 have been written and 4 of these have been fully tested and are working correctly. Two more programs are being written. Modules are of approximately equal size and complexity.

Determine the Value of Work Complete, the % Work Complete and the % Budget expended. When do you think the project will finish, based on this data? (15 mins)

Q12.3

Using the data from Q12.1 and Q12.2, how is the project doing relative to planned delivery dates? How far ahead or behind schedule do you think it is? Would you change any plans at this stage? (15 mins)

Q12.4 (Requires Project Management Software Package)

Update your project plan for the Organizer project in the software to reflect the planned and actual durations and effort as detailed in Q12.1 and Q12.2. If your tool has the facilities, calculate the VWC and budget conformance. (30 mins)

Q12.5 (Requires Project Management Software Package)

If your package supports this, export the data regarding planned performance and actuals to a spreadsheet package. Call this up in the spreadsheet and ensure that it is usable. (We will use this later for derivation of graphs, etc.) (30 mins)

ThoughtWell Books

Q12.6

The design stage is nearing completion. The following deliverables were planned, and columns shown indicate their status. Q.A. = Y indicates that the deliverable has been quality checked:

	Q.A.	Estimated Effort Persondays	Actual Effort Persondays	Estimated Complete Date	Actual Complete Date
Feasibility Study	Y	30	24	June 15	June 13
Requirements Specification	Y	60	75	June 30	June 30
Design Specification					
Physical Database Design	Y	18	30	July 10	July 8
Performance Requirements		12	9	July 10	July 10
Prototype Screens		9	12	July 12	
Design Programs					
Branch Programs	Y	30	42	July 30	July 30
H.O. Programs		39	36	July 30	
Test SpecificationY		15	21	July 20	July 25
Programming & Unit Test		180		Aug 30	
Implementation		60		Sep 15	

The date now is July 15. The project began on June 1.

Determine the value of work completed and actual expenditure. Project the likely end date of the project. Do you think the project will be over or under budget? By what percentage? What will happen if you lose Lars Bontsen, who is assigned to complete the H.O. programs? (Another analyst programmer with similar skills is available to be assigned.) (45 mins)

13 *Reporting*

Accuracy and Consistency

Accurate reporting is essential to let those not directly involved in the day-to-day activities of the project obtain a clear picture of progress: what is being delivered, and what problems are being encountered. A report should be done at the end of each phase, but possibly more frequently, as shown in figure 13.1. Consistency is vital to allow proper comparison across projects. A senior manager will not only be evaluating one project, but several. If each project manager uses a different format for the report, and the phases are not consistently named, it is like comparing apples and oranges through distorting spectacles. We could get three project reports as follows:

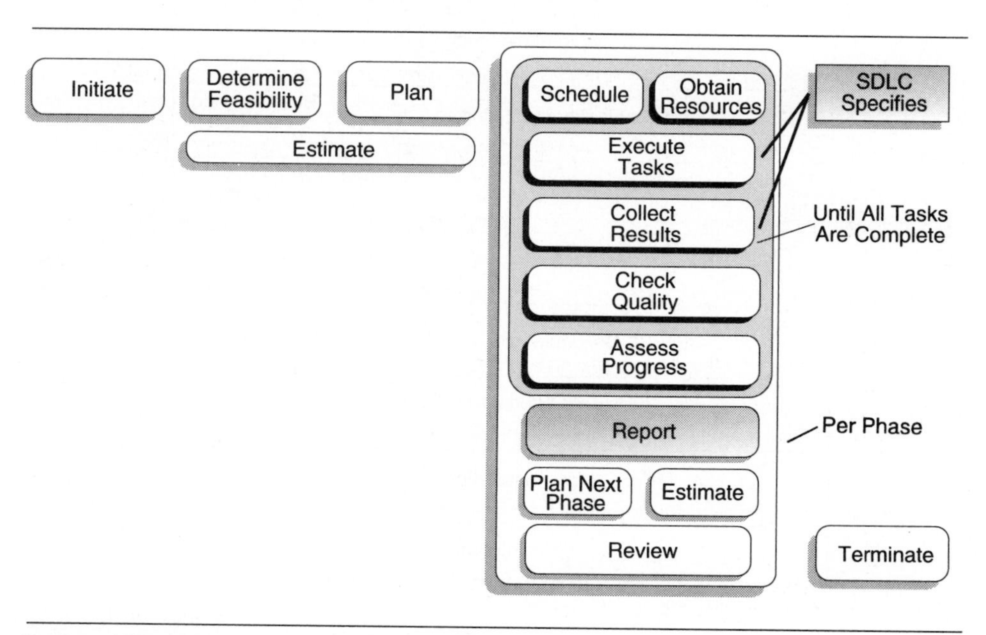

Project Lifecycle *Figure 13.1*

Project 1	"We are nearing the end of prototyping and have started formal data modeling"
Project 2	"We have finished functional modeling and data modeling, and are busy prototyping the user interface"
Project 3	"We are 80% complete on the unit testing phase"

What is the poor user to make of this? Are we behind, on schedule, or performing well?

A further problem is how to achieve consistency of reporting across a wide variety of project types. We have already discussed the concept of configuration management in previous chapters. This is a key tool in achieving consistent reporting. By having consistently named phases for a variety of project types, nontechnical managers and users can relate to where we are in a project. They will be able to get an accurate impression of progress on a variety of projects being run concurrently, even if these include a mix of development, package implementation, and technology deployment projects. For convenience, we repeat the configuration management diagram with the phase naming and baselines as figure 13.2. Note that the name of the phase and the baseline remain the same for all project types, but that the detailed activities within the phase may change.

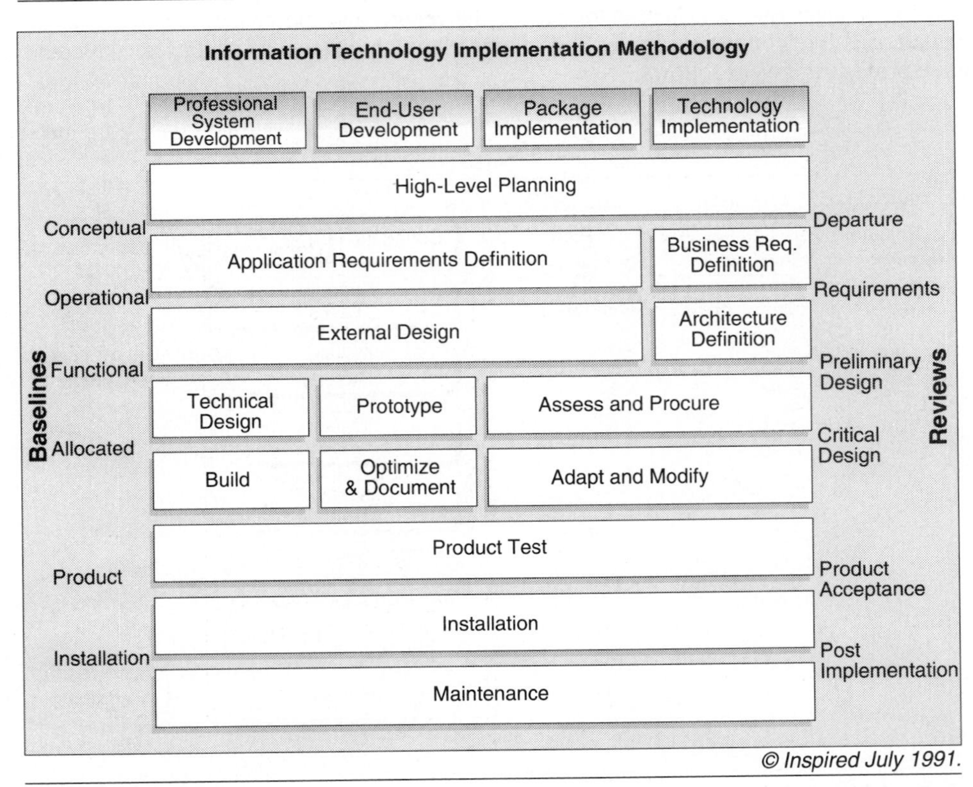

Alternative Lifecycles *Figure 13.2*

Now we might have three project reports like this:

Project 1	"We have delivered the conceptual baseline and have completed 40 percent of the deliverables for the operational baseline"
Project 2	"We have delivered the conceptual baseline and are 78 percent complete with the operational baseline"
Project 3	"We have delivered the allocated baseline and are 40 percent complete with the product baseline"

As I.T. professionals, we might be uncomfortable with these vague-sounding terms, when we would rather relate to process charts, file designs and so on. But users and senior managers find those to be incomprehensible, and would far prefer consistency. If you want to change the names from those used by the IEEE, fine, but make sure that you are consistent within your own organization.

Frequency

Reporting should be neither so frequent as to be a major overhead and irritation, nor so seldom that the team loses focus and the sponsor develops an ulcer wondering what is going on. Generally, the following guidelines are suitable:

- *The project plan and actuals should be updated and reviewed* by the team and project manager once a week
- *A formal report should be presented* to I.T. and user management once a month. This may take the form of documentation or a presentation, depending upon the organization culture. If it is verbal, make sure that the figures are recorded somewhere
- *A report to the steering body* once a quarter. This should include a presentation with time for explanation and questions, as well as formal documentation

You may want to adjust these frequencies to suit the circumstances. If the project is very risky and critical to the organization, reporting intervals can be shorter. If the project is routine, the project manager is very competent, and management is relaxed then maybe, just maybe, intervals can be longer. In all cases where formal reviews are to be held, participants should be provided with suitable documentation and an agenda ahead of time. This allows them to come to the session conversant with the facts, and prepared for discussion of problems, considering suggestions from the project manager, or asking penetrating questions.

Format

The report should be as concise as possible, but still convey all relevant details. We recommend the following structure:

- Cover page containing the name of the project, project code or identification, the date of the report, the name of the person making the report, and a project organization organogram indicating the reporting structures of the project. If anything in this structure has changed since the last report it should be highlighted, either by shading/color or a vertical sidebar. Further information should include the last completed baseline, and the total value of work complete calculated as set out in chapter 12.

- A status graph with time on the horizontal axis, and resource units (or money where non-personnel expenditures are significant) on the vertical axis. The time axis should be labeled in months, as well as have the phases indicated. Three lines should indicate:
  - The budgeted expenditure (which is also the planned rate of delivery of value) as a cumulative figure against time
  - The actual rate of expenditure derived from collecting actual resource consumption (and other expenditures if significant) graphed cumulatively against time
  - The value of work complete derived from summing the value of 100 percent complete and quality-assured deliverables as detailed in chapter 12, graphed cumulatively against time
- A Gantt chart showing the following:
  - All phases completed to date summarized to one line each
  - The current or imminent phase broken down and showing tasks at a summary level of approximately three weeks duration. Use your discretion, and do not show more than 20 tasks
  - Future phases shown as soft estimates

  Make sure that for all future tasks and phases you show the range of the estimates, and not a single figure. If management insists on a single figure, use the top of your range
- A single page listing current problems and concerns, together with the best approaches currently known for their resolution, initials of who is following up, and the expected resolution date
- A page containing the list of action items arising from the last review, and the status of these items.

A project plan in this format is a very powerful instrument, particularly when consistently used across the organization and managers become educated in analyzing it effectively. One can quickly identify projects which are running into trouble, others which are doing well, and what the problems might be in those which are not delivering according to plan.

Example

The sample project report shown in figures 13.3 to 13.6 illustrates the format discussed above.

MegaDodo Corporation

Personnel Records On Database Project

PROD

Progress Report as of	**June 1995**
Prepared by	**Stella Hayes**
Last Baseline	**Functional**
Work Complete	**37%**

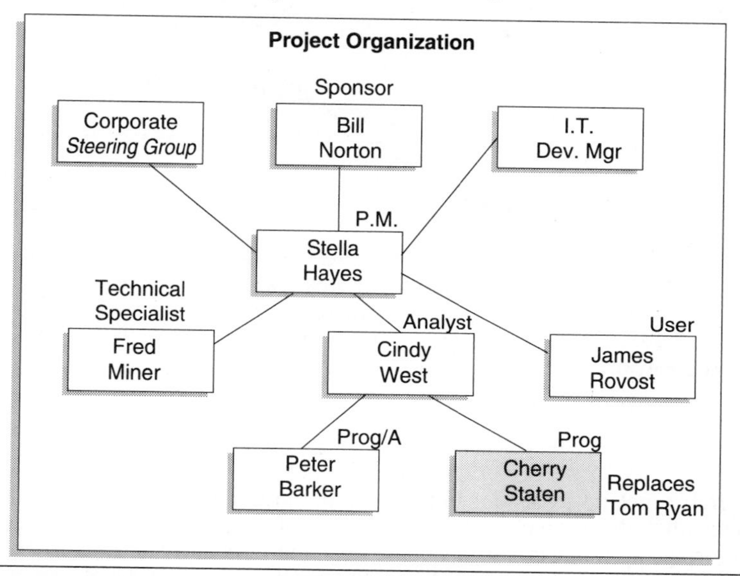

PROD Progress Report *Figure 13.3*

In figure 13.4, observe the following:

- The project expenditure is under budget, as can be seen in the Actual Expenditure line on the first graph

- Value of work delivered is behind schedule, and below the expenditure. This indicates a problem situation - the project is not delivering at the planned rate

- The project lost a staff member (Tom Ryan) who was replaced by Cherry Staten - this can be seen from the project organogram on the cover page (figure 13.3). Looking at the status graph, you can see that up to the middle of the requirements phase, budget, actual expenditure and actual delivery were all very close together. This indicates a project proceeding according to plan. At the middle of requirements, the expenditure rose suddenly untill the end of the phase, and at the same time the delivery rate (work complete) fell off. Since the end of the phase, expenditure has stabilized again, and the value of work complete is again going up to meet the budget line. What happened? The loss of a staff member meant that a new person had to be drafted into the team. There was a period of overlap where we carried the expenses for both people (Tom and Cherry). During this period, the output produced also fell as a result of the effort devoted to handover and the learning curve which Cherry

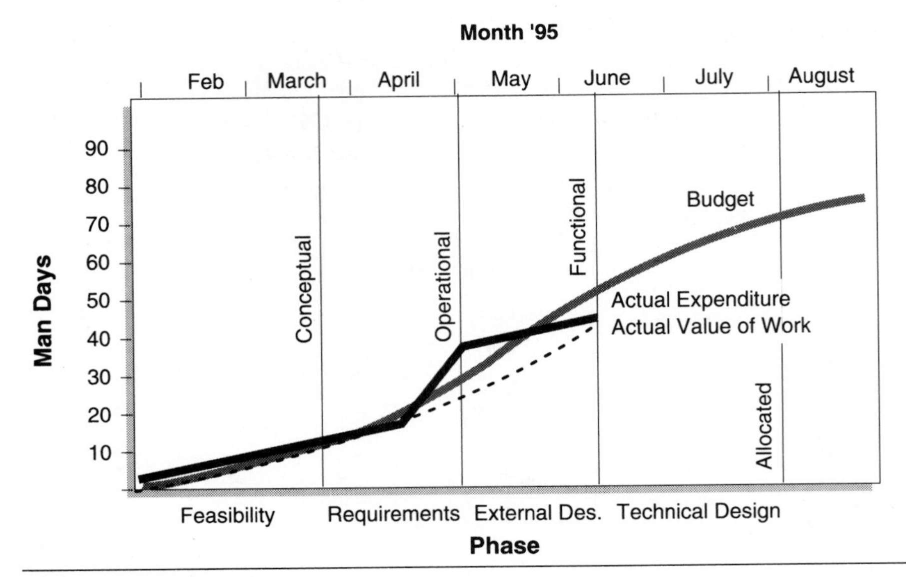

PROD Status Report *Figure 13.4*

incurred to get up to speed. Judging by the graph after the end of requirements, it appears that the project is getting back on track. We should remain cautious in monitoring, as there is still a chance that the delivery rate will catch up with the actual expenditure, but not with the original budget. This would result in delivery within the budgeted costs, but late

- Notice that all future tasks and phases in the Gantt chart are shown with both a minimum and maximum time. The earliest the project could be completed is in the final week of November. The latest the project should finish is the first week in January. You will see that the review markers in figure 13.5 have used the pessimistic schedule, so as not to create false hopes in the minds of management or sponsors. If the reviews occur earlier than these dates, that will be a bonus

If desired, more detailed information about resource assignments and spreadsheets for budgets and expenditures can be attached as appendices for those readers who would like to probe further; the information shown will give a very concise, easily interpreted view of the project in a minimum of space and time. Using this format consistently across all projects makes it very easy for management to track project progress, and to focus attention on the areas requiring it.

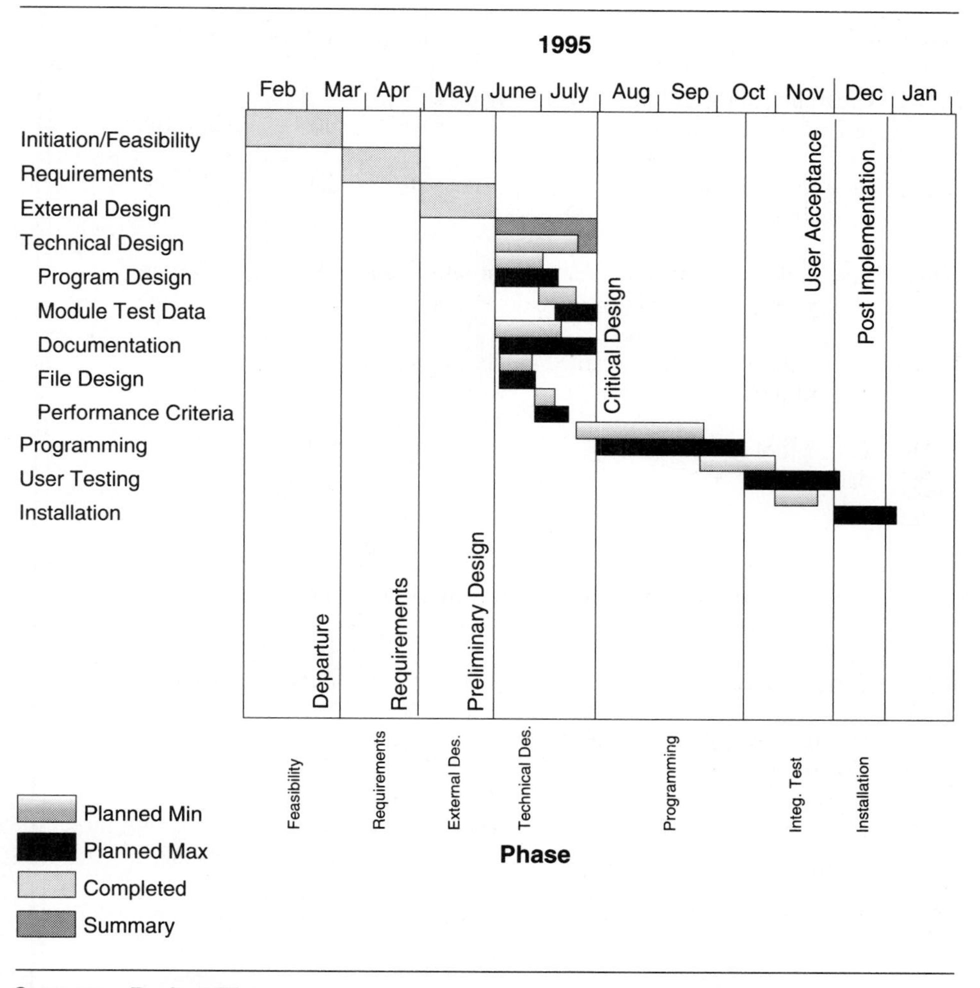

Summary Project Plan *Figure 13.5*

PROD

Critical Problems as of June 10 1995

Problem		Action	By When
JUN.1	Clearance not yet obtained from Tax office of Document Designs	JR	End June
JUN.2	Programming Standards Documents not yet available	SH	20 June
JUN.3	Screen design constraints for compatibility with all existing field terminals to be documented	FM	20 June
JUN.4	User sign-off still not obtained on tax calculation algorithms	CW	20 June
JUN.5	CASE tool still unreliable - vendor examining data	SH	15 June

Resolution of Review Issues as of June 10 1995

Issue	Description	Action	Status
FR.1	Fields on screen ENQ-EMP1 not reflected in data model	CW	Done
FR.2	Naming of items inconsistent on Report BEN-MTHLY and screens	CW	Done
FR.3	Missing performance constraints for batch runs	FM	Awaiting Vendor Data
FR.4	Budget to be prepared for new equipment required	SH	Done

Problems and Issues Report *Figure 13.6*

Case Questions

MyWay Organizer

Q13.1

Using the planned and actual data from Q12.1 and Q12.2, prepare a graph similar to figure 13.3. (20 mins)

Q13.2 (Requires Project Management and Spreadsheet Software)

Using the spreadsheet data which you exported in Q12.5, use the spreadsheet package to generate a graph similar to that in figure 13.3. (20 mins)

Q13.3

Prepare a summary Gantt on the status of the project for inclusion in a report to your senior management. You may do this manually, with the project management package, or with the spreadsheet data. (20 mins)

ThoughtWell Books

Q13.4

ThoughtWell senior management has become concerned with the state of the project. They have heard from a programmer that there are difficulties with something called TCP/IP which "doesn't seem to be working". Also, they are worried about how their staff will operate the system, which currently appears cryptic to them. They have also seen Lars Bontsen disappear off the site. They have asked for a full report on the status of the project. They would also like to know explicitly about any problems which may delay implementation. They are particularly nervous since it appears that Lars has joined one of their competitors. (Your instructor may have minutes of recent project progress meetings to assist you.) Your report should allay their fears, but should not compromise any facts or hide any difficulties. Assume the project is at the point detailed in Q12.6. (40 mins)

14 *Change and Configuration Management*

Project Scope Control

A great many I.T. projects fail when judged by our success criteria: delivery of desired results, on time and within budget. A large proportion of these failures are due to the scope of projects getting out of control. This can occur because we did not understand at the outset what it was we were undertaking to do, or because we allowed the specifications to grow or change during the project lifecycle. We have already dealt with the first of these causes in the chapters on initiation and project design. Two of the techniques discussed were context diagrams and technical environment diagrams. They scope the project from an application and technical perspective.

Studies show, however, that even if this initial scoping is done carefully, projects tend to grow. One study examined the number of function points calculated from projects at the requirements phase, and then re-counted these for the completed systems. The results indicated that the scope of projects had, on average, *increased fourfold.* This is an obvious problem for a project manager who is trying to stay within budget and deliver on time. The effect is called "creeping featurism" and is attributed to the fact that users ask for more and more to be included in the system as they see the potential for automation. In some cases they try to take advantage of the system under development to do their jobs. They will encourage systems analysts to put in facilities to handle every exception and circumstance that they can remember in their experience on the job. While this may be good for completeness, it can make systems unnecessarily complex and inflate project costs out of all proportion. It does not make sense to spend a month's analysis and programming effort putting in a function to handle an exception that only happens every two years, and can be handled manually in a few minutes.

An allied problem is that of including things outside the scope of the original system. The scenario goes like this: Projects A, B, C and D are considered during planning and feasibility. Projects A and C are approved. The users of B and D realize that their pet systems and functions will not be automated since their projects were not chosen. However, projects A and C are running - why not try to persuade them to include the desired functions? And this, of course is exactly what they do. Gullible analysts and project managers are too obliging, and projects A and C become larger and larger, eventually attempting to do everything that was in A, B, C and D in the first place. Chances are that they will fail, since we would not have left B and D out of our plans at the feasibility stage if we thought that they could be delivered within the same deadlines with the same resources as A and C alone.

Remember the effect of increasing size and complexity on project effort. It is like an expanding balloon (figure 14.1), which gets ever bigger, until it eventually bursts, allowing the benefits in the basket to fall out. Far better to get a simple system in on time to handle 90 percent of the cases, than to wait forever for one that will do everything.

An Expanding Balloon *Figure 14.1*

Configuration Management

We have seen in earlier chapters that configuration management concepts are useful to:

- Structure our projects
- Manage different kinds of projects under one consistent framework
- Report on projects in a consistent way, using consistent terminology

Configuration management, linked with change control, also gives us a way to manage the scope of our projects to prevent the problems described above.

We are now going to examine the configuration management concepts in detail, and relate them to scope management and change control.

Configuration management is a discipline that has grown up in the aerospace and military contracting industry. It is primarily aimed at managing large, complex engineering projects, with a research component. Many of these projects have significant software components, and the electronics and software engineers frequently have to work very closely together to develop hardware and software in parallel. People might be programming for a machine which has not actually been built yet. This means that interfaces and specifications where things overlap must be very carefully designed and managed. Configuration management does just this. More recently, the Institute for Electrical and Electronic Engineers (IEEE) in the United States has published documents detailing a Software Engineering lifecycle based upon configuration management. This is the basis for our use of configuration management. Configuration management is also fundamental to quality management, which we will tackle in the next chapter.

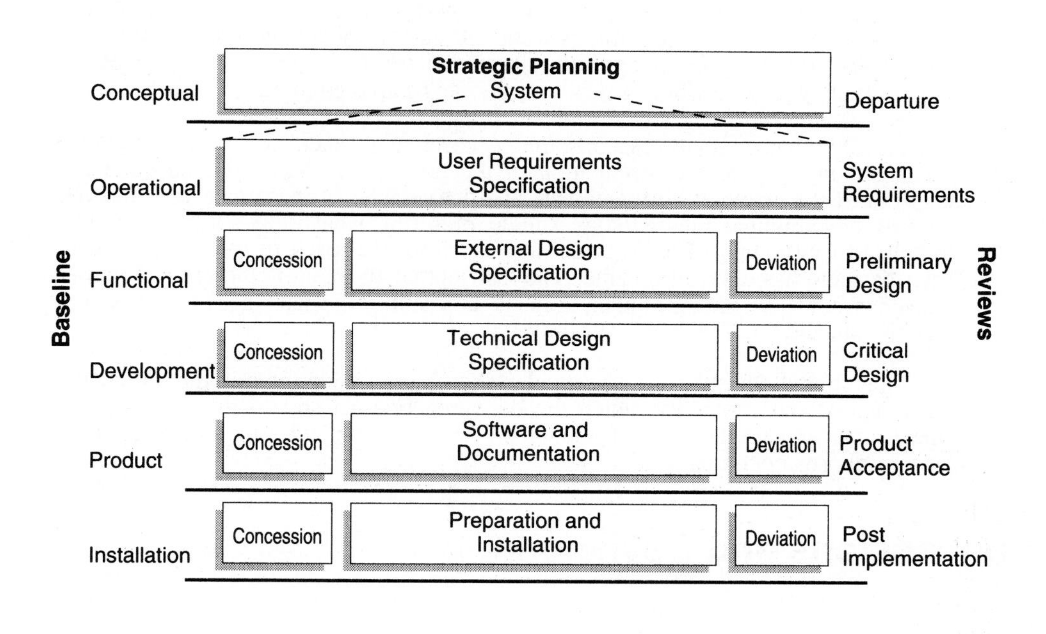

Source: ITIM based on IEEE

Configuration Management *Figure 14.2*

Referring to the configuration management lifecycle diagram shown as figure 14.2, consider the following:

- There are a number of *phases* in the lifecycle. The strategic planning area is considered to be outside the lifecycle, hence the dotted line. It is significant because this is where projects are assumed to be born. They arise out of a corporate planning process or business need

- There is a formal review following each phase. These have specific names. The *departure* review deals with the parameters from which the project departs: the initial vision, and tacit agreement with the client organization. This is followed by the System Requirements review, which, as its name suggests, reviews the requirements specifications. These specifications form the basis for the management of the project scope from this review onward. The Preliminary Design review checks the output of the external design phase. This will include detailed specifications of exactly how the system will behave in operation. Next we have the Critical Design review, so called because this is a critical point in the project where we make the transition from design (paper-based) activities to building and construction. After this point it will be very difficult and extremely expensive to change the design.

The next review is the Product Acceptance review once the product has been built according to the design. This is concerned with the client inspecting and testing the product and accepting it for installation in production. Finally, we have the Post-Implementation review, which is a learning opportunity for the project organization. This is where we look at how things went and gather knowledge for the future. In our adapted lifecycle, an opportunity is provided for this to occur at every review, thus shortening the delay before new insights can be applied to other projects.

- Associated with each review is a *baseline.* A baseline is the collection of deliverables which are in place at the termination of the prescribed phase. The baseline is a snapshot of the status of all documents and deliverables at that point. A competent systems development methodology should prescribe what deliverables should be completed by the end of each phase, and the form that each of these should take. Each baseline has a specific name. The Conceptual Baseline is the project charter, derived from the strategic plan, and is a vision of what the project should accomplish, and within what constraints.

  The Operational Baseline is of particular significance, since this is the anticipated functionality and capability which the delivered product should have. It is called Operational since this is what the client wants to be put into operation upon successful project completion.

Concessions and Deviations

Down the left-hand side of the diagram, you will see concessions, after the Operational Baseline. A concession is something which was present in the Operational Baseline specification, but which the team has found it could not deliver, and which the user *concedes* may be left out. An example would be that the original specification calls for a system which will automatically scan all source documents with an accuracy of 99.9 percent. We may find that this can be achieved with current Optical Character Recognition (OCR) technology for typewritten text of a certain quality. We find, however, that there is no similar capability for handwritten text available commercially. We discuss this with the user, and he concedes that we can alter the specification to stipulate that only typewritten documents will be accepted for scanning input.

Deviations are shown on the right-hand side of each phase after the Operational Baseline. These are items where the specification or design is *deviating* from the Operational Baseline. The user does *not* concede that this is acceptable. These represent problem areas where the design is not meeting requirements, and they must be redressed before the next review is reached. An example here would be: The Operational Baseline specification calls for a system which can process 1000 transactions per hour. Estimates at the Preliminary Design review indicate that the current approach will handle only 600 transactions per hour. This is unacceptable, and a deviation is recorded. Since the feasibility of the system and the calculated benefits are based upon achieving 1000 transactions per hour, a suitable approach must be found before the Critical Design review is reached. If this can be done, then the deviation will be closed.

The combination of concessions and deviations tend to guard against inadvertent changes in project scope. If we are unable to deliver something which the sponsor wanted, we are forced to negotiate a concession. This makes sure both parties are fully informed about the impact of the change. On the other hand, if we have not included everything that we should have, and it is possible to deliver the required features, then deviations alert us to this and provide a discipline to ensure that these issues are addressed before the termination of the next phase.

Estimating the Impact of Change

You will recall that we have built a product model during our project design. This contained an estimate of the size and effort of each deliverable. We can use this to assess the impact of requested changes as we proceed with a development project. For maintenance projects, or where the requested change is to a completed deliverable, the product structure map with actual sizes and efforts recorded can be extremely valuable in terms of assessing the impact and effort associated with a proposed change.

The PSM diagram (figure 14.3) shows the configuration for a co-resident personal computer "executive helper" product which provides diary, calendar and address-book facilities. It is a shrink-wrapped product sold through retail outlets. Note that documentation is measured in pages (pp) while software is measured in function points (fp). Both types of deliverables have the number of mandays (md) spent on their creation recorded next to them. These kind

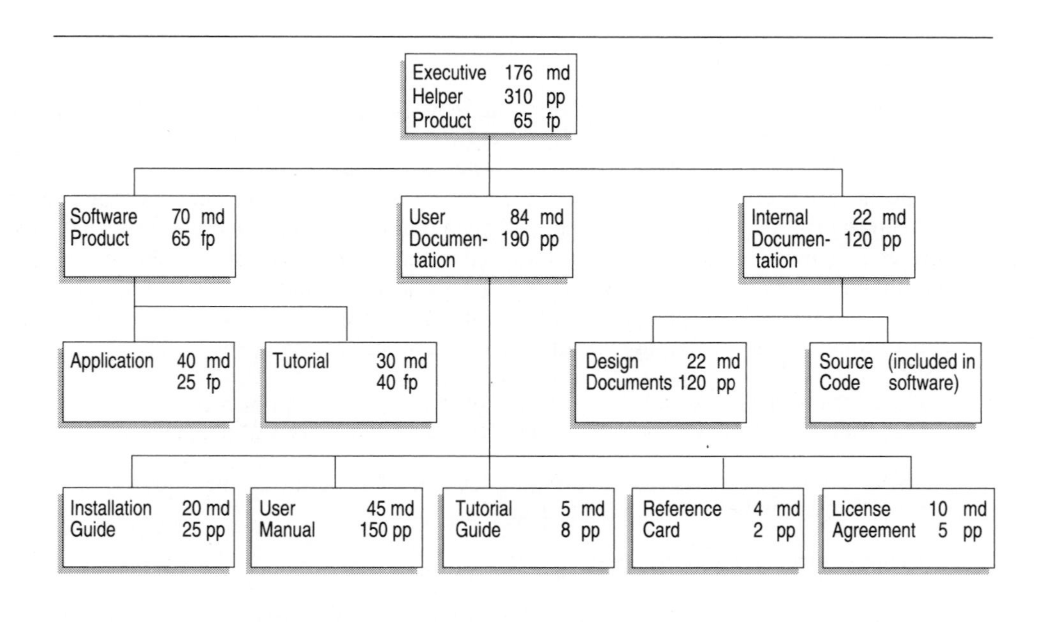

Product Structure Model *Figure 14.3*

of figures can be extremely useful when we come to consider the impact of a change. Let us assume that we require an update to the user manual, which will affect some twenty pages. A small change to the software component of the tutorial is also required, estimated at about 10 percent of its current functionality. We could then derive a realistic estimate for the effort involved in the proposed change as follows:

User manual: 20 pages @ 45/150 mandays = 6 mandays

Tutorial Change: 10% of 30 mandays = 3 mandays

Total change estimate = 9 mandays

Please bear in mind that the above estimate may need to be adjusted for all the usual factors. We are assuming that the team skills and size are the same, that the same time pressure will apply, and that the complexity has not changed markedly. A further complication can be that even a small change to software may require the entire suite to be retested thoroughly. This can introduce an overhead out of all proportion to the size of the change. This effect can be countered by the following strategies:

- *Structured Analysis and Design, or (better still) Object Oriented Design.* These have the effect of localizing functionality so that the effect of changes is constrained to one very small area of the total system. They also facilitate testing of the independent modules
- *Regression testing,* whereby all test data is kept in a test bank, together with known correct results. It is thus possible to rerun all tests automatically with minimal effort

Change Control

Systems tend to be living entities. They are born, they mature, they grow old and eventually die. Any successful system will typically undergo significant changes during its lifetime. The one thing that we can be sure of, is that things will change. What we need to do is to control the negative effects that change can have, while realizing the benefits that it can bring. Changes can occur after the product has gone into production, in which case they are normally seen as maintenance activity, or they can occur while we are busy with the project. The likelihood of changes occurring during the project is greatly increased in volatile business environments, if we have not done the early specification phases very well, or if the project is of long duration. We can thus minimize the amount of change by:

- Using strong analysis and design techniques which elicit all necessary perspectives (e.g., data, function, technology, user) during the early phases of the project
- Using techniques such as prototyping and JAD to achieve high levels of user involvement in the specification process
- Keeping project durations short
- Using spiral or simulation lifecycles such as those discussed in chapter 8, which allow for progressive refinement as we proceed

Despite the above, we will still encounter a certain amount of unavoidable change. This should be handled in such a way that it does not allow our project to get out of control. In the following paragraphs, we will explore a suggested change control procedure to minimize negative effects whilst ensuring minimal delays. This procedure is illustrated in figures 14.6 through 14.9. Forms supporting this procedure are shown in figures 14.4 and 14.5.

First, we must distinguish between requests for change and actual changes. The former is a formal request made by a team member, a user, management or other party. The second is a change approved by the project manager (or more senior management, depending upon the impact) that will be carried out by the project team.

CHANGE REQUEST	Serial No: 93.123
Originator	Date
Project Name	Project ID
Description of Change Required (attach fuller description if necessary)	
Business areas affected	
Projects or technical areas affected	
Business Impact Assessment (attach fuller description if necessary)	
Area ______	Area ______
Area ______	Area ______
Technical Impact Assessment (attach fuller description if necessary)	
Team ______	Team ______
Team ______	Team ______

Change Request Form *Figure 14.4*

CHANGE REQUEST TRACKING
Date Received
Business Impact Complete
Technical Impact Complete
SUMMARY
Business Impact
Technical Impact
Estimated Effort
Estimated Duration
Estimated Completion
Estimated Cost
DECISION
Reject ☐ **Proceed** ☐ **at Priority** ☐ **Refer Back for Info**
Reasons/Remarks
Signature of Change Approver
Work Completed on
Signature of Originator

Change Tracking Form *Figure 14.5*

The requested change is documented using a Change Request Form (CRF), shown in figure 14.4, and routed to the project manager (if during development) or the appointed change controller (usually in the Q.A. area if in production). Each CRF is numbered uniquely to facilitate tracking. The person receiving the CRF determines the business areas likely to be affected by the change, and routes a copy of the request to the necessary parties for an impact assessment. This could involve, for example, Stock Control and Distribution. These individuals assess the impact that the change will have and return this information to the project manager or change controller. The technical areas affected by the change are determined, and a copy of the CRF, with the business impact comments, is routed to the

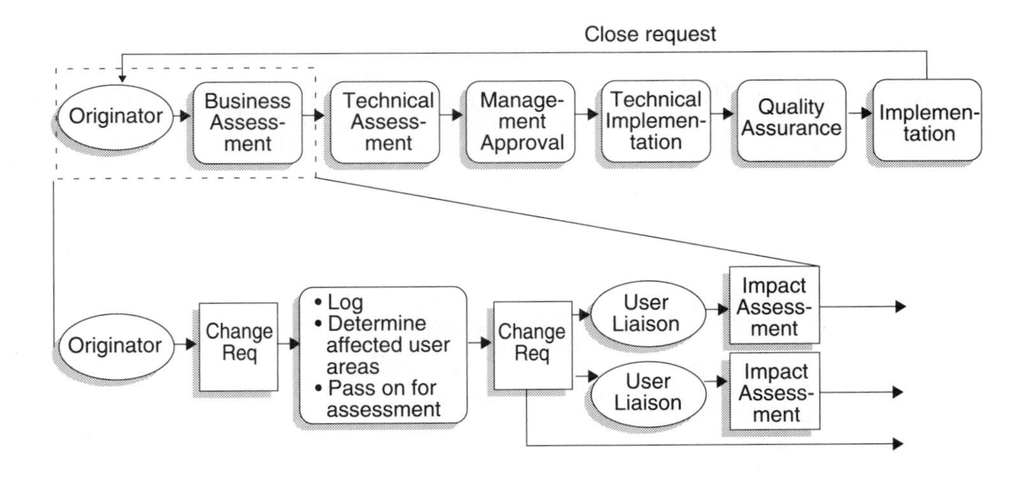

Change Control 1 *Figure 14.6*

technical areas, or other affected project teams. They in turn do a technical impact assessment, and estimate the effort and cost of implementing the change.

If the change can be accommodated without affecting the delivery date or the budget of the project as committed to management and the sponsor, then the project manager can decide whether to proceed with it or not. If he authorizes it, then the team or other necessary resources begin work on incorporating the change. If he decides not to make the change, the originator of the CRF is informed, and advised of the reasons why the change will not be made at this time. The originator may accept this, or escalate the request to higher levels of management. If the originator accepts the reasons why the change will not be made and does not wish to proceed further, then the change is "closed".

If the change is going to affect the project deadline or budget, the project leader must discuss this with the project sponsor and I.T. management. A joint decision is then made and documented. As before, if the change will not proceed, the originator is notified. If the change is to proceed, the project plan is adjusted to accommodate this, and work commences.

Some organizations put in place authorization levels based upon costs for the various levels of management. For example, a project manager may have the discretion to accept changes up to a cost of $10 000 in impact, but would have to refer the change to a Systems Manager above this figure. The Systems Manager might handle changes up to $25 000 in impact, but would pass larger amounts on to the Development Manager.

Where changes affect multiple project teams, these would automatically escalate to the next level of management. The CRF would be duplicated to all project managers affected for their input and comments. These would flow back to the responsible manager. Once all the

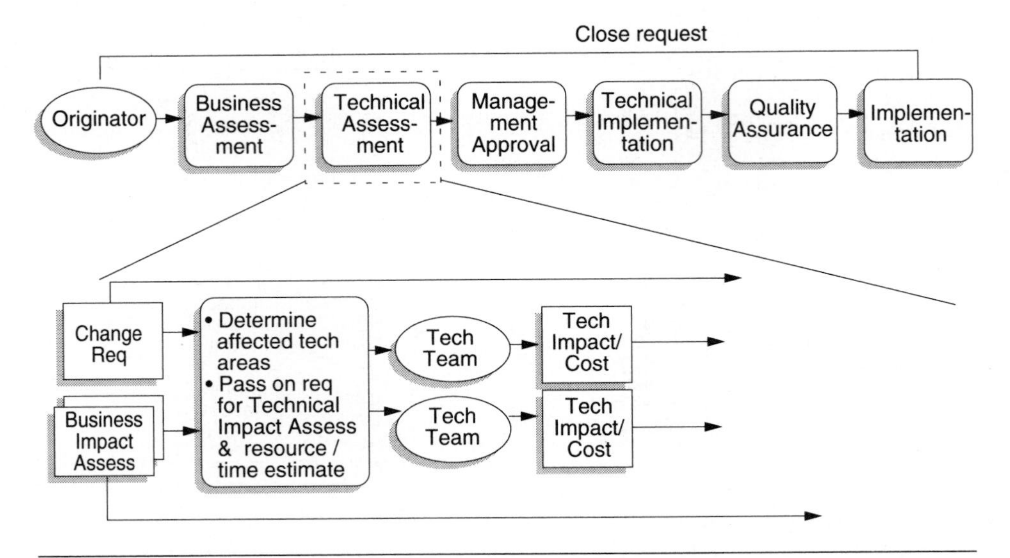

Change Control 2 *Figure 14.7*

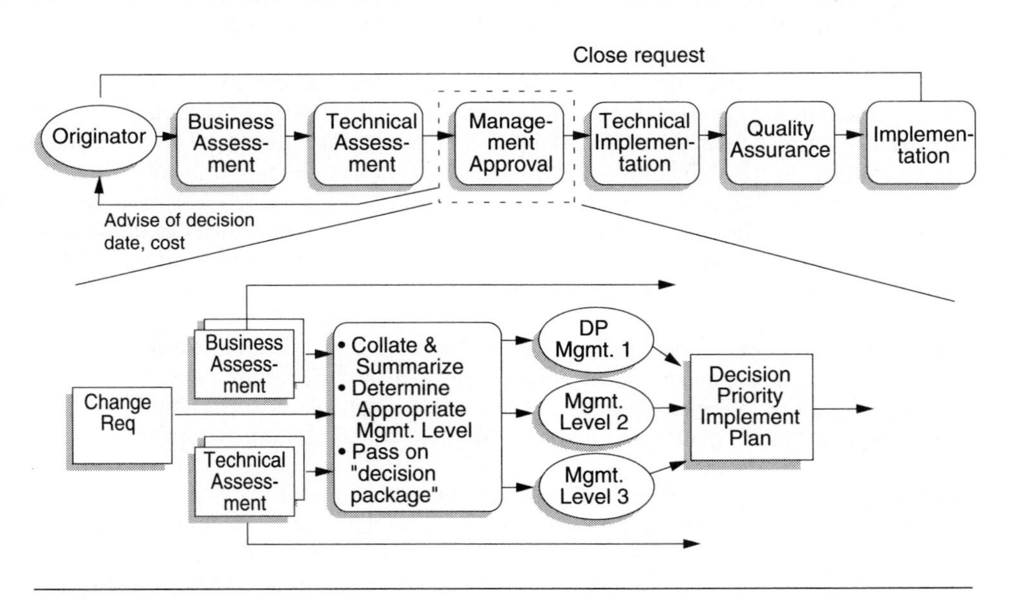

Change Control 3 *Figure 14.8*

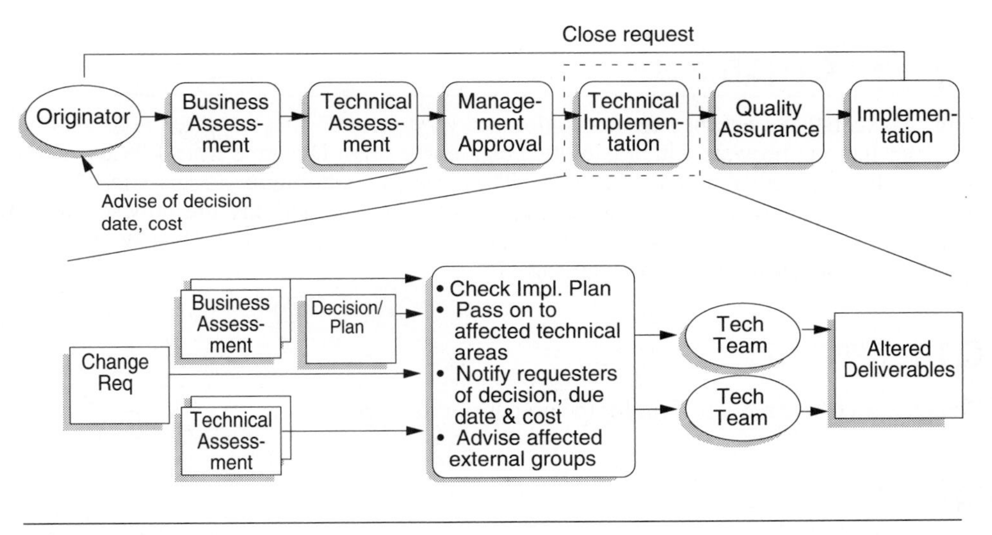

Change Control 4 *Figure 14.9*

figures were in, the manager could consult with advisers, and then reach a decision. Each change is also prioritized to indicate its level of urgency relative to other requests.

Two clients we have worked with have adopted similar systems and provided automated e-mail facilities to support the rapid flow of requests and information. This is very effective, eliminates delays, and allows those working on the same change request instant access to all the comments made by all parties so far.

It is also possible to integrate change management with problem management. The latter is the processing and resolution of problems which arise in the production environment. Some of these will result in changes being required for their resolution, e.g., a software error that is only discovered under large production loads. Using a single system to handle both simplifies management and increases responsiveness.

Case Questions

MyWay Organizer

Your managers have returned from a conference very excited. They have acquired the rights to use an Internet browser which is written as a fairly small Dynamic Link Library module and can be embedded into applications. They see an opportunity to give the MyWay Organizer a strategic advantage in the marketplace by incorporating the browser into the package in a seamless way. The package allows browsing of the World Wide Web and also basic receipt and sending of e-mail messages.

Q14.1

Complete a change request form for the proposal. (10 mins)

Q14.2

Assess the scope implications referring back to the context diagram and technical environment models completed in Q2.2 and Q2.3. Comment on the degree (%) of change to the original specification. (15 mins)

Q14.3

Your management wants to know if you can incorporate the facility into the product without missing your end date. They have said they are willing to provide you with two further resources if necessary. What is your response? (10 mins)

ThoughtWell Books

Q14.4

The ThoughtWell project is in imminent danger of missing the implementation date due to communication difficulties with the Branch POS equipment and servers. You have already reported on the status of the project to the client. It has been suggested that we might phase the implementation, and only install the system for head office initially, with branches placing their orders by telephone via a data entry operator at head office. The question now is: What proportion of effort can be saved if we follow this route, and how will it allow us to bring forward the implementation date. Determine your answer by revisiting the function point estimate done in Q7.12. Prepare a motivation to management regarding whether to phase the project in this way or not. You may suggest other alternatives if you wish.
(1 hour)

15 *Quality Management*

Definition

Ask anyone if they know what quality is, and they will almost certainly say "yes". Ask them to define it, and they will probably have a lot of difficulty. They will mutter words like "good", "well made", "durable" and "lasting". Ask them how we can measure quality, and you may get blank looks. So just what is quality, and how do we measure it?

Ask several people to name a quality watch and you will get replies like "Rolex", "Buren", "Chris Weill", "Seiko", and even "Casio". Now a Rolex is certainly a quality watch if you can afford it, and want a watch that will withstand extreme conditions. You could also live on the proceeds for a month if you pawn it. It has status. A Seiko is a well-made, accurate, fashionable timepiece at a more affordable price. For many people, this would represent quality. What about the Casio? If your needs are more a light, practical watch with a stop-watch for timing sports, cheap enough so that you won't be devastated if you lose it, then it is certainly a quality watch for you. So quality, like beauty, is in the eye of the beholder. We might define it as "fitness for purpose" - in other words, an article or product which meets the user or client requirements. This is exactly how the quality guru Philip Crosby defines quality:

Quality is conformance to requirements.

In the I.T. context, systems can have widely varying requirements. Some will have to process vast volumes with high reliability (e.g., a banking system), some will need to provide extreme ease of use, regardless of hardware consumption (e.g., an Executive Information System), while still others will need to be extremely flexible to serve changing business needs. Some aspects typically included in system requirements are:

- *Functionality* - what must the system be capable of doing?
- *Output* - what must the system produce?
- *Performance* - on what level of equipment must the system be able to run, and what performance should it deliver in this environment? What volumes of data and transactions or inquiries should the system be capable of handling?

- *Reliability* - what are the consequences of failure, and how often can this be tolerated?
- *Maintainability* - how easy should it be to alter the system?
- *Security* - how should access be controlled?
- *Operability* - how easy should the system be to operate?
- *Cost* - what is affordable to the enterprise?
- *Efficiency* - how efficient should the system be? What resources can it consume?
- *Interoperability* - is it necessary for the system to interface with or interact with other systems?
- *Portability* - how difficult is it to move the application to another environment?
- *Reusability* - how easy is it to identify and utilize reusable components of the product in subsequent projects?

All of the above should be quality goals to a lesser or greater degree, depending upon the project and the system under consideration.

Measuring Quality

If quality is so individual, how can we measure it? The answer is deceptively simple. If quality is conformance to requirements, then we can measure the quality of something by the degree to which is does not meet those requirements. Normally, this can be translated to money terms, and is called the Price of nonconformance (PONC).

The measure of quality is the price of nonconformance.

Just what is a non-conformance? It is anything which causes the product (system) not to perform as desired (as specified in the requirements). In system terms, PONC would include:

- Cost of rerunning a batch job which aborts because of incorrect data
- Loss of income because of a statement run not going out on time
- Effort to locate and correct a bug in production software
- Cost of removal of a virus from a computer, and the cost of restoring data to an operational state
- Damages awarded the client of a medical practice because of incorrect treatment resulting from incorrect patient history details
- Computer run-time over and above that budgeted for because of an inefficient system

Obviously, the lower the PONC, the higher the quality. At a quality level of 100 percent, the price of nonconformance would be 0.

So Why Worry?

Why should we worry about quality? The answers are very simple. The box shows some extracts from trade and other journals which give you a feel for the "state of the nation" when it comes to the I.T. industry. With information systems increasingly being deployed in mission-critical roles, we simply cannot tolerate the low quality levels which we have had in the past. One commonly heard quote is "If we built buildings like we build software, the first woodpecker to come along would destroy civilization". We tolerate a cost of nonconformance in our industry which is greater than 60 percent. In most industries, it is regarded as unacceptable if the figure exceeds 1.5 percent.

The State of Affairs

It seems dBase IV has done poorly because its 450,000 lines of code contain as many as 100 bugs, say outside developers. Ashton-Tate says the glitches number 44. Still, the flaws make functions such as file sorting worthless. "This program is nothing but a stick-up", says Denis Bellemare, a Montreal immigration lawyer and dBase IV buyer. "It's so bug-ridden I can't use it".

Business Week July 17, 1989

In 1981, a 1/30 second timing difference caused by a program change created a 1/67 chance of the space shuttle's five on-board computers not synchronizing. This was not detected during thousands of hours of testing, but caused a launch abort on the pad.

Paraphrased from Design News Feb 1988

In 1986 a Therac 25 machine administered allegedly fatal doses of radiation to two patients after a software problem caused the machine to ignore calibration data.

Paraphrased from Datamation May 1987

A software error was partly to blame for the information leading to the decision of a US aircraft carrier captain to fire upon a civilian airliner in the Gulf.

Newsweek story

Consider figure 15.1 which details software spoilage, derived from work by Tom De Marco. Similar figures were obtained by the IEEE. We are all aware that over 60 percent of effort in the average large installation is devoted to "maintenance" activity. We should also be aware that around 50 percent of the development costs are typically expended in testing and debugging software. Now we realize that some of the maintenance is to add new features and facilities, or to respond to changing business requirements. We also realize that we will have to spend some effort in testing, regardless of how good our product is. However, if we count only the proportion of effort spent on maintenance which is not adding new functionality, but fixing errors, and the portion of avoidable test/debug effort, we arrive at a software spoilage figure of 55 percent. This represents the proportion of effort which is not delivering any business benefit - it is totally wasted. With the high cost and scarcity of good I.T. skills, this is surely unforgivable, especially when we look at the growing application backlogs which most organizations face.

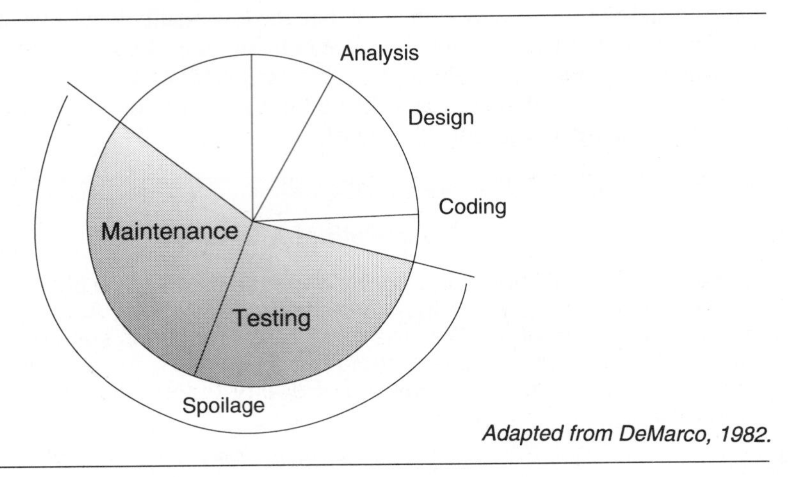

Adapted from DeMarco, 1982.

Software Spoilage *Figure 15.1*

The situation in most organizations, viewed from the perspective of senior management, is illustrated in figure 15.2. Despite the declining cost of hardware power, and the vast computing and storage capacities we are able to purchase at relatively small cost, the overall I.T. budget keeps increasing. The money goes into people and software-related activities. Approximately 60 to 70 percent of this is spent on maintenance and avoidable effort in the testing phase. This leaves about 30 to 40 percent available for delivering new functionality to the organization. In many mature shops, this is much worse and falls around the 20 percent mark. There is also a striking correlation between the length of time an organization has been computerized and this number. The more development we do over time with poor quality, the more burden we place on the maintenance activity, further reducing our ability to do new things. If this is allowed to continue, eventually we will spend all our time fixing things, and never produce anything new. Small wonder then that the backlog in some large shops is estimated to be 5 to 7 years of work for the entire systems department. At 30 percent of their capacity, this will take 15 to 25 years. Clearly, by the time the systems department gets around to some applications, the organization probably does not want them any more. The organization may even be out of business because it could not respond to a competitive thrust, or to comply with new legislation or circumstances. From a business perspective the situation is untenable: we keep paying more and more for less and less. It is not surprising that many organizations are outsourcing their I.T. needs.

How can we turn this situation around? The answer lies in doing things properly. We frequently see client organizations which never have time to do it right, but always have time to do it again. One company we are familiar with is embarking on the third attempt at a key business system. This process has lasted five years, and cost tens of millions. They still

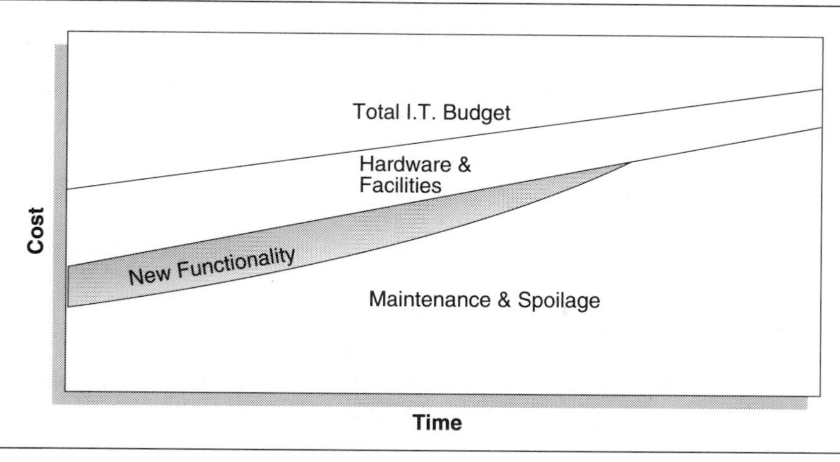

Ability to Deliver New Functionality *Figure 15.2*

do not have a solution. When we suggested some 4 years ago that they should take a short break to review their methods, establish standards, do some training and design the infrastructure carefully, we were told, "Forget it, we don't have time!"

If we can create a culture and set of capabilities that does do things right, preferably the first time, then we can turn the situation around. This is a long-term view and requires the support of senior management. Referring to figure 15.3, this scenario works as follows: We initially spend some money and time to educate our people about the quality philosophy. This impacts short-term productivity. Next, we begin using the best techniques and methods we can find, with the express purpose of delivering the highest quality possible. As we improve the quality level of our output, we will find that the spoilage component declines, releasing more productive resources. Gradually, over a period of several years, we can reverse the situation to where maintenance and spoilage are a small component of our

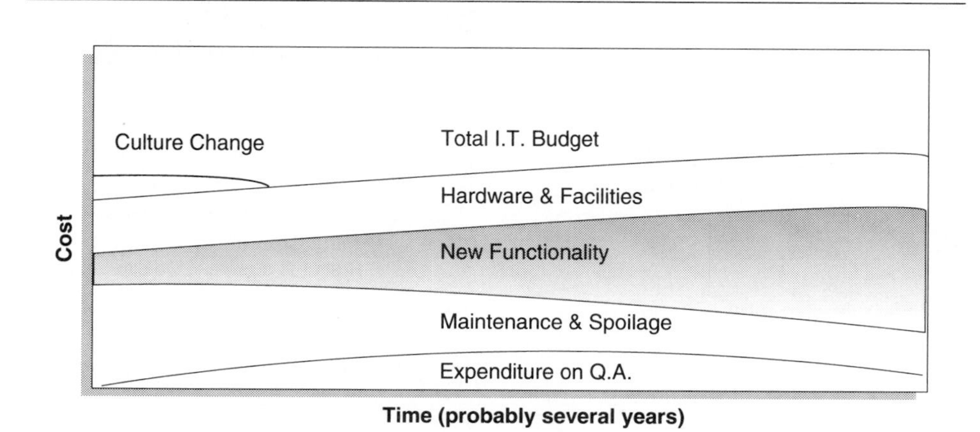

Turning Productivity Around *Figure 15.3*

workload, and we have great ability to add value to the business. If you are skeptical, stop ten people at random and ask them which country is a quality leader in electronic goods. Most will say Japan. Now stop ten other people and ask them which country has the highest productivity per capita. Again, most will say Japan. How can it be that they are both quality and productivity leaders? As we will see shortly, quality is the key to high levels of productivity.

Quality Management

Quality Management is a total approach (also called Total Quality Management [TQM]) embracing the necessary things to achieve quality at every level. It is a common myth that quality is the responsibility of the "workers". Quality is the responsibility of management. Total quality is the responsibility of senior management. There are three main components: Quality Environment, Quality Assurance (QA) and Quality Control (QC).

Quality Environment

A quality environment includes the organization's approach to quality - its culture and the infrastructure provided to support the achievement of high quality. Consider figure 15.4. It is the responsibility of senior management to create a quality-enabling environment, and to inculcate a quality-conscious culture. Unfortunately, this is more difficult than it sounds. There are many organizations where management pays lip service to quality, but demands unrealistic deadlines from staff with inadequate training or tools. This is a sham. True quality culture is created by example. When employees see senior managers behaving in a way consistent with the total quality philosophy, then they too will feel free to behave in this

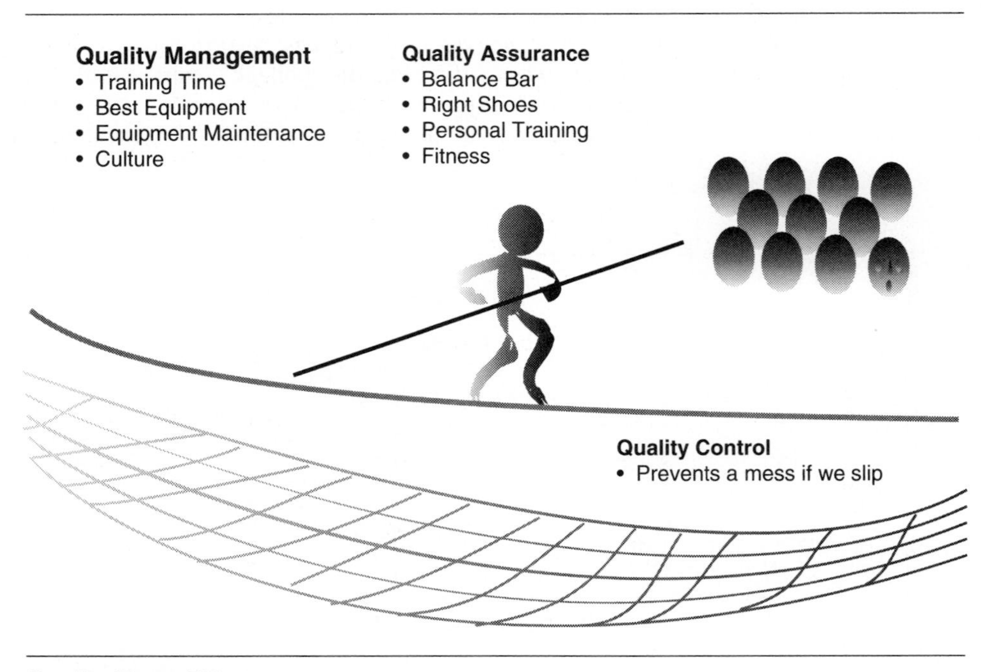

Quality-Team Effort *Figure 15.4*

way. It extends to things like providing a quality working environment, integrity and honesty in dealings with staff, measuring results not busy time, adhering to standards, establishing the quality expectations, funding the creation of measurement and monitoring programs, and training the entire organization in the requisite concepts and techniques.

We need to be very careful here. There are organizations we know which have a formal quality policy, plaques on the wall, graphs of performance, a Quality newsletter, and a training program, but still, when the chips are down, decide to breach the principles when making project decisions. They will put a system into operation because of user pressure, even though they know full well that it has not been performance or reliability tested, for example. There are others who have no fanfare, no formal program, but where quality is evident in everything that managers and workers do. Formality is a necessary evil when we need to change the culture of a large organization. The point we are making is that, to work, the commitment must be real. Token adoption of quality programs will yield token results. The follow-through is all important. The General Manager (I.T.) of a large insurance company we know tells the story of how he knew the quality improvement program was working when a junior team member, requested to implement an untested change to a production system by a board member, had the gumption (and presumably the faith in his management) to say, "We don't do things that way here."

To sum up, the actions management must take to launch a quality program include:

- Publicly declare the philosophy, and the corporate commitment to it
- Implement company-wide training in the concepts and principles involved
- Set up a measurement program to establish the current quality level
- Identify problem areas, look for solutions which will prevent recurrence of problems and implement them
- Keep monitoring to see if the solutions are indeed improving the quality level
- Ensure that staff at all levels, and suppliers, are involved
- Involve internal and external clients in requirements definition
- Set an example
- Keep doing it

Quality Assurance

Quality assurance is the second major component, and is the responsibility of middle management. It includes everything which we do to ensure that things are done right the first time, including:

- Development of standards
- Training in techniques and tools
- Use of proven methods with defined deliverables

- Provision of tools to assist in proper performance of tasks
- Assignment of correct skills to tasks
- Prevention activity - finding the root causes of errors and eliminating them

Quality Control

Quality Control is the final safety net. It is the last line of defense. Its purpose is to catch the odd nonconformance that occurs despite QA before it goes out the door and is seen by the client. Quality control detects errors; it does not correct them. Correction should be performed by the people responsible for producing the product. Quality control is an operational level activity. QC includes activities such as:

- Inspections
- Testing

Ideally, we want the people performing QC to find nothing to report. Achieving quality is a team effort: Management must establish the culture and infrastructure to allow high quality; Project Managers must do everything in a way that assures quality is built in, and Inspectors must ensure that if we slip up, that this (rare) fault does not reach the client.

Note that in terms of the project lifecycle, what is a QC activity for one phase may be seen as QA for the next. For example, we may inspect the requirements document. This is a quality control on this deliverable. At the same time, this is preventing poor input to the design phase, thus assuring quality there. Elsewhere in the book we have seen the relative cost of errors, depending upon where they are discovered in the lifecycle. It thus makes sense to have frequent QC/QA activities early on in the lifecycle. These will increase the proportion of effort expended on these phases of the lifecycle, but the overall project effort will be greatly reduced. In work which we have done in productivity analysis with clients, we have seen a strong relationship between those using good methodologies which emphasize requirements analysis and design, and high delivery rates. Those who have a high proportion of lifecycle time spent in the early phases are very likely to be the ones who demonstrate the highest overall productivity.

Total Quality Costs

From the foregoing, we can see that the total expenditure on quality will not only be the price of nonconformance, but also the cost of the training, culture change and assurance activities. Thus the total quality costs are defined to include PONC and the Price of Conformance (POC). POC includes cost of:

- peer walkthroughs
- inspections
- developing and implementing standards
- training in quality philosophy
- setting up and running a quality measurement program

The cost of quality (COQ) = Price of Conformance (POC) +
Price of Nonconformance (PONC)

An obvious goal is that the COQ must be less than the value of the benefits achieved by the quality program over a period of time.

Two Philosophies

Early ideas about quality centered on an approach where finished products were inspected, and defective ones were sent back into production for repair (so-called *rework)* or discarded (known as *scrap)*. This is the *appraisal* school of thought, depicted in figure 15.5. Adherents to this approach believed that there was an *economic level of quality*. This would be a point where the total cost of quality would be at a minimum, and that this would occur somewhere below 100 percent quality level. The thinking was that to achieve very high quality levels, you would have to increase testing and inspection to find all the faults, and that you would incur increasing costs of rework or scrap to correct the problems. While this is intuitively appealing, it has been proven incorrect. The problem lies with the assumption that we must continually correct products with the same fault, thus incurring the rework expense over and over again. This should not be necessary if we follow a different approach, known as the *prevention* approach.

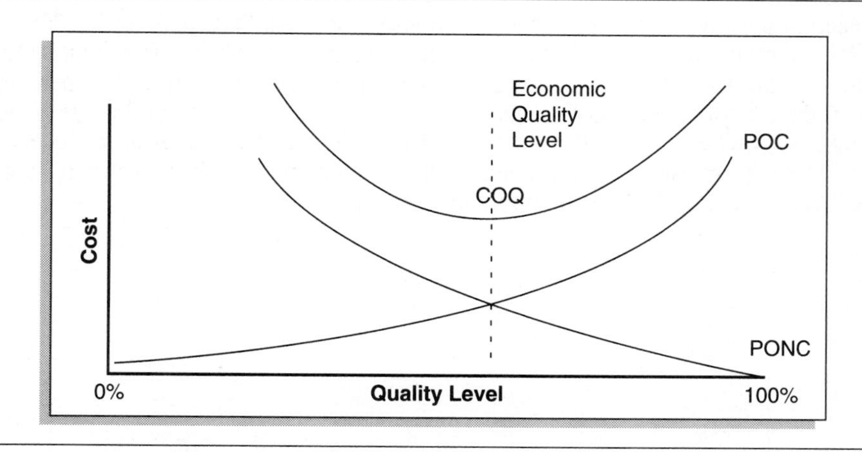

Quality-Appraisal Philosophy *Figure 15.5*

A Quality Scenario

During system testing, a 4GL reporting program is found which performs very poorly. It is producing correct results, but accessing the database in a very inefficient way. We could just fix the problem and continue testing the rest of the system. What we do under the prevention philosophy is this:

- Correct the program and continue with testing
- Look for other programs exhibiting the same problem
- Correct the unit testing procedure which allowed the problem to get this far
- Discuss the problem with the responsible programmer to determine why the program was designed in that fashion

  (We discover that the programmer believed that it was the correct way, based upon knowledge gained from a language course given by the 4GL vendor)
- Advise all our 4GL programmers of the correct way to tackle the problem
- Liaise with the vendor and get the training course fixed

This seems like an awful lot of trouble to go to because of one poor performing program. It is. The secret lies in the fact that we are tracking down and fixing the root cause of the problem once and for all. We should never encounter exactly the same problem again. Contrast this with the appraisal approach where we would continue to find similar programs, and fix each one in turn. We can sum it up like this:

- In appraisal, the effort to correct the problem is small, but this will be repeated many times in the future
- In prevention, the initial effort is high, but we will not repeat this effort

The prevention approach thus changes the relationship between POC and PONC. The COQ will not be at a minimum below 100 percent, but *at 100 percent*, as shown in figure 15.6. In appraisal, expenditure continues to rise as we approach a quality level of 100 percent. In prevention, the rate of expenditure on quality is controllable. It will affect the rate at which quality improves, not whether it does. It is also independent of the number of products being produced. The key is that we cure problems once at source, and this expenditure is never repeated. This leads us to declare two important principles:

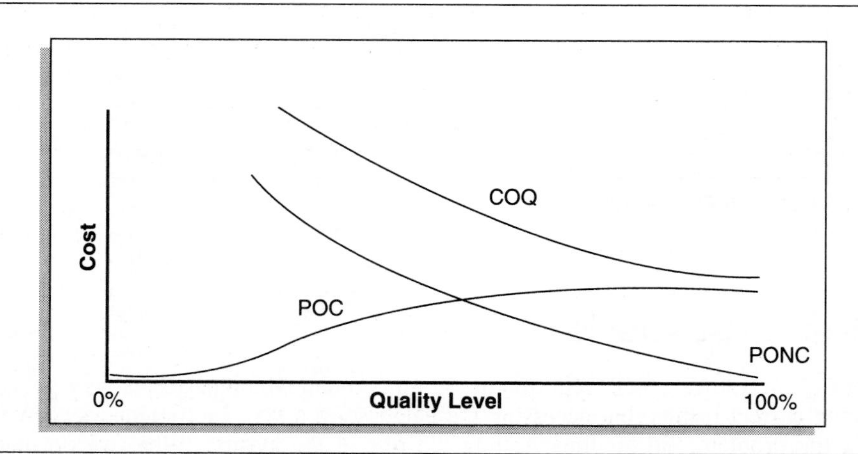

Quality-Prevention Philosophy *Figure 15.6*

The system of quality is prevention.

The quality standard is zero defects.

The latter is often a contentious statement. People think that we are talking about perfection, and argue that this is not attainable. This is not what zero defects means. It means *no deviation from requirements*. The requirements might state something like: "There should be no more than one hour of downtime per month". If the system has 20 minutes downtime per month, then it is performing at a zero defects level - it has met or exceeded the specification.

A Quality Model

A useful way to think about everything we do from a quality perspective is as a process with inputs and outputs as in figure 15.7. Inputs come from suppliers, which may be internal or external. Examples include a systems analyst providing a data model (internal), or a vendor providing a user manual (external). Outputs go to clients, internal and external. An internal client might be a designer who will use the specification we have created. An external client might be the ultimate recipient of a software product and its documentation where we are a software house. The requirements for the deliverables from our process come from our clients. We must ensure that the outputs we produce conform to their requirements. This means that we must know what good quality output looks like. We must know the content, the format, and the timing constraints, etc., that our client dictates. Good methodologies assist greatly here by identifying what form deliverables should take. We also need to understand the process, techniques and tools that we will use to produce the required output. This too is the province of good methods, assisted with tools to automate the process, and training to know how to use both.

We need good inputs to produce good outputs. If I am a designer expected to produce good file structures, I will need a data model as input. If this is erroneous, my output will be too. There are thus requirements that I will have of my supplier. The data model should be complete, in a recognizable form, include volume information, and indicate how data groups

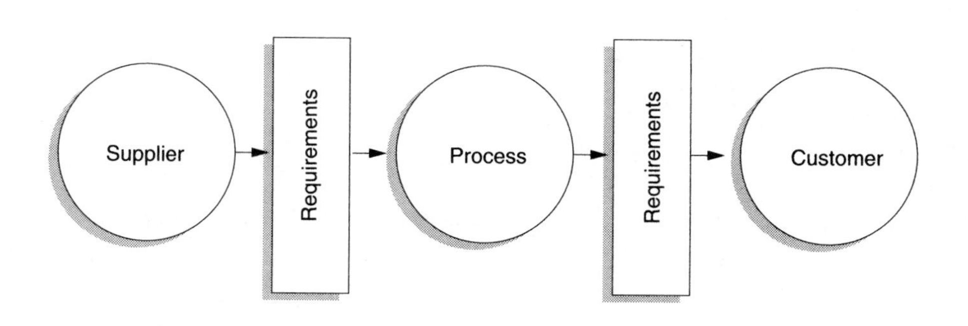

Quality Model *Figure 15.7*

will be accessed. It must also include the detail of data item types and sizes. Without this information, I cannot produce a good design. I should ensure that my suppliers are made aware of what constitutes good quality from my perspective.

Obviously, with the SDLC and PLC, there will be long chains of activities and associated deliverables. A competent methodology will have specifications for the tasks and the deliverables. Where these are lacking for your particular project, they should be created at the outset when you design your project.

Figure 15.8 shows what the model looks like to describe the file design process we have mentioned. Having a comprehensive model in place, coupled with the concept of statistical process control described in the preceding chapter on measurement, allows us to continually improve quality.

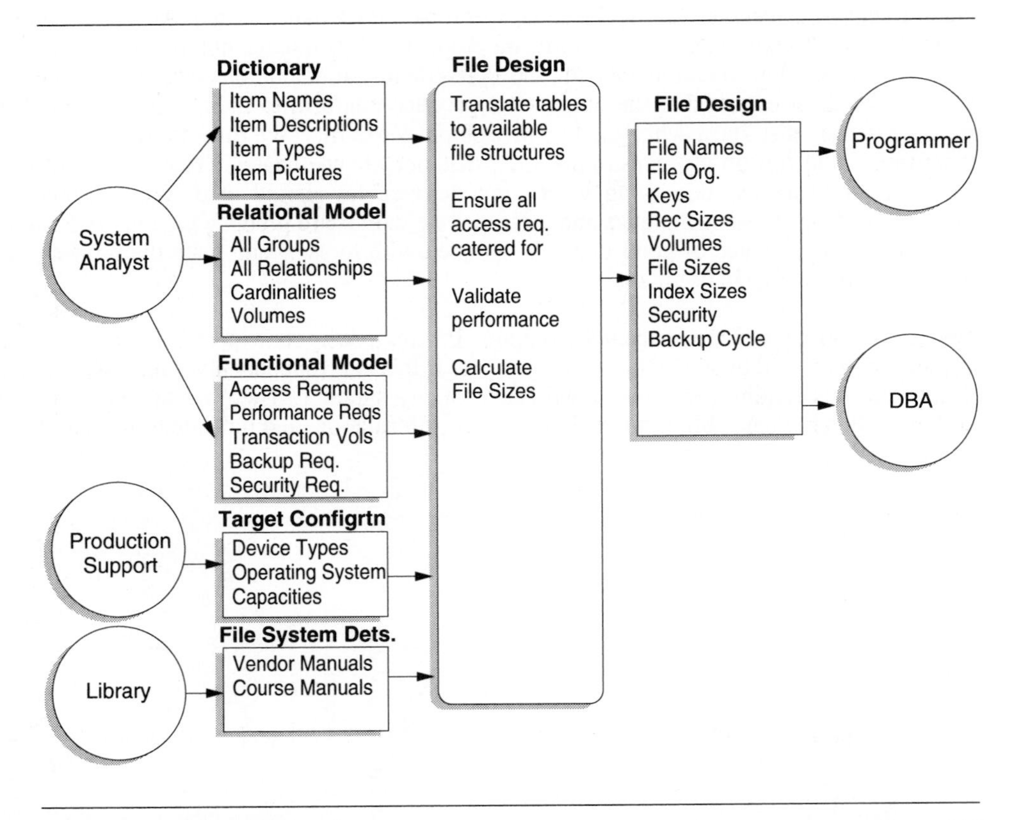

Quality Model for File Design *Figure 15.8*

Innovation versus Kaizen

Many will argue that these techniques work only in certain cultures, e.g., the Japanese culture where organizations and employees have very high loyalty, often extending to lifetime commitment to each other. It is interesting to reflect that virtually all the quality techniques that we attribute to Japan were first described in the United States. The Japanese just took them more seriously. There are two major ways of achieving improved processes and better quality output. The first is *innovation*. This is the favored approach in Western culture. It relies on adoption of new techniques or breakthrough technologies which radically change the way in which the task is approoached. An example would be the use of a code generator instead of manual writing of code. Innovation is attractive because it can offer order-of-magnitude improvements, and apparently does not require that much effort or discipline from the adopter. This, of course, is deceptive. There is significant effort involved in learning how to use the new technology and adapting all the surrounding processes. It is also risky, since there is usually no guarantee that the new approach will work. Indeed, the history of the I.T. industry is full of examples that promised much, but have not delivered in the vast majority of cases, e.g., 4GL's, CASE. Innovation can be used to very good effect, though, if we are currently at very low quality levels (as most installations are) because we are simply not using the best "state of the practice" techniques which are proven to work. In this case it is an innovation for us, but has relatively low risk because it is proven elsewhere. Examples in this category might include the use of prototyping, user involvement in the lifecycle, and using query tools for ad-hoc requests.

Far harder to sell to the Western mind, but extremely popular in Japan, is the concept of continuous, small improvements, or *kaizen*. This relies on experience with the task, and actively looking for opportunities to do it just that little bit better the next time. It relies heavily on worker participation and detailed knowledge of the tasks performed. Quality circles are one way of facilitating and promoting the approach. Kaizen dovetails with the statistical process control approach previously covered.

There is disagreement as to whether the Japanese management techniques can work in Western organizations. The American automobile industry was seriously threatened in their local markets by Japanese competition. The major automakers sought refuge in government restrictions on imports, instead of looking at their product quality. The Japanese response was to set up plants in the United States. These used Japanese management and quality approaches, but relied exclusively on American workers. The result was that these plants now dominate the domestic American market. American workers responded very well to the approach, and often added their own tendency to innovation as a bonus.

The conclusion we can draw from the above is that both approaches are useful. Innovation tends to be unpredictable, risky and disruptive, but it can yield spectacular results. It is best employed where quality levels are still fairly low, or in a controlled way in a stable environment. Kaizen is slow, predictable, nondisruptive, but yields steady improvements. It is best used at high quality levels where no easy gains are available. It relies heavily on worker participation, communication and a disciplined approach to work. The last of these is probably the reason we use it so seldom.

If hindsight is the only exact science, why are we so sparing with it?

In the next sections we will examine ways to improve our quality levels.

Quality Improvement

There are a number of things that we can do to improve the quality of our work, products and services. They include:

- *Well-defined lifecycle.* Only when we have a lifecycle which is understood and under control can we be in a position to measure our performance on an ongoing basis, and to implement on a wide basis those changes which really yield benefits. We frequently rush in and make wholesale changes based upon no sound facts whatever. A defined lifecycle also facilitates the accurate definition of deliverables and standards for these. In addition, we build examples of good work and skills in our team members

- *Tasks associated with well-defined deliverables.* It is insufficient to define only the tasks. We must also understand the products and deliverables that are to be produced. These should be specified in detail in terms of both content and presentation

- *Quality standards for each deliverable.* Each deliverable should have an associated quality standard. This can be brief, but should be enough to ensure that we can evaluate a deliverable and see whether it meets requirements

- *Informal walkthru's should be encouraged.* These encourage people to make what they do visible. If I have to show my work to my colleagues, I am far more likely to check that it is rigorous. It is also frequently true that we find our own errors when we start to explain how we have done something to somebody else. Teachers are familiar with the phenomenon that you think you understand something until you have to explain it to someone else

- *Formal walkthru's* have similar benefits to informal ones, but are more resource hungry. They should be saved for critical deliverables, or review points

- *Code inspections.* These seem rather old-fashioned, and are criticized for being expensive and time consuming. However, Yourdon has shown that organizations using this technique alone improved productivity by 38 percent. How is that for innovation? Work by Jones shows that they can remove up to 85 percent of all code defects without testing being done. If we relate this to the amount of effort expended on testing, maybe we should dispense with testing and just have inspections?

- *Reviews and audits of deliverables* as we have built into our lifecycle, are fundamental to catch errors early and prevent them escalating in later phases of the lifecycle. A $10 error caught at the requirements phase can save $10 000 down the line. Reviews should be formal and documented. Time must be scheduled into plans to respond to the issues they raise

- *Defect seeding,* as discussed by Tom Gilb, is an interesting approach to determining if our testing and defect removal process is working. The idea is that we knowingly introduce some errors to code before the testing process. We then see if the testing finds these errors. In large software products, the technique can give us valuable feedback on the effectiveness of our testing and correction procedures

- *Rewards for Quality.* The old management maxims apply: You get what you measure and what you reward people for. If we want quality, and quality improves productivity, and this saves money, then we can reward the people who deliver quality. Rewards do not necessarily need to be financial, although they can be. As we will see later, I.T. personnel are frequently motivated by other factors. The reward could be sending the person on a seminar they would like to attend, buying them a book they have been coveting, or giving them an upgrade for their PC. Key factors are to make it public, and make the selection of recipients fair - this requires clear policy and guidelines, as well as objectives and measurement criteria

- *The importance of training and mentoring* cannot be stressed too highly. Many shops we see are in the mode of "we can't stop to sharpen the axe, we're too busy trying to chop trees." Cutting training may save you schedule time and money in the short term. In the long term it will cost you productivity, motivation, and ultimately your best performers. Training need not necessarily be formal, but it should definitely be an objective and receive priority. Don't forget mentoring, which is much neglected, but vital. Many subtle skills can only be transferred effectively in this way. A course may be great to learn about the latest release of software, but this will not teach the new analyst how to handle a particularly tricky user in group sessions

- *Separation of duties.* In the financial world we have the concept of joint signatories. In auditing practice, we make sure that the person who draws the check is not the same person who signs it. In systems, we let the same people who wrote systems take them live. This allows shoddiness where we mentally say to ourselves, "I know that the calculation routine is a bit slow, and that the print routine really isn't tested thoroughly, but the users want it now and I can fix it later." Of course, we never do get around to fixing it before it fails in production. This attitude is also a major cause of poor documentation. Writing documentation is never fun. If the only person we are writing it for is ourselves, we are hardly likely to be very motivated. All systems and changes, before going into production, should be formally accepted by the personnel who will run them (operations) and the team that will maintain them. The latter should be separate from the development group. This will ensure that this group will not accept anything from the developers unless they can prove that they have done a thorough job. The accepters must not fix any errors, though, or this will encourage developers to use them as a quality assurance resource.

We will expand on several of these in the following sections.

Bugs or Defects?

The first "bug" was reportedly discovered by Admiral Grace Hopper of the U.S. Navy, working on an early machine using electromechanical relays. The machine had displayed a puzzling fault which was traced to a moth jammed in a relay. This was removed and taped into the log book with the message "I found the fault, it was a bug," or something similar. This was a real bug, and it crept in there all by itself. However, ours are not real bugs, and they don't get there by themselves: we put them there. They are also very costly. De Marco estimated the cost of software *defects,* as we will call them from now on, at $7.5 *billion* in 1982. Bugs are cute, defects aren't.

We should not call them bugs, since this removes the responsibility from the creators. We should refer to them as defects, which is what they are. A defect is a deviation between desired and observed results. We are often told that it is beyond human ability and the state-of-the-art to produce software without defects. De Marco tells a lovely story about an engineer who wrote his first system after teaching himself to program. The system ran without any hiccough for many years, and no faults were ever found. When asked how he had managed to do it, the engineer replied, "I didn't know that errors were allowed." We should be more like him.

The vast differences in ability we discussed with respect to productivity of developers extend to their ability to produce error-free code, and their ability to find and remove errors. We can expect an order of magnitude difference between our best and worst performers as highlighted in figure 15.9.

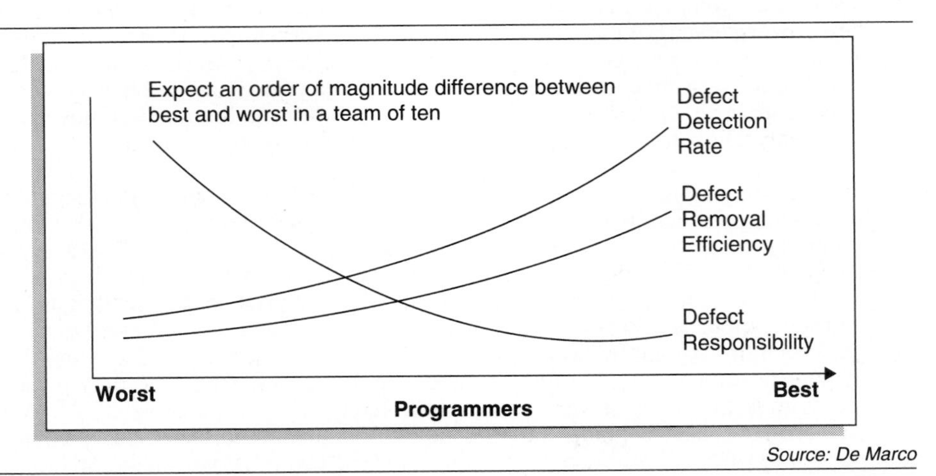

Source: De Marco

Differences Between Developers *Figure 15.9*

Start Testing Early

It is a common fallacy that we can only test when the first programs are complete. Nonsense. We can test the moment the first deliverable is produced. We can test a project plan for the reasonableness of its assumptions, to see if leave and training have been included, and to make sure that resources are not routinely overloaded. We can test a functional model to see if all the necessary data to support it is in the data model. We can test a user document to see if it is comprehensible, and that the readability level is not too poor. How? Just present it to your peers for a start!

Adversary Teams

We have mentioned the problem of not having a formal hand-over procedure, and the need to introduce separation of duties. One extreme way to do this is to have a testing team whose job it is to try to make the product produced by the developers fail. An example of this was the notorious "Black Team" at IBM. They were started as a normal testing group, but quickly took it upon themselves to develop a "mean" image. They started dressing in black, and delighted in breaking code submitted to them for acceptance testing. It became a challenge, of course, for the developers to see if they could beat the Black Team, and their quality shot up. Members were rotated through the team so that it didn't become a personal issue between individuals.

Looking at it more generally, and considering figure 15.10, construction teams should produce their work and submit it to the testing team for acceptance. Both construction and test teams have a copy of the requirements specification. The test team will try to find any defects they can. They then return their diagnosis and comments to the construction team.

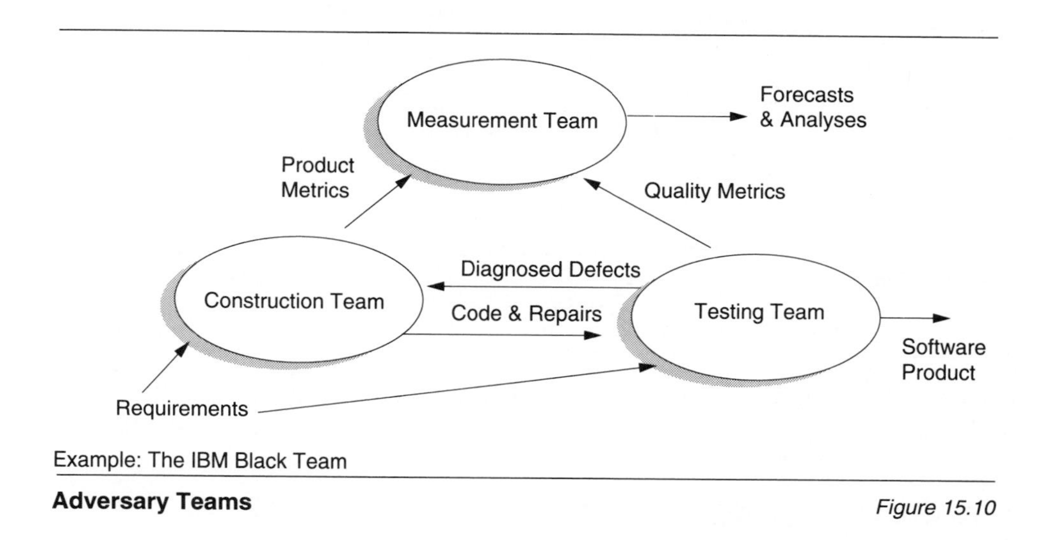

Adversary Teams *Figure 15.10*

The construction team makes the necessary repairs, and resubmits the product. When the test team is satisfied, the product is accepted for implementation. A separate measurement team maintains information about the software development process. The construction team will provide them with measurements of application size, effort expended, etc. The test team will provide the metrics team with information regarding product quality. The measurement team will build up a database of information. This is useful in monitoring productivity and quality, and in assisting with estimating and process improvement.

Myers's Findings

In his research into the defect density of code in large systems, Myers found a relationship that shows that the likelihood of finding errors in a portion of code is proportional to the number already found there. See figure 15.11. Simply stated, those areas where early tests locate defects will probably continue to produce more defects with further testing. Those where early tests show no or few defects will tend to exhibit this behavior during continued testing. This is termed the "cockroach theory". If you see one cockroach in a kitchen, it is unlikely to be the only one in the place. There is probably a nest somewhere. If conditions are right for one, there will be more. A practical application of this is in assessing modules for maintainability. If we analyze the defect density of a total system on a module-by-module or subsystem-by-subsystem basis, we can identify those areas with high defect densities, and those which are relatively defect free. An analysis of IBM's IMS database system in this way localized 57 percent of errors to just 7.3 percent of modules. We can then decide whether we can fix these by a restructuring or inspection process, or whether they should be

"The probability of the existence of more errors in a section of a program is proportional to the number of errors already found in that section"

- IMS - 57% defects in 7.3% of modules
- Defects' presence is not consistent
- Cockroach theory
  - Kill the nest

Myer's Findings *Figure 15.11*

rewritten. This approach can greatly reduce maintenance effort, and simultaneously increase quality significantly.

Monitoring Defect Removal

Having found defects, we need to remove these. In doing this, we run the risk of introducing new defects. We need to track the number of defects found and the number corrected. These should produce a graph with a characteristic shape, as shown in figure 15.12. This indicates that defect discovery proceeds up to a point, and then levels off. Defects corrected should climb to meet this line. After this, no more defects should be discovered. If the number of defects discovered plateaus and then rises, this is indicative of problems in the correction process - it is introducing new problems.

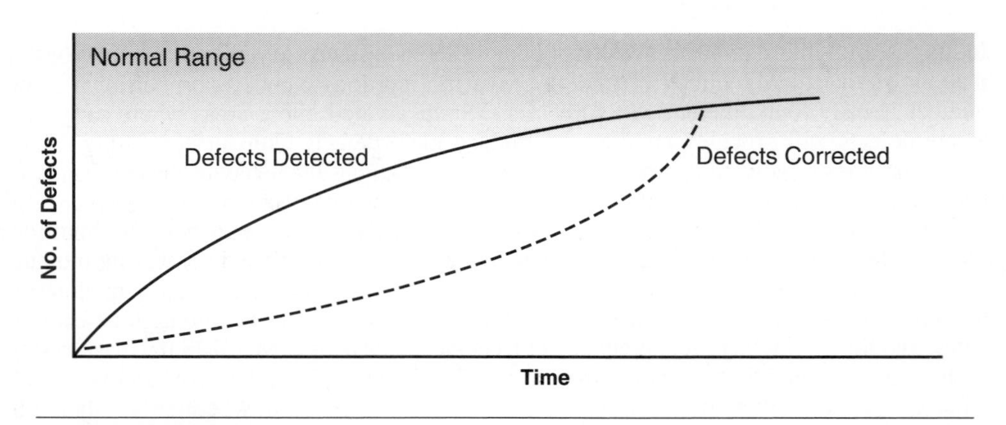

Monitoring Defect Removal *Figure 15.12*

It can also be useful to put a band on the graph to indicate the normal range for defects in products of similar size to the current one. This should have one standard deviation either side of the mean. If our product falls outside these ranges, we should be wary. If defects are very high, then we should review our production process. If defects are unusually low (below the band), it may be that we have really got the development process producing very high quality. On the other hand, it is far more likely that our testing process is at fault and that we are not finding all the defects present. Of course, over time, the defect rates should fall and the mean and thus the "normal" band will move downward.

Problem Incidence as Reliability Indicator

Once systems are in production, it is useful to monitor their production performance. This can include aspects of performance and reliability. Performance can be monitored in terms of batch run times, and response times. Any nonlinear increase could be an indicator of impending capacity problems, or a need to reorganize files.

Reliability is normally measured in number of incidents per unit time. An incident is any occurrence in the production environment which requires corrective action. It could be a program "abend", incorrect figures on a report or a corrupt record on the database. These may be documented using Problem Incident Report (PIR) forms. Monitoring the PIRs for a system over an extended period will give us an indication of its reliability, and the quality of the maintenance being performed on it. See figure 15.13. A system which is improving in

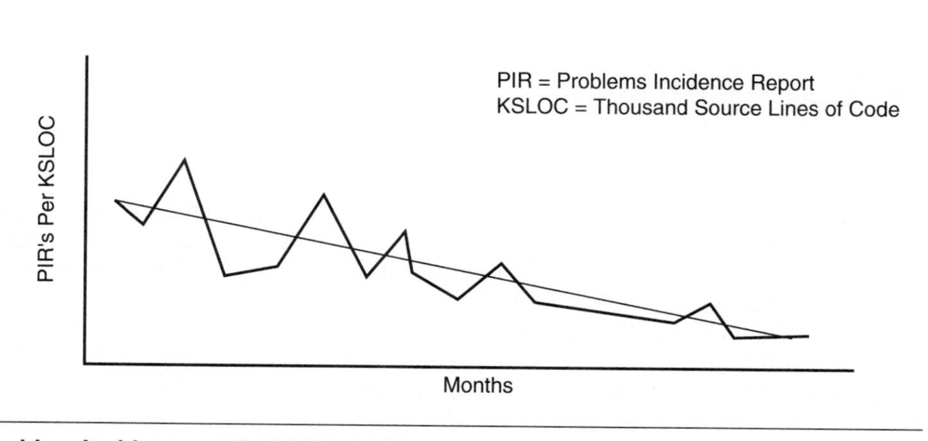

Problem Incidence as Reliability Indicator *Figure 15.13*

quality will have a downward trend in PIRs, one being maintained badly, an upward trend. We need to balance these with usage patterns, e.g., an increase in PIRs may be experienced when new facilities are exploited in production. Normally we will use a simple regression to assist us in seeing the overall trend, as opposed to the valleys and peaks. To normalize them for system size, we might express the trend in PIRs per 1 000 function points, or some similar measure.

The Productivity Link

If we do it right, what can we achieve? We would like to cite two examples. The first is Hitachi Software in Japan. This is a very large software house producing system software. They ran a quality improvement program over some 4 years. The results are astounding, given the figures we have presented earlier in the chapter. Hitachi were unhappy with its PONC, measured as a cost of fixing any defect in software delivered to clients as a proportion of revenue. The company began its program with the figure at 1.48 percent! By the time the progress was reported, this had been reduced to a figure of just 0.08 percent as shown in figure 15.14.

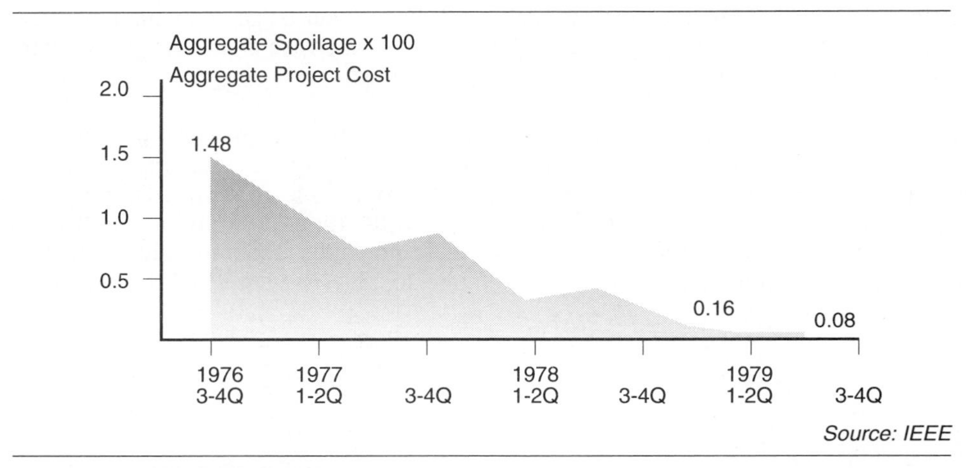

Quality Improvement at Hitachi *Figure 15.14*

An American example comes from Computer Sciences Corporation, an aerospace contractor which builds complex software for embedded systems in missiles, spacecraft and airliners. The company mounted a program to improve quality. They used statistical process control concepts as we have outlined. Over a period of some four years, they were able to reduce the error delivery to less than half of the previous norms as shown in figure 15.15. An amazing byproduct that they had not counted on was a *threefold increase in productivity*. These results support our earlier contentions that the major problems and challenges which we face are managerial, rather than technical. If a vendor were to offer us those sort of benefits from some brand new gizmo, we would buy it today. Similar results are available to us if we take the trouble to learn the concepts, plan a suitable strategy for our organizations, and stick to making it work.

Reusability

One way to achieve high quality and productivity, available since the inception of computing, but which is only now beginning to be widely exploited is *reuse*. If we can create functional, modular, high-quality software components, then we can reuse them in the same way that hardware engineers use integrated circuits in a variety of devices. An example

• Computer Sciences Corporation

• Six Projects - 2.7 Million lines of code

• Card, Clark and Berg, 1987

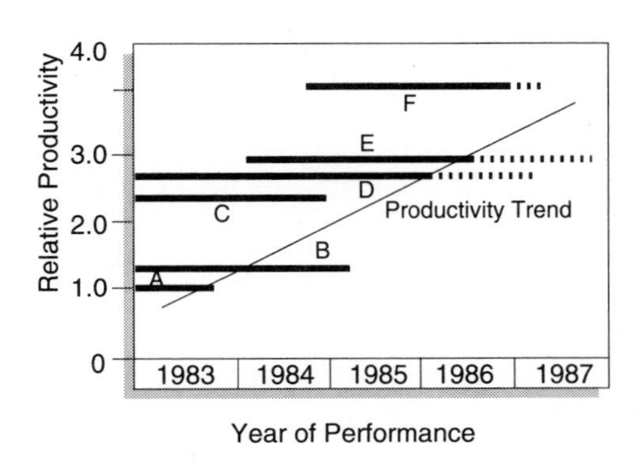

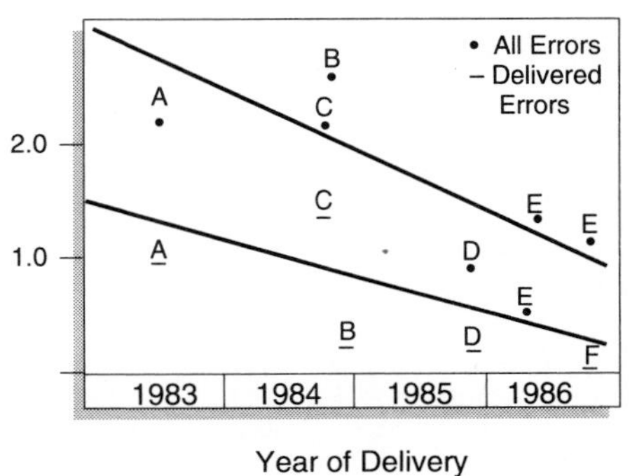

Process Improvement Results *Figure 15.15*

would be a timer chip which could be used in a digital watch, an alarm clock, a VCR, and a personal computer. The application is different each time, but the designer of each device can use the standard component in each case. All we have to worry about is what connections it needs, what signals it accepts and what it will give back. We can design software in the same way. One good example of this over the years has been the Fortran scientific function libraries. These items of code have been reused by millions of developers in widely varying applications.

Until recently, developing reusable code required very high skills and was technically difficult, mainly due to lack of tool support. With the advent of object oriented techniques and development tools, we should see reuse becoming common. The advantages are great. If we do not have to design, code and test a module, the savings are enormous. If the module has been used by thousands of other developers before us it is likely to be extremely reliable. If it is highly used, the developer can take the trouble to optimize its performance - something we could not justify for our single application. The benefits for quality and productivity are apparent. Already large class libraries for technical environments (e.g., Windows™, Presentation Manager™, Motif™ user interfaces) and application areas are being marketed. We may finally see improvements in productivity of software development on a par with what the hardware engineers have delivered.

We should make reusability part of our strategy and reward developers for producing reusable components and for using existing components. It is vital that these components be of the highest quality if we are to reap the benefits, so careful attention must be paid to quality throughout the development process.

Case Questions

MyWay Organizer

Q15.1

Below is the Entity Model developed for the MyWay Organizer. Using the criteria given in chapter 4 for this type of deliverable, critique the model. (20 mins)

Q15.2

Using work breakdown and network planning techniques, devise a detailed plan for the testing of the MyWay organizer product. Include reviews and other QA activity in your plan. (20 mins)

Q15.3

Using the Quality Model presented in this chapter, develop a description of the process, suppliers, clients and criteria for the task "Cross Platform Test" in the Organizer project.
(20 mins)

Handover Trust

Q15.4

Mr. Renfrew has asked you to propose an organization structure and job description for a quality assurance function within Information Systems at Handover Trust. Remember that this function will need to establish standards, provide assistance in achieving quality results to a variety of project teams, conduct walkthroughs and quality audits, and mount other activities to ensure that quality does improve. Detail in your proposal:

- Your proposed organization structure (show relationships of the function to other groups and projects)
- Draft a mission statement for the function
- List the five most important objectives for the function
- What do you recommend as the first order of business to which such a function should apply itself?

(40 mins)

Q15.5

A standard project plan has been suggested as the basis for planning parallel projects undertaken in future. You are asked to critique the structure of the plan with particular reference to quality assurance and configuration management. Amend the plan where necessary.

(30 mins)

Indentation is indicative of work breakdown hierarchy:

- Feasibility Phase
  - Justify Economics
  - Consider Technical Options
- Specify Requirements
  - Model Data
    - Collect Documentation
    - Analyze Data
    - Draw Entity Model
  - Examine Existing System
    - Interview Users
    - Draw Data Flow Diagrams
  - Define Logical System
    - Develop Logical DFD's
    - Design Reports
  - Design User Dialogues
  - Design Online Screens
- Design System
  - Database Design
    - Normalize Data
    - Draw Relational Model
    - Add Physical Access Requirements
    - Calculate Sizing
    - Specify Physical Schemas
  - Map Functions to Modules
  - Define System Interfaces
  - Program Design
    - Define Program Requirements
    - Determine Access Maps
    - Design Algorithms
  - Design Batch Flows
    - Specify Batch Process
    - Define Job Control
- Write System
  - Code Programs
  - Test Programs
- Parallel Test With Old System
  - Prepare Data
  - Move Software to Production
  - Run Test
  - Verify Results
- Take System Live

Q15.6

Consider how you might implement a version of the "adversary team" quality assurance approach in the ThoughtWell Books situation. Could you make use of client personnel? What are the advantages and disadvantages that you foresee? (30 mins)

16 *Project Documentation*

The Need for Documentation

There is ample evidence to indicate that between 60 and 70 percent of the effort expended on systems projects occurs *after* the implementation of the system, i.e., in the maintenance phase - recall our discussion of the Putnam-Norden model. Some of this is adaptive maintenance to meet changing requirements, some is corrective maintenance to fix defects, and some is perfective maintenance to fine-tune the system in production. Regardless of which type of maintenance we are to perform, good, accurate, up-to-date documentation of the system is essential if we are to be efficient. It is also vital to ensure that our changes do not introduce unanticipated problems. Good documentation is vital to realize the full return on investment from a system. Documentation should thus be a major component of the product which our project sets out to deliver.

The Traditional Trap

Producing documentation is a lot of work. It is normally not a job that creative systems people enjoy; consequently, it is avoided like the plague. It is normally tucked away somewhere at the end of the project plan, and assigned to the most junior resource, who can't protest too much. Of course, this person knows the rationale behind the system inside out and is an expert communicator and technical writer, so we won't have any problems, or will we? We frequently find that this isolated task of "document the system" is planned to occur in parallel with integration testing, and preparation of the production environment. Teams are under enormous pressure at this time, and, you guessed it, the documentation suffers. We would feel a supplier of electronic goods was negligent if we were sold a product that could not be fixed by technicians because there were no schematic diagrams, yet we frequently leave our own users in exactly this position.

As we have pointed out, good documentation is vital to realizing the benefits and potential of a system over the long term. This means that it should be done by competent, senior people. But how do we persuade them to do it?

Painless Documentation

The most painless way to produce documentation is in small, manageable chunks. How do you eat an elephant? One bite at a time! Secondly, we should not see documentation as a separate task - it should be spread evenly throughout the project, and spread over several team members. It should be a painless byproduct of the activities we normally perform, not something discrete. To achieve these aims, we can:

- *Use automation* in the form of CASE tools, and other much simpler tools, such as word processors, graphics packages, data dictionaries and so on. This takes the sting out of documenting for technical people. It gives them some technology to play with, which they like. It also makes revisions much less labor intensive, and therefore documentation tends to stay much more current. (Have you ever had to redraw a 200 box diagram by hand to include the user's latest idea? - You soon start discouraging him from having ideas at all!). Even better, is if the tool is integrated with the environment in an active way, e.g., an active data dictionary. This means that if you want your change or definition to be usable in the final product, it has to go through the documentation tool: there can be no difference between the documentation and the reality. See figure 16.1.

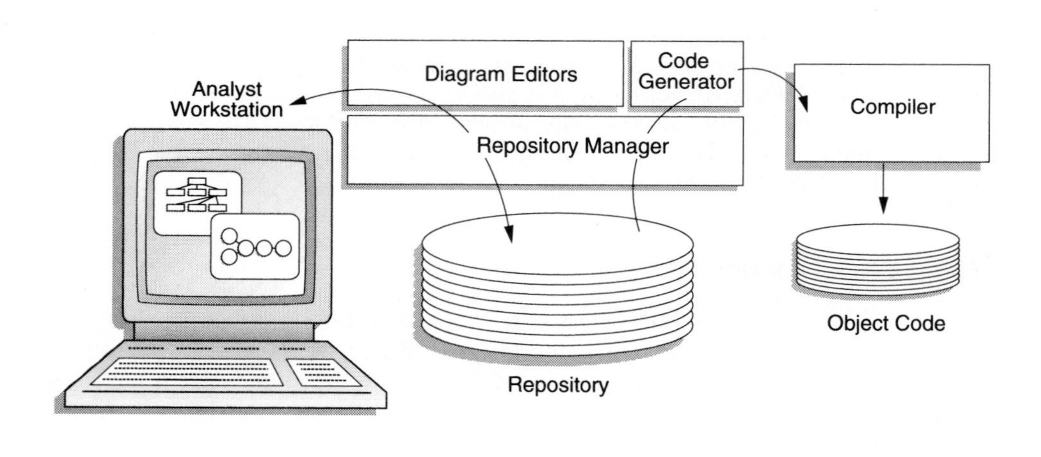

A Typical CASE Tool *Figure 16.1*

Maintenance and changes can be further simplified, and reliability increased, by using a principle called *transclusion.* This is attributed to Theodore H. Nelson, the inventor of the Hypertext concept, now at Autodesk, Inc. Transclusion is the inclusion of components in documents by reference, rather than by copying. Thus there is *one correct copy,* which can be changed once, and which will then be available in all its scales, orientations in all documents which reference it. How we have desired that capability in our systems for so long! It can become a reality for documentation with the emergence of compound electronic documents, hypertext and techniques such as Object Linking and Embedding (OLE) in Windows™. Similar techniques have been available in other environments (e.g., Apple and Unix) for some time.

To facilitate the reusability of diagrams and figures, they should be prepared as structured graphics, not bitmaps. Bitmaps are usually created using scanned images or paint programs. They are stored as a color value for each "bit" (dot on a printer or pixel on a screen). While simple, they suffer from a loss of resolution when scaled up or down. This is normally seen as a ragged edge, normally called "jaggies". See the accompanying examples. Structured graphics, by contrast, are stored as a mathematical description of the collection of objects or shapes making up an image. They are more complex to create and are normally produced in one of three ways:

- Specialized design and drafting packages as used in Computer Aided Design (CAD) or CASE tools (figure 16.1)

- Structured art packages, such as Corel Draw or Harvard Graphics

- Vector tracing a bitmap image to determine the predominant shapes and edges, interpreting these into a structured image

Structured graphics can be scaled to any size without loss of resolution, as shown in figure 16.2

• *Provide secretarial support to the team.* It is incredible how many bad but extremely well-paid typists we see on project teams masquerading under the title of senior analyst! Organizations are short-sighted if they do not provide much more productive (and much cheaper) secretarial support to project teams.

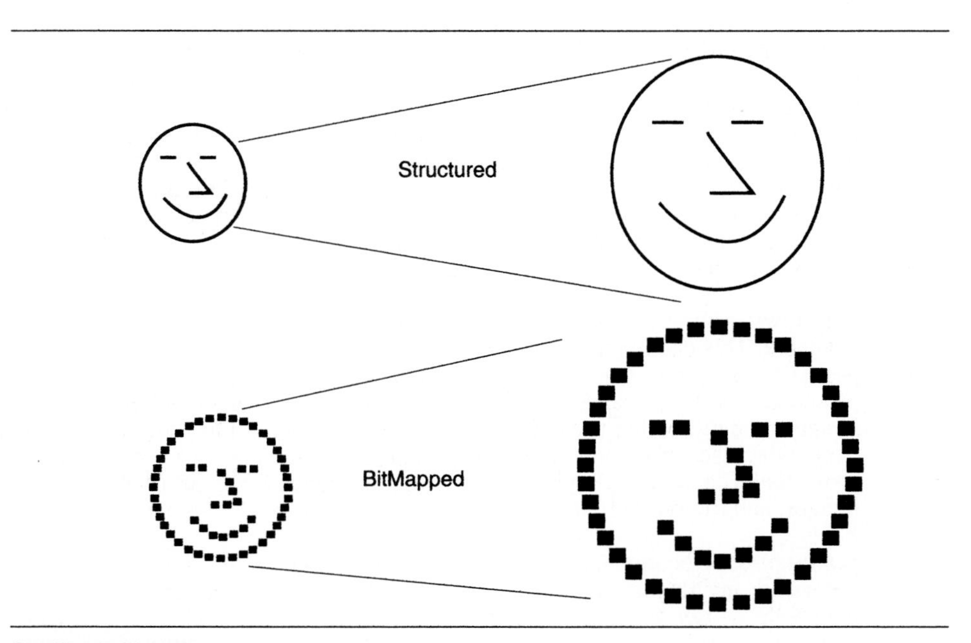

Scaling of Images *Figure 16.2*

- *Having a good librarian* that keeps track of things. This can be a secretarial resource, but could also be a technical person who would provide support to several projects. This function can enhance communication across projects, and sometimes identify reusable components, thereby realizing huge savings. It is a good role to rotate relatively new employees through, as they will gain a broad perspective on the organization's systems

- *Using technologies such as scanning*, CD-ROM storage and automated indexing to reduce the bulk of documentation, make it easy to back up, and easy to access. This will also simplify impact analyses in future. An apocryphal story tells us that when the Patriot missiles were deployed in the Gulf war, that the missile system needed three trucks to deliver, and the documentation to go with it required seven! Shortly afterward, the documentation was all reduced to about a dozen CD-ROMs

Documentation Principles

There are some important principles we should follow in planning and producing our documentation. They include:

- *Communication medium.* This should be appropriate to your environment and intended audience. Online documentation is fine for system developers used to browsing it in this form, but may be useless to a business analyst or user manager. For users written, linear documents with an introduction, body and summary are best. Some of the new hypertext authoring packages provide "linearizing" facilities which not only allow you to set up a hypertext machine-readable document, but to automatically produce a linear version from this. Make use of devices such as diagrams and tables to aid clarity and explain the structure of the document where necessary. Ensure that all printed material is of good quality, preferably laser printed. No one wants to pore over scores of pages of faint text off an aging dot matrix printer with a worn-out ribbon.

  For presentations, overhead transparencies normally work best. They are relatively easy to prepare, they are reusable, and they can be shuffled or extracted quickly to suit the audience and the presentation (even on-the-fly if necessary). Be careful to keep the size of lettering quite large (say an 18-point font) so that they do not become too cluttered or detailed. When designing them, it is a good idea to divide an A4 sheet into quarters, then pencil your content for each slide into one quarter. The limited space ensures that you do not try to cram too much onto a single foil. Overheads can be prepared quickly on computer with the aid of a graphics package (e.g., Corel Draw™, Powerpoint™) and a good printer.

  Pre-drawn artwork can be included from "clip-art" libraries which are widely available. This can liven up foils if used carefully. Preparing your foils on a graphics package also gives you the option of reusing the images in computer-controlled presentations. These can be accomplished with a device which projects the computer image through a video projector (for example, the BARCO™ system) or by using a panel which accepts computer output and projects the image via a standard overhead projector. The mouse, keyboard, or a remote control can be used to change the images, and to reverse if necessary

- *Conciseness.* Documentation should be as long as necessary, but no shorter. Before preparing any document, think about the objective it must achieve. An example would be a user manual. Here the goal might be: "To enable a user with basic computer literacy to understand the system adequately to install it successfully, use all of its features, secure application data, and respond correctly to messages and prompts generated by the system". This sounds simple, but to write such a user manual is quite a challenge. Where possible, use diagrams of a type known to the target user to reduce bulk and convey information accurately. An analyst could make good use of an Entity Relationship diagram, for example. Use consistent references on diagrams to act as an index to more detailed information. The Entity model might be supported by detailed record layouts, for example

- *Clarity*. The meaning should be clear and not dependent upon the perspective of the reader. Be careful to define uncommon terms in the text and in a glossary. Get a nontechnical person, uninvolved with the project, to read the document and see if he understands what it is you are saying. We frequently develop a vernacular within a project team that will not be easily understood by outsiders

- *User friendly*. User-friendly documents are well structured. They allow the reader to gain an overview quickly, and to find relevant information without going through reams of paper. They provide for different levels of readers by making use of introductions (which more experienced readers may skip), as well as providing "jargon busters". These are panels which explain terms and concepts not familiar to all readers. Those who know them can look at the title of the box, and skip these parts easily. We should provide a table of contents at the beginning, as well as an index at the end. The former aids the reader in discerning the structure of the document and which sections are of interest to the current purpose. The latter helps us to find information on a particular topic. References should be provided to allow the reader to pursue issues in greater depth, or to find background information.

  Serif fonts in point sizes from 10 to 12 should be used for body text. (This is set in 10 point Times Roman, a serif font). Serifs are the little "tails" on the edges of letters which aid the flow of the eye across the page. Non-serif fonts in point sizes from 15 to 18 should be used for headings. In this text, headings are set in 15 point Triumvirate, a sans-serif font. These are very crisp and clear, attracting attention. They can be more tiring to read for any length of time, though, so should be avoided for long passages. Italics and bold styles can be used to attract attention, and for emphasis.

- *Best knowledge*. Each document should be prepared by the person with the best knowledge of that particular aspect of the project or system, assisted, if necessary, by someone who can write or document well. For example, the context diagram might be the responsibility of the project leader, the Entity Relationship model the systems analyst, and the user manual that of the analyst/programmer who developed the prototype and user interface design. We should never assign documentation to a junior person without the necessary in-depth knowledge to do a good job. Remember, the documentation is an integral and very visible part of the total product your project will deliver.

- *Consistency*. It is vital that all documentation be consistent. This applies to technical content (especially naming of parts of the system, data items, etc.); structure; presentation and language.

- *Accuracy*. Equally vital is the accuracy of the documentation. One of the authors once spent two very frustrating weeks trying to nondestructively change a municipal billing system, written in assembler, with little success, only to discover that the impressive documentation was hopelessly inaccurate with regard to memory location usage. Using automated tools (e.g., CASE) tied to a dictionary or repository can greatly facilitate keeping the system and documentation in step. Even if your installation does not have these facilities, some simple, home-grown utilities can go a long way in preventing problems.

- *Don't spoil the ship for a ha'p'orth of tar!* If we are going to put a lot of effort into doing the documentation right, reward your team and encourage their pride by getting it printed and bound professionally. This can be expensive, but not compared to relation to the overall effect on the morale of the team. It also sets a benchmark for later projects to achieve, thus lifting the standard of documentation generally.

Martin-McClure-Odell Notations

For systems projects, we recommend the diagramming standards developed by James Martin and Carma McClure, and now extended for Object Orientation by Martin and James Odell. You will find that these are very consistent, usable, comprehensive, and are well supported by a variety of CASE products. They include techniques for process models, data models, structure charts, object models, state diagrams, event models and program structures.

Meetings, Decisions and Minutes

All project meetings should be minuted. These can be just "action minutes". They do not need to show the flow of discussion and bargaining which has taken place, just the decisions taken. Be sure to record:

- The date and time of the meeting
- Who was present, who sent apologies, and who was absent
- Any decisions taken
- Any actions which must be pursued as a result of the meeting (be sure to identify responsible parties and expected date of completion)

Distribute these minutes as quickly as possible after the session.

PBM of Documentation

We have reproduced a Product Breakdown Model of the documentation structure for a mainframe development project, where a package was considered, but in the end not accepted. This is an indication of the extent and coverage of complete documentation for a project. In some cases, the item mentioned could be very brief (a single paragraph). It should nevertheless be considered.

Documentation
- Technical Documentation
    - System Charter
    - Technology Charter
    - Business Requirements Definition
        - Context Diagram
        - Conceptual Model
        - Technology Review
        - Formal Business Rules
    - Technology Requirements Definition

        - Functionality
        - Capacity
        - Operability
        - Reliability
        - Security
        - Connectivity/Compatibility
    - External Design Specification
        - Functional Model
            - Functional Decomposition Chart
            - Function Narratives
    - Data Model
        - Entity Model
        - Data Item Definitions
        - Normalized Relations
        - Relational Model Diagram
    - Behavior Model
        - Technology User Interface Standards
        - Dialogue Definitions
        - Batch Flows
        - Prototype Screens
        - Prototype Reports
        - Interface Definitions
    - System Consistency Matrix
    - Logical Access Maps
- Invitation to Tender
    - Requirements
        - Function Model
        - Data Model
        - Operational Requirements
        - Technical Requirements
    - Procedure for Submission
        - Submission Process
        - Submission Format
    - Supplier Short List
- Package Evaluation
    - Selection Criteria
    - Package Functional Mapping
    - Package Data Mapping
    - Package Technology Mapping
    - Package Throughput Assessment
    - Package Technical Quality
    - Package Supplier Support Assessment
    - Package Economic Assessment
    - Package Recommendations
- Technology Evaluation
    - Selection Criteria
    - Technology Functional Mapping
    - Capacity Evaluation
    - Operational Evaluation
    - Reliability Evaluation
    - Security Evaluation
    - Connectivity/Compatibility Evaluation

- Summary Technology Evaluation
- Technology Recommendations

Technical Design Specifications

- Product Structure Map
- Physical Database Design
- Run Unit Specifications
  - Process Graphs
  - Process Narratives
  - Physical Accesses
- Bridging and Conversion Strategy
- Test Specification
- Performance Prediction

Physical System Documentation

- Program Listings
- Control Language Listings
- Data Definition Listings
- Sample User Interfaces
  - Input
  - Output
  - Input/Output
- Parameter Data

Test Documentation

- Test Plans
- Test Cases
  - Test Conditions
  - Test Data
  - Correct Test Results
  - Actual Test Results

Installation Documentation

- Physical Distribution Map
- User/Location Matrix
- User/Function Matrix

Project Control Documentation

- Project Charter
  - Project Definition
  - System Charter
  - Feasibility Report
- Project Plan
- Project Actual Measured Performance
  - Time Reports
  - Deliverable Completion Record
  - Expenditure Record
- Project Progress Reports
- Quality Assurance Documentation
  - Review Reports
    - Departure Review
    - System Requirements Review
    - Preliminary Design Review
    - Critical Design Review
    - Product Acceptance Review
    - Post-Implementation Review
  - Requests for Change

        - Impact Analyses
        - Change Notices
    - Project Meeting Minutes
    - Project External Correspondence
        - Intra-company
        - Extra-company
- Operational Documentation
    - Facilities Guide
        - System Overview
        - Batch Processing Runs
        - Processing Run Control
        - Hardware/Software Environment
        - Backup, Recovery and Restart
        - Reports
        - Data Structure, Volumes and Growth
        - Program & Control Language
        - System Errors
        - System Failure
        - System Maintenance
        - Performance Monitoring
        - Data Capture
    - User Guide
        - System Overview
        - Basic Organizational Documents
        - Reports
        - Screens
        - On-line Procedures
        - Computer Equipment
        - Management and Control Procedure
        - Departmental Structure
        - Training
        - Run Requests
        - Back-up and Recovery
        - Document Storage
        - Liaison
- Training Package
    - Student Materials
        - Course Notes
        - Workbook
        - Tutorial
    - Instructor Materials
        - Presentation Materials
        - Course Time Schedules
        - Lesson Plans
        - Case Study Background
        - Data
            - Installation Procedure
            - Data Files
            - Hard Copy Content

Summary

Documentation is an integral part of the total product you deliver. It determines the ease with which the product can be used, installed, modified and repaired. These activities consume more than half of the lifetime costs. Documentation is thus an extremely important activity. Do it well.

Case Questions

MyWay Organizer

Q16.1

Define a Product Structure Map for the contents you believe should be in the shrink-wrap package in which the Organizer is planned to ship. (15 mins)

Q16.2

You want to include a large number of captured screen images in your user manual. What kind of graphics are these? What will happen when we scale them? Do you have any suggestions for how we should proceed? (10 mins)

Handover Trust

Q16.3

Your suggestions for the standard lifecycle incorporating quality assurance tasks have been approved by management. Your next task is to develop a structure for a standard set of documentation which all projects will use for planning, tracking and reporting. Present your answer in the form of a PSM. The structure should ensure that projects are thoroughly planned, that adequate information is collected for proper tracking, and that progress is consistently reported to management across a variety of project types. (30 mins)

17 *Communication*

Introduction

You may have met good project managers who are good technicians. You may also have met those who are good managers. We believe that you have to be a good technician and a good manager to be effective. The difficulty is that many of us have long experience in technical areas but are new to management. Although the next chapter will deal with the management of people, this chapter focuses on the skills many of us have already been exposed to - human communications. If you are in the property industry the most important factors are location, location and location. In a project the most important factors are communication, communication and communication! This is reflected in the highlighting of all the

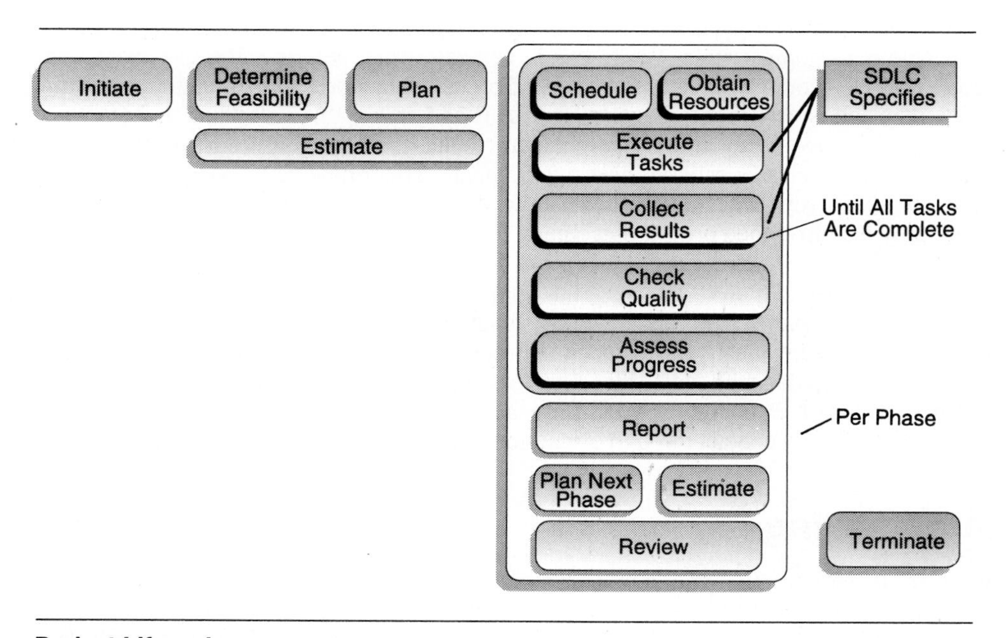

Project Lifecycle *Figure 17.1*

boxes in the project lifecycle diagram (figure 7.1) - communication is vital at all stages.

While traditional writing and formal speaking skills are important, today's challenges demand skills in developing and maintaining relationships. These relationships depend on open communication established by probing for information, presenting views and discussing individual behavior.

It is becoming increasingly difficult to be a successful project manager. Projects are becoming more complex and project team members are more demanding. Traditional management approaches have been replaced with those where the establishment of mutually satisfying relationships between manager and subordinates prevail. This is the key to successful project management. To develop these relationships, the project manager must be a skilled communicator - skilled at getting information from others; skilled at expressing views without creating problems and skilled at discussing behavioral issues. This chapter is about these skills and how they fit in the project context. Remember that skills are easy to learn but more difficult to apply.

Communication Styles

Developing an appropriate communication style is a critical step in becoming a project manager. Traditional managers tend to have a one-way style where they direct staff using specific instructions which are not negotiable. The I.S. industry has many managers who are (or think they are) technically superior to their staff. These managers always seem to know the best way to solve problems. Subordinates are not able to develop their own approaches when working with them. A preferred approach is a consultative one where subordinates are allowed to develop and test their own approaches. In this way, more initiative can be used and staff have a greater feeling of commitment to the task.

As a project manager, you may feel that the consultative approach should be the only applicable management style. While this is generally true, bear in mind that there will be occasions when you will have to use a more direct style in a pressured situation to get things done. Your job is certainly to develop your staff and ensure that individual and team needs are satisfied. However, you are measured on results: producing an information system on time, on budget and that satisfies user requirements. The following interview with a project team member (opposite page) shows how a consultative management style does not necessarily lead to a good project outcome.

Our view is that a project manager has to get things done through people. Creating a productive environment for people to work in is important. So is a good consultative approach. But there must be a balance between the task, the individual's and the team's needs. Being popular with staff (like Andrea), but unable to influence senior management, will not produce the required results. Influencing people at all levels in an organization is key. This is done by contact through discussion, interviews, presentations or meetings.

Interviewing

Most project managers who have been systems analysts will have developed strong interviewing skills - especially probing users for needs and problems. A project manager will have to conduct other types of interviews, mainly in staff selection, appraisals and performance discussions. These are discussed in detail in the next chapter. These interviews differ from a probing interview in that there are no second chances and a bad interview will inevitably lead to a deterioration in performance or relationships.

The best approach to any interview is to plan it. You should determine beforehand what you want to accomplish and the interview structure. A typical structure could include some introductory focus on the reason for the meeting, followed by the specific discussions, then ending with some clear concluding remarks which include individual action and timing. These should be documented to ensure understanding. Setting aside adequate time and providing a location without distractions are important. This is especially important in large companies where open-plan offices are the norm.

"I have known Andrea for 5 years and worked for her for the last 2. I got to know her as if she was a family member - in fact, I have never worked for a more caring person. She always had time to listen to my problems - even when they were nothing to do with work. In the last project we were working on, we were 2 months behind schedule. Everyone was working overtime to catch up, even weekends. Even under this pressure, Andrea still found time to discuss individual problems. Despite the long hours, we still could not catch up even though Andrea was always there. When Andrea explained that we needed more staff, the management did nothing about it. Eventually, a new project manager was appointed who seemed to get everything she wanted. The project was completed ahead of schedule and everyone was excited with their contribution. But, we all still miss talking to Andrea."

Negotiating and Influencing

Project organization structures often reduce the ability of the project manager to control project team members. Project managers have to spend time negotiating with superiors, co-workers and subordinates. As hard bargaining does not work well in the project context, a climate of trust and cooperation has to be developed. Negotiation takes place between two parties because there is a conflict of interests. Typically, a project manager will have to negotiate for user resources, computer time and time of computer specialists such as network designers and database administrators. Because pressure for resources leads to conflict, we have to find ways to deal with the situation. One way is to ignore the conflict, hoping that it will go away. Another approach is to get your own way irrespective of the consequences. Alternatively, you might resolve the conflict by giving in to the other party. Compromise is another approach, and in this case both parties at least get something. The last alternative is problem solving where a solution is sought that can satisfy everyone's interests.

All of the above strategies have their place in negotiation. If you don't care what happens - withdraw. Giving in to other parties is often used in projects where managers or users demand deadlines. This is realistic when dealing with a customer but it is essential to re-negotiate project scope or resources if time is so important. Giving in to unrealistic demands only delays problems. Compromise can be used as a quick solution but is inferior to problem solving. The latter demands considerable participation and trust. For an agreement to be advantageous to both parties, there must be a climate of total honesty.

Negotiation is a process. First of all you have to define what the conflict is about. Identify the situation or events that led to the negotiation. Think about the situation you would like to achieve and some of the likely alternatives. Then try and project how the other party will respond. Once you have done your homework, schedule a mutually convenient meeting. At the meeting:

- Introduce the problem
- Agree on the problem structure
- Search for alternative solutions
- Select the best alternative

If a deadlock is reached where the two parties cannot agree, sleep on the problem and try again the next day when new ideas may be forthcoming.

Given that problem solving is a powerful negotiating tool, you should have good problem-solving skills. These include being able to focus on the problem being discussed without being diverted to other issues; searching for areas of common interest and emphasizing areas of agreement. Negotiation often takes place within the organization's procedures and practices. Projects sometimes create novel situations and this is where company politics comes into play and decisions get made by the "old boy's network". The network is built on personal relationships and a good project manager will try to build long-term relationships through normal daily activities.

Presentations

The presentations you give are crucial to the project's success. Most of them will be to senior managers. The presentations will require careful planning and execution, demanding a lot of your time. As most of these presentations are "selling", the structure of the session will be slick, the slides will be constructed using the best graphics package and you will have rehearsed your presentation to ensure success. Because meetings at this level include a fair amount of company politics, make sure you know the audience and their position with regard to the project.

Meetings

Most meetings will be dealing with output from the SDLC and PLC activities. However, there are others related to team development and new techniques like Joint Application Development (JAD) sessions. The Status Meeting is a regular project meeting, perhaps weekly, and is attended by the entire project team. The team reports on project progress and discusses problems. The status of the project is then assessed.

Input may be verbal or written, depending on the complexity of the project. Although people claim that these meetings are a waste of time, it is amazing how many issues affect everyone. The ideal time for a status meeting is on Friday afternoon. This gets people working to that deadline in midweek, whereas a Monday meeting will ensure more weekend work to make the necessary progress. It is also remarkable how quickly people can get through an agenda when the weekend beckons.

Review meetings use a lot of people time. They should therefore be planned and run efficiently with a set agenda that has been distributed in advance; they should be in a good venue free from interruptions; and they should have a strong chairperson who keeps to the agenda. Minutes should be taken with clear action points assigned to individuals and progress on these items must be followed up.

Technical reviews relating to planning, design, coding, testing or documentation are carried out using a walkthrough technique or code inspection approach. These are discussed elsewhere but it is important to note that this is not a management appraisal of the individual but a peer review of a technical product. Its primary objective is to detect and correct technical problems.

Management reviews are carried out from time to time to keep appropriate user managers aware of progress and problems in the project. A Project Steering Committee normally consists of the project manager, the user project coordinator, functional managers and a senior user manager. This committee typically meets monthly to receive information about the project. Milestone reviews are built into the project plan and meeting them calls for a celebration. The review takes place in two sessions - one for the technical team to discuss the current position and to plan for the next phase, and the other where the users are brought in to share in the celebrations. These sessions are important to maintain morale and to get everyone excited about the work still remaining. There are several milestone meetings required in a project and these have already been specified in a previous chapter.

The traditional SDLC provides for interviewing to determine user requirements. This is a slow process and does not necessarily lead to user concensus. Bringing the appropriate users together in a formal session called a JAD session has gained considerable popularity. These sessions can be very effective when combined with the use of CASE tools. Because of the growing importance of JAD, it is discussed in a separate section.

Joint Application Development

A JAD session is a facilitated, team-based approach to solving business problems. It comprises one or more meetings. Conflicts are resolved and consensus is reached through brainstorming meetings. Thus, in a JAD meeting, the development of user requirements is carried out in real-time in contrast to the traditional and slow approach of individual user interviews. JAD works because the right people are involved who can make decisions by contributing to a decision-making process and reach consensus on solutions with their peers. This ensures better executive commitment and improved user input.

The concept originated in IBM Canada in 1979 and is now a standard published by the world-body GUIDE. The technique involves the right people getting together at the right time to make decisions about a project and produce specific deliverables. JAD sessions are structured in that they have specific objectives and are led by experienced facilitators. The technique goes far beyond the standard management meeting. A unique feature of a JAD session is the use of CASE tools. CASE automates the use of modeling techniques generally used in systems analysis and design. The "automation" of the meeting increases productivity and produces more rigorous deliverables.

JAD can be used to:

- determine I.S. strategy
- prioritize projects
- determine system requirements
- do data modeling
- produce a technical design
- choose technology
- do a technical review
- produce an implementation plan

The participants in a JAD session include managers, users and I.S. staff. The group should be able to understand the business and technology while having the authority to make decisions.

Typically the JAD membership should include the following (shown in figure 17.2):

- An executive sponsor who is a senior executive and who can make project decisions regarding resources and user selection
- User managers who take final responsibility for the system and are involved in the entire project
- Users who understand the business in detail and will eventually use the system
- The project manager and/or leader who is responsible for the project
- I.S. professionals who are part of the project team
- The JAD leader (facilitator) who runs the session and ensures action is taken on outstanding issues
- The JAD scribe who documents sessions, often using an automated CASE tool
- Observers, who cannot participate in the session unless requested to do so, but attend to learn about the JAD process

In some organizations some of the above roles are combined. This has been found to be detrimental to the process. For example, if the project manager and JAD leader are the same person, there may be a bias because the project manager is often an I.S. professional. The JAD leader also needs a different set of skills.

The JAD process works in three distinct phases as shown in figure 17.3. Session preparation is followed by the JAD session proper leading to post-session work.

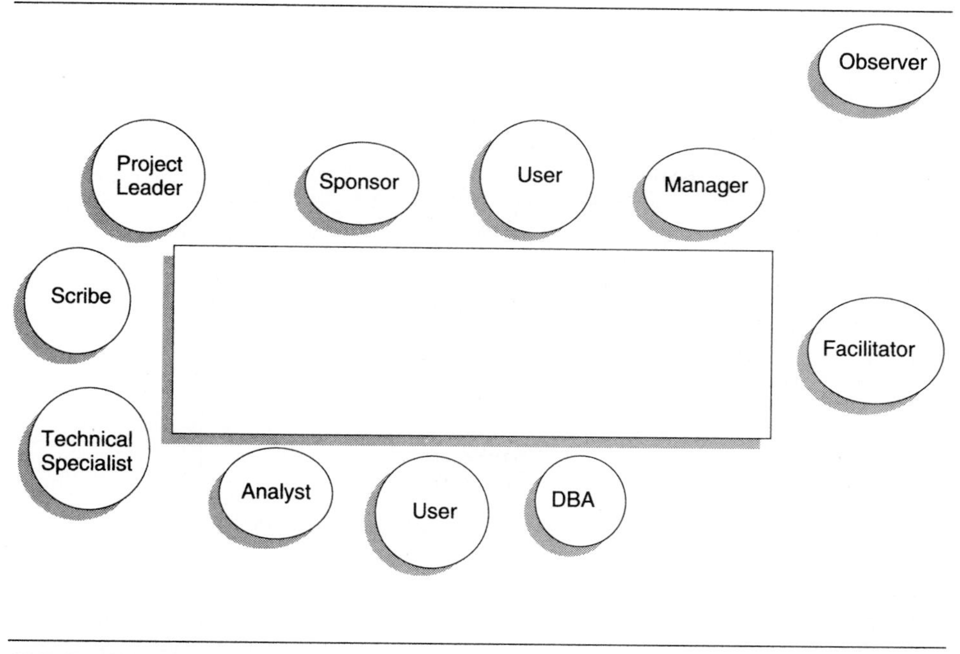

JAD Participants *Figure 17.2*

Session Preparation

JAD sessions involve a lot of expensive time. If we get 5 to 10 senior professionals in a room for a day, that could easily cost about two month's salary for a team member. It is vitally important that sessions go well. To ensure this, pre-session planning is carried out by the JAD leader and the scribe. Session preparation consists of:

- Reviewing JAD documentation
- Developing JAD controls
- A review by the executive sponsor
- Interviewing the participants
- Providing JAD training (where necessary)
- Reviewing the agenda

Running the Session

The JAD session normally starts with introductions and the executive sponsor introducing the purpose of the session. The JAD process is then reviewed, followed by a presentation of

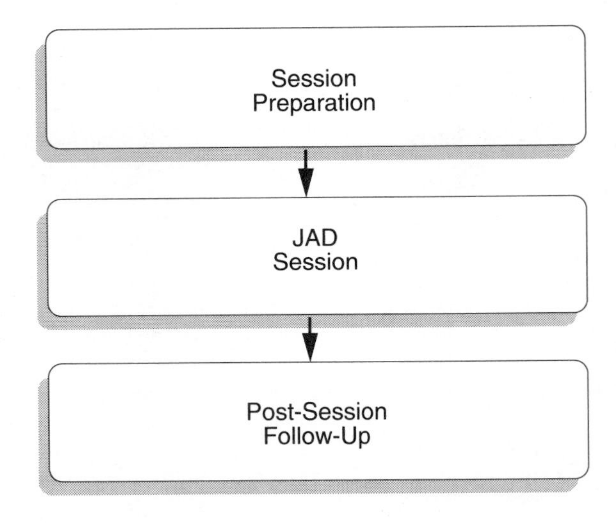

JAD Process *Figure 17.3*

the agenda after which the session gets under way. In the session, the JAD leader ensures that:

- There is only one conversation at a time
- Every participant is given a chance to contribute
- The subject focus is not lost
- Consensus is reached
- All ideas are considered

At the end of the session the leader reviews the results, determines future activities, decides when deliverables will be available and closes the meeting. Post-session work comprises a review of the session followed by the completion of the deliverables using members of the project team.

Good JAD sessions are run by leaders who can influence people without being dictatorial. They should be people-oriented, with a humorous touch, while being persistent enough to get results. JAD is not a panacea and can fail if incorrectly used. To be successful, there must be senior management commitment, the session must be well prepared, and the right participants must attend.

A successful JAD session requires:

- executive support
- company-wide use of JAD
- a well-organized session
- a strong JAD leader
- the right participants
- a good outcome

JAD succeeds because of everything being visual and the JAD leader continually seeking clarification and commitment. Companies are reporting up to a 45 percent reduction in time (not effort) in the SDLC phases from planning to external design. Successful JAD sessions happen because of good group dynamics. The effectiveness of a group relates to how well the facilitator and project manager build the team.

Team Building

A project team consists of a group of people with divergent ideas. Thus disagreements will occur from time to time. These should be appraised in terms of group think to enhance group performance. Conflict can lead to creativity. When a team realizes this, they will enhance group creativity. Once opinions are encouraged and discussed, there will be increased consensus and a strong commitment to decisions. A good project manager will resolve conflicts through confrontation, negotiation and resolution. This gives team members a feeling of belonging and ownership of the project's goals. This in turn leads to an increased responsibility by each individual to achieve these goals (discussed further in the Motivation section in the next chapter). This participative approach gets far better results than managing through coercion, persuasion and directives.

Case Questions

MyWay Organizer

Q17.1

At your previous company, you used JAD techniques extensively with clients involved in development projects. This is a fairly new concept to your software house, since most of their development to date has involved products conceived in-house. Structure a presentation to management detailing the key elements and benefits of JAD, as well as the critical success factors for its implementation. (20 mins)

Q17.2

The above may be physically presented via overhead foils to your study group at the discretion of the instructor. If you undertake this, we suggest a presentation not exceeding four overhead foils and completed in around 10 minutes. (foil preparation time, about 10 mins)

Gleam Stores

Q17.3

Implementation of the Gleam Stores system necessitates training of a vast number of people at widely distributed locations around the country. Consider what media and format would be most appropriate for a cost-effective, but foolproof, training program. Your suggested solution should specify who will be trained, to what level, how much time this will entail per type of person, what medium(s) you will use, what trainers or supervisors will be involved, and the overall approach and philosophy to be followed. (30 mins)

Handover Trust

Q17.4

Handover senior management is quite keen on the idea of JAD, but middle managers are very nervous, particularly because of all the other changes going on in their lives! Nonetheless, you are going to use the technique with a group of managers in the Claims Processing area. Prepare a presentation, which you will walk through with each person individually, telling them about the process, how they can participate, what is expected of them, and what the advantages are over other approaches. Also try to anticipate any fears they may have and allay these. If possible, divide into groups and role-play these sessions. The interactive session should take approximately 10 minutes. (Preparation, 30 mins)

ThoughtWell Books

Q17.5

Given the problems which have been occurring on this project, personal relationships are strained. There has also been less of a team feeling with the departure of Lars. Carefully consider what team-building activities and approach you could use to:

- Get your own staff working as a cohesive, motivated group
- Build common goals, objectives and understanding
- Build a larger team among yourselves, the sponsor, and other ThoughtWell personnel affected by the project

(30 mins)

18 *Managing People*

Introduction

The management of knowledge workers is one of the most critical areas facing business today. In first world countries, the number of knowledge workers now exceeds any other job category. These workers have a disproportionate effect on all aspects of society as it is they who determine what will be accomplished in the foreseeable future. Differentiation between organizations will depend largely on the contribution of these professionals. The job they do is not routine. Difficult to specify and unlikely to be totally automated, the job is highly dependent on the situation and the problem at hand. It demands judgment, ingenuity and creativity.

Techniques used to manage people in routine jobs are often inappropriate when managing professional activities. For example, elaborate job descriptions are often developed but these have little effect on the individual's performance and can lead to political infighting. Approaches such as Management by Objectives (MBO) and Quality Circles are installed at great expense, then quickly forgotten, or worse, lead to disastrous results. Despite these potential failures, managing professional activities does need structure. Understanding the nature of management provides us with the knowledge and tools we require to manage people. Good management can improve the overall performance of the professionals we manage.

The changing nature of the Information Systems (I.S.) industry makes managing I.S. personnel more complex. The industry has specific problems relating to technology and scarcity of personnel. Being a new industry, we have not yet adopted many management techniques used in other disciplines.

As a project manager, you will have to deal with all managerial activities relating to people, teams and tasks. To manage people, it is useful to define the major process and interactions shown in figure 18.1.

Before recruiting or selecting staff, the project environment must be clearly defined and job descriptions and job specifications completed. Suitable staff, from inside or outside the organization, are then selected to fill the vacant positions. Once hired, newcomers need to be socialized into the organization, after which specific objectives and action plans are set for them. Having completed a task, the individual must be given feedback on his performance.

This feedback is kept and used in a later, formal performance appraisal. We must also ensure that project-team members are highly motivated and that their career development is not neglected.

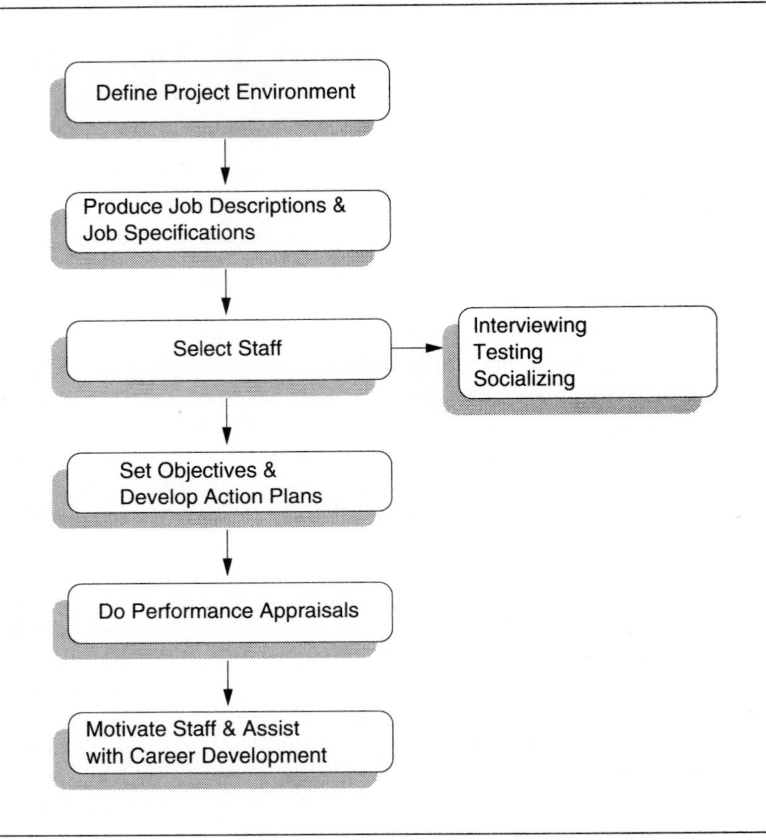

People Management Process *Figure 18.1*

Job Design

Jobs are the basic building blocks of organizational structures. A manager decides who does what job and the level of authority given to each individual. Jobs provide income, meaningful life experiences, self-esteem and often regulate our lives. We see ourselves and others in terms of jobs. Thus we see ourselves as systems analysts or programmers rather than Paul Simon fans or football supporters. The performance of organizations and individuals depends on how well management is able to design jobs. The new concept of business process re-engineering includes job design within the redesign of organizational processes.

Analyzing a job involves looking at the skills, abilities and responsibilities required by an individual to do the job, as well as the specific tasks that make up the job. A project manager can use the Work Breakdown Structure to define the tasks and then derive the people skills required to complete them. Job information can be described in a job description and a job specification (the terms "position description" and "position specification" are also used). This job information assists in recruitment and selection of staff to ensure skills are optimally matched to the job requirements. This information is also used to establish levels of seniority in organizations and pay scales.

To perform a job analysis, a description of the job and the description of the person required to do the job are developed using the characteristics typically shown in figure 18.2.

Job Analysis
Procedure for obtaining facts about a job

Job Description
Statement that provides personal information about:

- Job Title
- Responsibilities
- Experience
- Technology
- Techniques Used
- Supervision
- Working Conditions
- Salary

Job Specification
Statement of the characteristics needed to perform the job:

- Education
- Work
- Experience
- Judgement
- Vision
- Creative Skills
- Communication Skills
- Ability to Work with Others

Job Analysis *Figure 18.2*

Many job advertisements focus strongly on the job description aspect. To build a team, it is important to ensure personality characteristics are also specified clearly. It is a sad fact that we hire people for their skills and experience, but fire them for their personality characteristics and behavior.

The characteristics of the job as designed can encourage or discourage job performance. This will depend on the views of the job holder - different people may have different values within the same job. For example, some people may view responsible and challenging work in a negative light while others regard it as highly positive. These aspects will be discussed later in more detail under motivation.

It is important to complete the job description and job specification before initiating a search for suitable candidates. This search could be either internal or external to the company. You must know in advance what the job offers and the type of person that best matches the job. An example of a typical job description and job specification can be seen in figures 18.3 and 18.4.

JOB DESCRIPTION

Position Title: Manager, Systems Development

Salary Range: $80,000 - 110,000 depending on experience

Duties and Responsibilities

1. Provide leadership and direction to three project leaders.
2. Plan systems development work in consultation with the I.S. Manager. This involves working closely with managers of departments requesting services.
3. Recruit and develop all staff with proper training and experience.
4. Chair a committee to discuss the latest tools and techniques in systems development and to recommend possible acquisitions.

Work Environment

The job is based at head office. The department is part of the I.S. Division (23 analysts, 70 programmers, and 47 support staff), equipped with two mainframe computers serving a network of 200+ terminals, extensive database management systems and 145 PC's distributed throughout the organization. This is a key position in the Division.

Job Relationship to I.S. Division

Two other managers report to the I.S. Manager. These individuals interact closely through weekly meetings. The S.D. Manager is expected to develop subordinates and to maintain a work environment conducive to high performance at all times.

Typical Job Description *Figure 18.3*

Staff Selection

One of the most important responsibilities of a project manager is hiring staff. Sometimes you will be able to choose staff, at other times staff, who would not be your first choice, will be transferred to your project. The capabilities, productivity, quality and the quantity of output achieved by the team is determined to a large extent by the team members. The organization will also have to live with your choice long after the current project. The apparently simple task of hiring is therefore critical. While it is possible for project managers to

JOB SPECIFICATION	
Education Level	A higher degree in Information Systems or Business.
Work Experience	At least 5 year's pertinent work history in the I.S. industry at a management level.
Communication Skills	Demonstrable oral and written skills at senior level.
Interpersonal Skills	Capable of motivating others through strong leadership and energy. Entrepreneur with vision in a changing environment while ensuring current problems are solved.

Typical Job Specification Sample *Figure 18.4*

influence the quality and quantity of work of those hired, you will be severely constrained by their capabilities and attitudes. Good staff selection procedures will enhance your chances of recruiting satisfactory staff; a poor approach will generally produce an undesirable outcome.

Two factors should be carefully noted when selecting staff:

- The quality of performance of an individual persists through time. Thus a poor performer would tend to remain a poor performer over his career path
- A relatively small percentage of those hired are responsible for a disproportionately large percentage of both the best and the worst performances

Research has shown that individuals judged as high performers in the early stages of their career continue to be judged high performers at a later stage. Similarly those judged as low performers tended to persist at that level of performance. There is very little migration from high to low performance over time. Only those people who were judged medium performers, either improved or worsened toward the extremes.

A small number of professionals account for the best and worst performance in a typical team. This is an example of Pareto's Law, which states that in social and economic affairs, many relationships can be expressed as a straight line on a double log scale. Simply put, a few of the X's account for a large percentage of the Y's. Managers know this as the 80/20 rule. For example, 80 percent of inventory costs can be attributable to 20 percent of the items. Managers in many professions report that 80 percent of the best output of their organization could be attributed to only 20 percent of their people. Conversely, 80 percent of the problems and disasters could be attributed to a different 20 percent of the staff! Most managers can easily recall the names at each extreme. Pareto's Law can be used to the project manager's advantage. A small improvement in the staff-selection process can achieve very large improvements in performance. The trick is to identify and hire "stars" and ensure that "disasters" are not hired. Overcoming mistakes in staff selection takes a very long time, and is very expensive in terms of damage caused.

Interviewing New Recruits

Before considering the interviewing process, consider the following interview originated by Tom DeMarco and Tim Lister. Imagine you are a circus manager interviewing a candidate:

Circus Manager:	How long have you been juggling?
Candidate:	Oh, about two years.
Circus Manager:	Can you handle three hoops and four hoops?
Candidate:	Yes - both.
Circus Manager:	Do you work with flaming objects?
Candidate:	Of course.
Circus Manager:	Do you work with knives, axes, open cigar boxes and floppy hats?
Candidate:	I can juggle anything.
Circus Manager:	Do you tell funny stories while you are juggling?
Candidate:	I do - everybody loves them!
Circus Manager:	Well, that sounds perfect - you have got the job.
Candidate:	Er - don't you want to see me juggle?
Circus Manager:	Oh - that's interesting, I never thought of that!

Hiring a juggler would be nonsensical without first seeing a performance; that should be mandatory. When you set out to select a designer or a programmer, it is easy to discard common sense. Ironically, the interview is just talking. If a person is being hired to generate and produce a product, then they may well have done this many times before. Examining a sample of the products previously produced to see the quality of the work is an obvious part of hiring - but is it ever done? There appears to be an unwritten rule when interviewing programmers and analysts that says that it is okay to ask the potential candidate about the work they have done in the past, but not to ask to see it.

Selecting I.S. personnel should follow a four-step process:

- Preparation of job description and job specifications
- Initial screening of candidates
- Company visits and interviews
- Final selection

A job specification and job description are required to initiate the job search. The manager must know in advance what the job offers and the type of person that will best match the job. The next stage is to carry out an initial screening of the candidates. The purpose of initial screening is to weed out unsuitable candidates. This is done by reviewing the candidate's CV (curriculum vitae) or application form and performing reference checking with previous employers. Note that reference checking cannot normally be done with the current employer, making current performance of the candidate difficult to determine. An initial interview may also be carried out to verify some of the information about the candidate's abilities and skills. This interview gives both the candidate and the company an initial view of each other. The outcome is a decision to continue further with the selection process or not.

On passing the initial screening test, the candidate is then invited to visit the firm for further interviews when the candidate will be asked to expand on their career plans, experience and motivation. Aspects such as problem-solving, oral communication, interpersonal skills and ability to cope are also probed. This is a more intensive screening process, allowing a number of managers to talk to the candidate before a final decision is made. Note that the company at the moment has a relatively clear idea about the job and the kind of person they are looking for. They have very limited information about the candidates for the job. Although aptitude tests, achievement tests, personality tests and other tools are more objective in assessing individual characteristics, the interview remains the most popular method of obtaining information. We know that interviewing is a good procedure for gathering factual information during systems analysis. However, it is severely constrained when trying to make judgments, because of its subjectivity. It is surprising that use of interviews as a means of gathering and assessing people has increased in the selection process to the point that few other techniques are used.

Problems found in interviewing include:

- Interviewers often make their decision on whether the interviewee is suitable within 5 minutes of the start of the interview. You will have to be more objective than that!
- The interviewer and interviewee may not cover all the important dimensions within the interview. The interviewer may not have a complete description of the job or the interviewee may end up not knowing the work conditions being offered
- When several managers interview one person, some questions can overlap while others are missed entirely. Planning this type of interview is essential
- The interviewer may allow one attribute to sway the evaluation. This is called the Halo Effect and occurs when an interviewer judges an applicant's entire potential for job performance on the basis of one characteristic (such as how well the applicant dresses or talks)
- The interviewer's judgment is affected by pressure to fill the position. This is especially true in I.S. where turnover is high and skills are scarce. This may cause a lowering of standards
- Scheduling interviews in rapid succession can produce poor decisions, bias and use of stereotypes in selection

There are several ways to overcome potential interview problems and to increase the validity and reliability of the interview:

- One approach is to use past behavior to predict future behavior. This should consist of examples of specific experiences and involvement, possibly with documented examples (e.g., specifications completed, programs written)
- Having several managers involved in the interview. The final decision will then be based on several varying perspectives

- An interview consists of a verbal component and a nonverbal component - body movement, gestures, firmness of handshake, eye contact and physical appearance. Some interviewers place more importance on the nonverbal than on the verbal aspects of the interview. Get help in this area from a trained personnel manager

- Useful questions that may be asked by the interviewer to assess the applicant's performance include:

  - How the applicant has performed in a similar capacity in the past

  - Why the applicant wants to change jobs?

  - What are the applicant's career objectives?

  - Does the applicant like working closely with other people?

  - Ask the candidate to rate his performance against his peers in his last company. If previous performance ratings are available, get them in writing. (Note that these questions are very predictable and could lead a well-prepared candidate to give a non-genuine response).

- Other more personal questions should be asked in an appropriate manner. For example, whether any physical defects may require adaptation of the work environment. Hobbies or interests could also have a direct bearing on the job. Ask about any courses taken which may enhance job performance.

Employment Testing

Testing is another important procedure to assess information about an applicant's aptitudes, experiences and motivations. Employment tests include any pencil-and-paper (or computerized!) measure used as a basis for an employment decision. The most common types of tests measure aptitude, achievement and motivation. Note that these tests must be valid and reliable to ensure correctness. The results are only one aspect of the evaluation process and should be used in conjunction with interviews and other information.

Aptitude Tests

These measure the potential of individuals to perform in a particular area. Popular tests administered to programmers include the computer programmer aptitude battery from IBM. Advantages claimed from such tests include improved accuracy in selecting employees and more objective means of selecting the best applicant. However, many of these tests have proven controversial. For example, there is no proof that a high mark in these tests equates to being a top performer. Also, in many companies tests are given too much weight and have replaced judgement. In the United States, many tests have been discontinued because they were seen as discriminating against ethnic minorities. Aptitude tests are used because of a lack of alternatives.

A good reason not to use aptitude tests is that we may be measuring the wrong thing. Programmer aptitude tests are designed to measure programming aptitude or whatever is required in the first position that the employee takes up. If that person should join the company, it is unlikely that he will be in the same position in several years' time. An

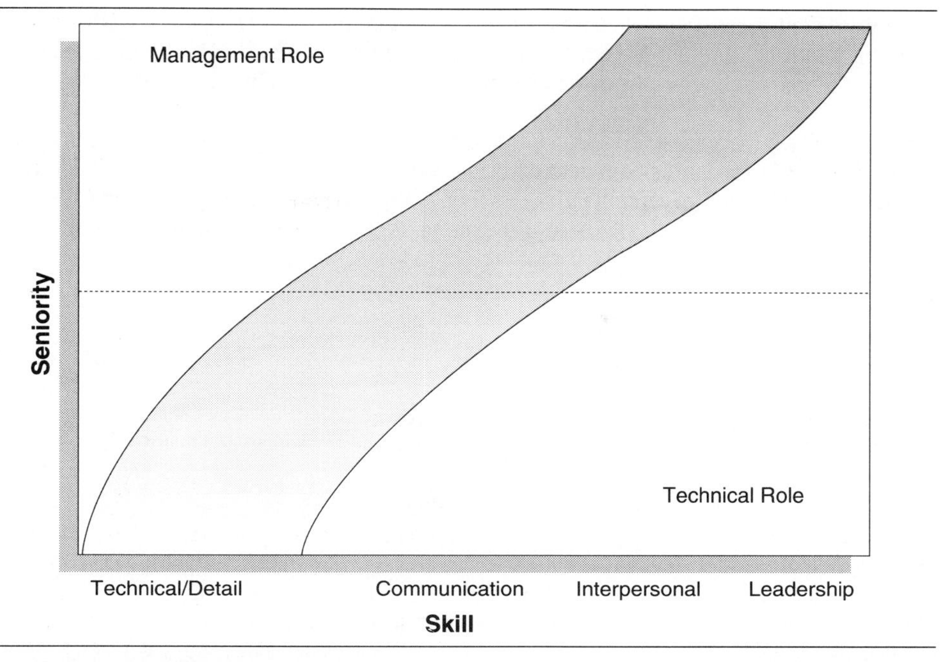

Changing Skill Requirements *Figure 18.5*

aptitude test, therefore, may provide the right type of person in the short term, but is less likely to predict longer-term performance. In a relatively short time, I.S. staff can develop from a technical to a management role with very different skill requirements, as shown in figure 18.5.

Achievement Tests

These are used to predict an individual's performance on the basis of what they know. These tests tend to be samples of the job which will be performed if the candidate is successful. Many achievement tests are pencil-and-paper oriented and tend to be less job-related because they measure facts and principles, not the actual use of them. Thus a candidate could be very successful in passing a test measuring knowledge of analysis, whereas they may well become poor systems analysts. Despite this problem, these tests are used widely in the professions: for example, the bar exam for admission to the legal profession and the final qualifying exam for admission as a chartered accountant. A variation of the achievement test is a recognition test widely used in the advertising and modeling industry to select applicants. In a recognition test the applicant brings a portfolio of their work to the interview. Note that these portfolios contain no clues regarding the conditions or circumstances under which they were produced. This approach could be useful in I.S.

Psychometric Tests

These are designed to measure an individual's motivation and categorize personality and behavior, as well as preferences for certain types of jobs. They are sometimes referred to as personality inventories. Typical examples are the Personal Profile Analysis (PPA) from Thomas International and the Myers-Briggs Type Inventory (MBTI). This type of testing is popular as it lends a behavior or personality dimension to the evaluation process.

Personality defines the characteristics that determine the way people think and behave. For example, the MBTI measures four related dimensions of behavior identified by Carl Jung. These are introversion/extroversion (I/E), sensing/intuitive (S/N), thinking/feeling (T/F), and judging/perceiving (J/P). Each of these dimensions is measured on a continuum shown in figure 18.6.

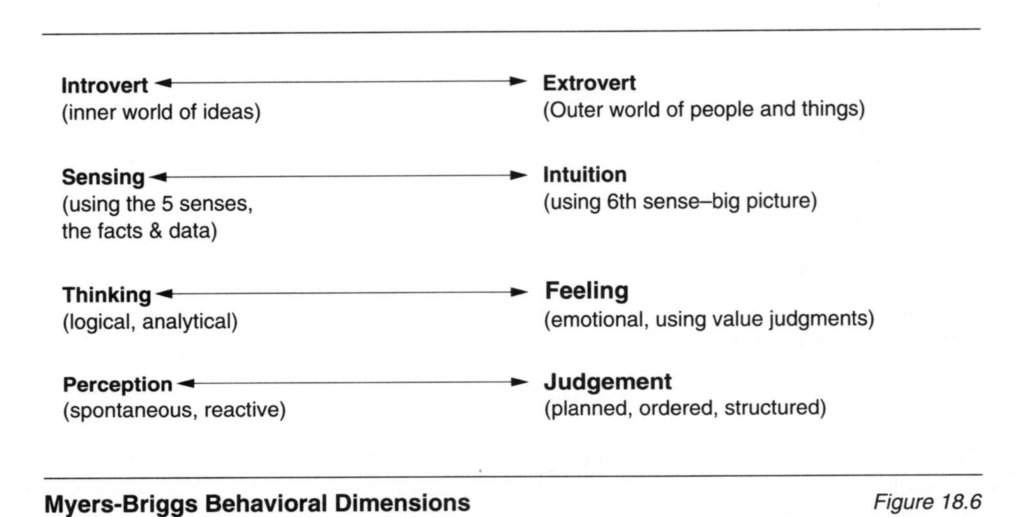

Myers-Briggs Behavioral Dimensions *Figure 18.6*

A person's degree of extroversion or introversion can range from extreme extrovert to extreme introvert. The personalities at these two extremes are very different: an introvert is generally concerned with the inner world of concepts and ideas whereas the extrovert is more concerned with people and the world at large. An individual's MBTI is measured in terms of all four dimensions. This is done using a structured questionnaire with questions similar to those in figure 18.7.

Thus, for example, an individual can be "typed" as an ISTJ personality. This means the individual's preferences are introversion, sensing, thinking and judging. Such a person would be serious, quiet, practical, orderly, thorough and responsible. Research has shown that the I.S. population has a different proportion of the various types than the general population. This is shown in the table which follows. Thus, an individual can be typed, but so can a project team or even all the analysts in a company. This information can assist managers at the project or the organization level to bring an appropriate mix of personality types together, or to know how to handle the mix they have.

Research using the Myers-Briggs instrument (the MBTI) shows that there is a greater predominance of the STJ personality type in the I.S. industry than has been found in other professional groups. See figure 18.8.

MYERS-BRIGGS TYPE INDICATOR
Sample Questions

Does it bother you more having things
a) incomplete b) completed

In doing ordinary things, are you more likely to
a) do it the usual way b) do it your own way

Are you more often
a) a cool-headed person b) a warm-hearted person

Are you more
a) firm than gentle b) gentle than firm

Myers-Briggs Type Indicator *Figure 18.7*

The STJ personality is more suited to working on self-contained tasks and will not relate strongly to teamwork. They will be less inclined to communicate with other people but will have a strong need for a job that has a high motivating potential. A suitable management style would be to ensure specific tasks are set (using the WBS approach) and regularly check the individual's needs concerning future work assignments.

The use of assessment centers is growing in popularity. With this approach, several job applicants attend a two- or three-day assessment at a hotel run by managers who are skilled in rating the candidates on a large number of characteristics. Events during the course of the assessment include management games, psychological testing, decision-making exercises, role-plays, group problem-solving exercises and written and oral presentations. Assessors measure applicants' ability, drive and motivation. The applicants may therefore score low on motivation but high on ability and the assessors will have to balance the two to gain a comprehensive view.

Affirmative Action Programs

Many organizations have mounted affirmative action programs to increase the proportion of certain groups. These programs, designed to ensure proportional representation, are referred to as affirmative action programs (AAP's). This approach can equally apply to handicapped workers as well as the socially or educationally disadvantaged. Because this practice is popular, the project team could be affected in terms of selection and productivity.

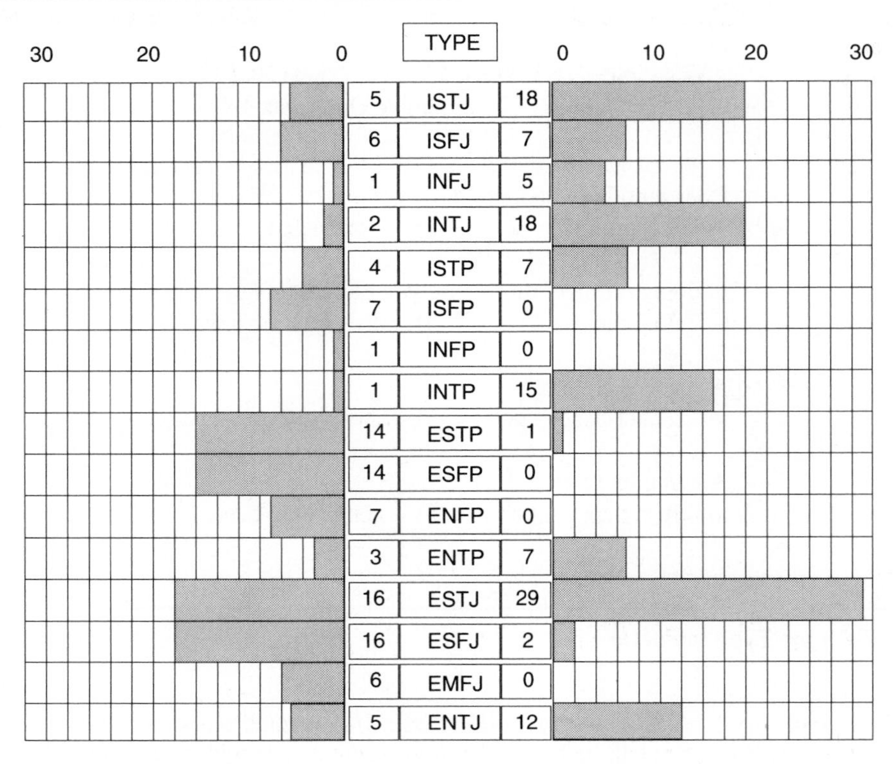

Distribution of Myers-Briggs Personality Types *Figure 18.8*

Interview or Audition?

It is said the I.S. business is more sociological than technological. That is, it is more dependent on people's ability to communicate with each other than their ability to communicate with machines. It is therefore reasonable that the hiring process should focus on communication ability. An interesting approach is to use auditions for job candidates. The interviewee is asked to prepare a ten-minute presentation on some aspect of past work. It could be about technology issues or a particular problem identified in a past project. The interviewee chooses the subject and a small audience is assembled, made up of the project team. Obviously the candidate will be nervous and should be put at ease as soon as possible. It is important to explain to everyone, including the potential employee, that the audition is to see the candidate's communication skills and to give co-workers a say in the selection process. After the audition each person comments on the suitability of the candidate. This feedback is invaluable in the hiring decision. The only restriction on the topic should be that the candidate speaks about something immediately relevant to the work of the organization.

The final decision must now be taken by the appropriate decision-makers. Once employed, the new employee should attend a company socialization program. A new employee must understand the appropriate values, norms and behavior patterns existing in the organization.

The selection process does not ensure that new employees know the values and culture of the organization. Socialization is the process by which people acquire the knowledge, skills and disposition which make them able members of their organization. It is facilitated by orientation training, informal after-hours social events, assignments to small teams and the appointment of a mentor. The socialization process assists in building a sense of belonging in a new employee. It also helps to increase performance and satisfaction in the job, thereby reducing the turnover and hiring costs. Furthermore, it reduces the anxiety that is often mentioned by new employees in the early days on the job.

Early job experiences are a major factor in socialization and the first supervisor is often a key figure. These supervisors serve as role models and the supervisor's expectations can have a positive influence on the new employee. This influence is called the Pygmalion Effect. If the supervisor believes that the new employee will do well, this belief will be conveyed to the employee who will try to live up to these expectations. After the staff member has been selected and socialized, objectives and action plans should be determined.

Setting Objectives

An objective consists of two elements:

- the statement of a deliverable
- a result standard

For example, the deliverable could be:

"to produce career development plans for a project team"

The result standard could be:

"within three months' time when all initial skills training courses have been completed and satisfactory marks, (>75%), have been gained"

The result standard makes the objective measurable and is expressed in terms of at least two of the following:

- Quantity (how much)
- Quality (how well)
- Time (by what time)
- Cost (at what cost)

Any objective can be measured by incorporating the result standard. It should define an end result and not an action. Thus the statement:

"talk to all relevant users"

is unsatisfactory because it does not indicate the expected result from the discussion.

Types of Objectives

Two types of objectives may be established for any position:

- *Maintenance objectives,* where a satisfactory performance standard has already been attained and it is important to maintain this standard

- *Improvement objectives,* which are established where there is a need or opportunity to improve the results over the current performance standard. An improvement objective may be achieved through solving a problem, through innovation or through constant small improvements

An example of a maintenance objective would be "to ensure that computer usage will not increase this year over the previous year." An improvement objective would be "to more effectively utilize computer time by reducing last year's utilization level by 10 percent this year."

Criteria for Objective Setting

Objectives should logically be set at senior management level and cascade down through the organization. Setting of objectives is the responsibility of the manager who, after involving the project team members, would specify the minimum performance standard acceptable. The team member is responsible for developing an action plan with the manager to meet these objectives. In this way, a high degree of participation can be achieved.

The number of objectives to be set per employee will vary. However, it is important not to set too many objectives. Typically, the number of improvement objectives should be somewhere in the range of 5 to 8, depending on complexity.

Generally, objectives should be difficult but attainable. They should call for extra effort or an improvement in methods. However, there may be rare occasions when a manager may want to set reasonably easy objectives for a subordinate whose confidence needs a boost. On balance, it is better to set levels that can be reached and exceeded than to set ones that can never be achieved. Don't challenge to the point of failure.

Objectives should be reviewed on a regular basis as conditions change. However, only in extreme circumstances, when it is clear that conditions have changed to the extent that the objective loses its meaning, should an objective's performance standard be lowered. Normally, if circumstances change, then the action plan (the method by which the objective was to be achieved) and not the objective itself should be altered. Only when all alternative methods and action plans have been considered, and it is clear that none of these will meet the objective under the changed circumstances, should a performance standard be lowered. On the other hand, if circumstances become more favorable, then the performance standard can be revised upward.

Thus:

- when conditions change for the worse, change your action plan.

- when conditions change for the better, upgrade your objectives.

Action plans need only be established for the major improvement objectives.

Developing Action Plans

Once objectives are established, action plans are then developed. The tasks of the action plan should correspond to those in the WBS. An action plan consists of:

- Action steps
- Deadlines
- Responsibilities

For example, if the objective is to reduce computer usage by 10 percent this year, based on last year's figures, the output from this objective would be a set of action steps (e.g., determining the specific usage of computer time last year and the requirements this year). These steps would be completed by a certain deadline (a specific date) and would be carried out by a specific person (e.g., a nominated systems analyst in the project team).

Performance Appraisal

Most large organizations have an annual evaluation system because performance evaluation is fundamental to organizational effectiveness. The process is often linked to salary increases. Evaluation is a consequence of the way large organizations are structured and jobs are designed. The assignment of individual responsibility to task performance makes the assessment of individual performance both possible and necessary. Appraisal identifies the results for which the person is responsible. To operate effectively, large organizations need information on how well jobs are being performed. Performance appraisal is therefore a system to measure and evaluate an employee's job-related behaviors and outcomes to discover how well an employee is performing on the job and how performance can be improved in the future. The following criteria should be used:

- Performance is measured by results.
- Performance is linked to behavior (the things people do that produce the results).
- Behavior can be either active or passive (the person can do something or do nothing). Most of the behavior discussed in performance appraisals is "on the job" behavior. However, some "off the job" behavior belongs in the appraisal if it affects results.
- Performance appraisals should always be restricted to behavior that matters. A rule-of-thumb question to be asked is "What difference does it make?" If the answer is "none" then don't evaluate it. Behavior such as chewing gum while working on a PC; wearing a bow-tie to work or whistling loudly in the corridor do not belong in a performance appraisal. You will certainly have to discuss these issues if they cause offense or are against company policy, but they should be resolved at the time of the incident, not saved up for the annual appraisal.

The Importance of Performance Appraisals

Performance appraisals should be used for the following reasons:

- Management development - provides a framework of future employee development by identifying potential and preparing individuals for future responsibilities

- Performance measurement - where the relative value of an individual's contribution to a company is established and specific individual accomplishments are evaluated
- Performance improvement - where continued successful performance is encouraged and weaknesses are measured and ways sought to eliminate these
- Compensation - where appropriate pay is determined along with performance. Ideally this should be done at a different time to the performance appraisal.
- Identification of potential - where possible promotion posibilities are identified
- Feedback - gives the employee your perception of performance achieved and how it fits the plan
- Communication - provides the structure for dialogue between the you and the project team members

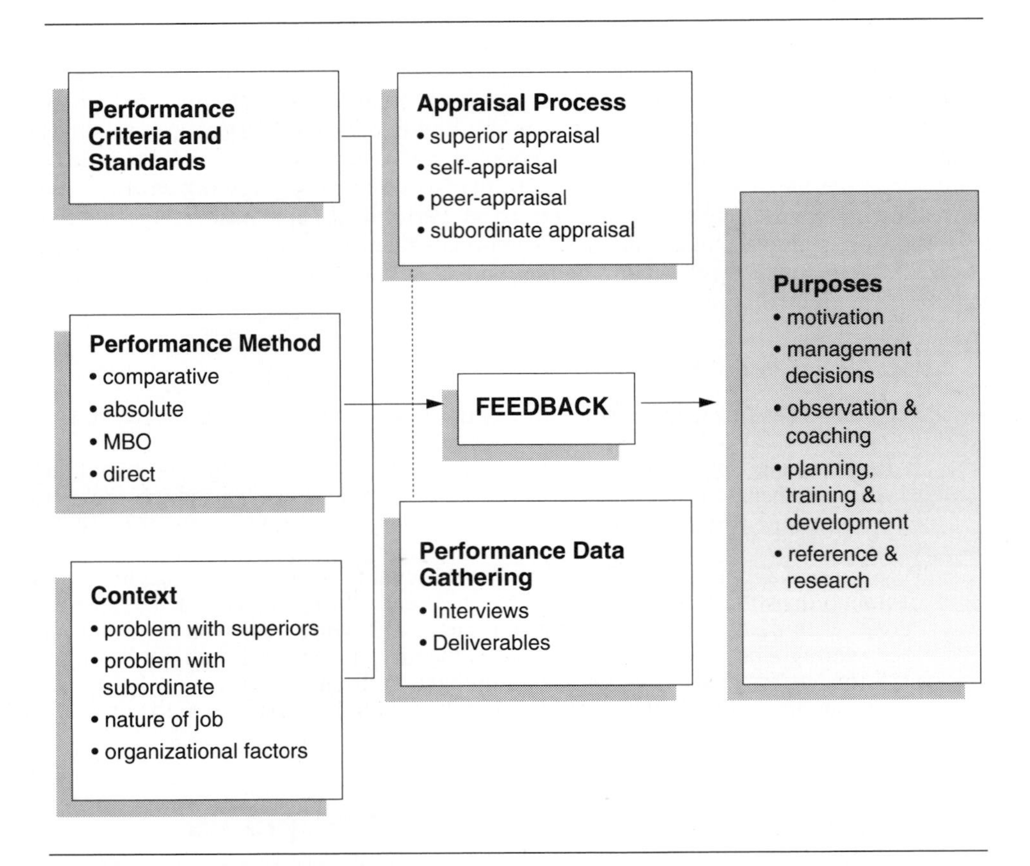

Performance Appraisal System *Figure 18.9*

The Performance Appraisal System

A general model of the performance appraisal system is shown in figure 18.9.

The performance criteria must first be determined using job analysis. The employee's contribution can then be evaluated based on his performance and the results specified in the job analysis. The job analysis will include several performance criteria that determine an employee's contribution and these will be measured in the performance appraisal.

There are four different kinds of performance appraisal methods:

- comparative
- absolute standards
- management by objectives (MBO)
- direct (or objective) indices

These are discussed in detail below.

Comparative Approach

In a simple comparative approach, the project manager ranks the subordinates in order from best to worst, using performance as the criteria.

A more sophisticated approach is the paired-comparison method in which each subordinate is compared to every other subordinate one at a time, using a single criteria (such as overall performance). The subordinate with the greatest number of favorable comparisons is ranked first and so on. This suggests that no two subordinates perform exactly alike. Although this may be true, many managers say that differences between subordinates are too small to distinguish performance.

Another method, the forced distribution method, was designed to overcome this problem. The manager is allowed to assign a certain proportion of subordinates to each of several categories on each factor. A common forced distribution scale is divided into five categories. A fixed percentage of subordinates is allocated to each category. The problem with this method is the project team may not conform to the fixed percentage. In fact, all comparative methods assume that there are good and bad performers in all teams. People in teams sometimes perform in a similar manner, making these approaches difficult to use. This is especially true in a project team where everyone may be performing at a high level. In this case, you will have to justify to management that your team has a skewed distribution of high performers. If you select and motivate well, you should have a skewed distribution!

Absolute Standards

The absolute standard approach allows managers to evaluate each subordinate's performance independently of other subordinates and often on several dimensions of performance. This approach is in common use and measures personality characteristics as indicators of performance. Frequently used traits include aggressiveness, independence, maturity and sense of responsibility. The appraisal ratings are quick and easy to complete and are very popular. Unfortunately, the results are sometimes very difficult to convey to subordinates, especially if they are unfavorable. In addition, the results do not help subordinates improve their performance.

Management by Objectives

Management by Objectives (or MBO) is a goal-setting method to measure objectives. The manager first establishes the goals for each subordinate during a time period. In many organizations, the manager and subordinates work together to establish these goals, thereby increasing commitment. Once established, the manager monitors performance over the time period. The subordinate knows what there is to do, what has to be done and what remains to be done. The manager then compares the actual level of goal attainment against the agreed goals. Reasons for goals not being met, or goals being exceeded, are explored with the subordinate. This helps establish possible training needs and also alerts the manager that organizational issues beyond the control of the subordinate may be affecting performance. The final step is to decide on new goals and action plans for the new time period.

MBO was originally designed to assist in the development of subordinates. It is especially useful in jobs where objective measures of performance are difficult. It is a valuable approach in project management since we mount projects specifically to achieve goals.

Direct Index

The fourth performance evaluation method is the Direct Index approach. This approach measures performance by objective impersonal criteria, such as productivity, absenteeism and turnover. For example, a project manager's performance may be evaluated by the number of team members who quit or by the absenteeism rate. Productivity can be measured by looking at both the quality and quantity of the manager's work through customer complaints, creation of system components and quality assurance. This is the most objective method but the measures must be carefully chosen. The approach can also disfavor valuable contributors in areas which do not contribute to the measures used. For example, if we measure development productivity, what about the maintenance programmer?

The Context of Performance Appraisal

Regardless of the performance methods used, there are several factors which may reduce the effectiveness of the appraisal system. Perhaps the most important is the relationship between the manager and the subordinate. The manager may encounter four categories of problems:

- the manager may not know what employees are doing and may not understand the work well enough to appraise them fairly. This is true of technical staff who report to user project managers, or when a manager has a large project team

- the manager does not have performance standards for measuring and evaluating the work. This may lead to unfair evaluations because of variability in standards and ratings. This is particularly obvious in large organizations when comparing evaluations of subordinates working for different managers

- managers may use inappropriate standards, allowing personal values or biases to replace the organization's values and standards. A frequent occurrence is when a manager evaluates his subordinate similarly on all dimensions of performance, based on evaluation of one dimension. This is another manifestation of the Halo Effect. Alternatively, managers may give all their subordinates favorable ratings. This is called an error of leniency. An error of strictness is the opposite of an error of leniency. An error of central tendency means all subordinates are evaluated as average. These three errors may occur intentionally or unintentionally. Some managers may, for example, intentionally evaluate their best performers as slightly less than excellent to prevent them from being promoted out of the group.

Many of the above errors can be minimized if:

- Each dimension being measured addresses a single job activity
- The manager appraising the job can observe the behavior of the individual first-hand
- Terms like "average" are not used on a rating scale
- The person appraising the jobs does not have to evaluate large groups of employees
- Training is given to all individuals who have to do appraisals
- The dimensions being evaluated are meaningful, clearly stated and important

The subordinates may also have problems with the performance appraisal. For example, they may not know what is expected of them, or they may not be able to do what is expected of them. Thus they may have the ability but just don't know how to apply it.

Apart from the manager-subordinate relationship, the nature of the job is also an important factor. In many jobs the quality or quantity of performance may be outside the subordinate's control. This is especially true in many project tasks which are highly interdependent and where many people are involved in the outcome. In these circumstances it is difficult to separate the individual's performance from that of the group.

As employees become more unionized, performance appraisals may disappear completely as unions have traditionally favored seniority to determine salary packages - similar to government employees. This is a negative trend where performance is not recognized or rewarded and will ultimately lead to lack of motivation and a drop in productivity.

The Appraisal Process

The appraisal process is a data-gathering exercise. The appraisal data can be generated by the manager or through self-appraisal by the appraisee. When the manager appraises the individual, the process is often seen as "one way", making the subordinate feel defensive. Self-appraisal makes a subordinate participate in the evaluation process and become more involved and committed to the outcome. Done in isolation, the self-appraisal approach will have biases and distortions.

A further method of gathering performance data is through interviews. This can be highly productive if the climate is correct and the individuals both participate in setting goals and determining job responsibilities as well as deciding on career development measures.

Feedback of Performance Results

Feedback is an integral part of any learning experience and is the process whereby the subordinate is told how to improve performance through an objective assessment of his present position. It should be given immediately after the performance appraisal itself and is therefore best discussed in the appraisal interview. Clearly, the manager will find discussing successes far easier than discussing failures.

Managers feel that negative feedback can lead to poorer performance rather than better. Often negative feedback is confused with criticism. The latter is evaluative, implying "goodness" or "badness", whereas feedback is descriptive. Feedback provides the subordinate with information and data which can be used to perform a self-evaluation. Feedback, given correctly, should not create a defensive reaction. However, where the appraisal system is tightly coupled to the salary system, defensive reactions will dominate and handling these discussions will require extra sensitivity.

Feedback is not always easy to provide but nevertheless effective feedback is essential. Effective feedback is specific rather than general. Specific instances must be cited rather than generalizations. In addition effective feedback is focused on behavior rather than on the person. It is important to refer to what a person does rather than to what that person seems to be. A manager may therefore say that a person talked more than anyone else at a meeting, rather than he or she is a loud-mouth. The former allows for the possibility of change, the latter implies a fixed personality trait. Effective feedback should be given to help, not to hurt. The interview should include information to share rather than merely giving advice. The person receiving information can then decide for himself on the changes required. Effective feedback should be given immediately after an incident and not saved up for a particular appraisal time. Effective feedback should be formalized to ensure clear communication.

No matter what the intention, therefore, poor feedback is often seen as "threatening" and can thus be the subject of considerable distortion or misinterpretation. Make sure yours is objective and honest!

Matching the Purpose and Method

The last step in the performance appraisal system is deciding on the required action. The five broad categories where action may be required are:

- Motivating subordinates to perform well
- Providing data for management decisions
- Helping in human resource planning, training and development
- Encouraging managers to observe and coach their subordinates
- Providing reference and research data.

Although all the methods mentioned handle all these categories to a lesser or greater extent, the MBO method relates more closely to the needs of the I.S. industry than the others.

Ineffective Performance

Employees do not always perform the way managers want them to. Sometimes this ineffective performance is a reaction to the work environment and sometimes it is the result of being prevented from reaching career goals. Regardless of the cause, telling people the truth about their ineffective performance is difficult and therefore often avoided. The biggest problem with ineffective performance is determining what it is. Ineffective performance can be caused by personal problems or be environmental. For example, there may be a general lack of motivation due to the wrong selection procedure or poor working environment. The internal work environment may also be at fault and this is linked to the management style of the project manager. Often the "nice guy" approach, i.e., the manager who likes people and does not want to hurt their feelings, is infinitely inferior to the manager who pushes subordinates to do their jobs better but lacks the personal warmth exhibited by the other manager.

Ineffective performance is difficult to handle and therefore the easiest way out is to ignore it. Some action must be taken to handle the problem and could involve several distinct steps, starting off with coaching and feedback. If no improvement is obtained, we may issue a verbal warning, followed by formal written warnings, eventually ending up with a dismissal.

Staff Motivation

Motivation comes from the Latin "*movere*", meaning "to move". A manager does not see motivation, just as a manager does not see thinking, perceiving or learning. All that a manager sees is a change in behavior. To explain these changes in behavior, managers make inferences about the psychological processes; thus motivation can be seen as an inner striving in terms of needs, desires, drives, motives and so on. It is an inner state that activates or moves. Managers often observe behavior and make inferences about motivation. If an employee displays the following type of behavior, they are considered to be motivated:

- Is regularly on the job
- Puts forward his best effort at all times
- Is always trying to improve performance on the job
- Directing his efforts toward accomplishing meaningful goals

Managers observe the presence of these aspects and make inferences about whether or not an employee is motivated. All behavior is in some way motivated as people have to have a reason for doing what they do. Human behavior is directed toward certain goals and objectives. Such goal-directed behavior relates to the satisfaction of a need - and need is a physiological, psychological or sociological desire that can be satisfied by reaching a specific goal.

An unsatisfied need or desire initiates motivation by causing a tension in the individual which leads the person to engage in behavior to satisfy that need and thus reduce the tension. A thirsty person needs water, and is driven to satisfy that need and is motivated by desire for water to achieve that need. Depending on how well the goal is accomplished, the inner state is modified either completely or partially. Thus motivation starts with an unsatisfied inner-state condition and ends with movement to release that unsatisfied condition with goal-directed behavior as a part of this process.

It is important to note that motivation and performance are two distinct concepts. One management concern is to ensure that employees accomplish significant work goals. Successfully accomplishing these goals is a result of a number of factors, including the employee's commitment to the job and his ability to do the job. It is essential for a manager to see that some of the factors are internal and others are external to the employee. Otherwise a manager can reach an incorrect decision about motivation. For example, if a manager notices a drop in performance and decides to increase pay to overcome this decrease in performance, the reaction may be incorrect, as the drop in performance may be simply due to domestic problems or sickness. Wrong diagnosis of motivational problems are common in the I.S. industry.

Motivation theories have evolved over the past decades. The early management theorists regarded money as the main motivator. Most behavioral scientists agree that human beings are motivated by the desire to satisfy many needs, including money. However, there is a wide difference of opinion as to what these needs are. For example, Maslow developed a theory of motivation called the Needs Hierarchy, shown in figure 18.10. This approach is based on two important assumptions: First, each person's needs depend on what the person already has. Only needs not satisfied will influence the behavior, as a satisfied need cannot influence behavior. Second, needs are arranged in a hierarchy of importance. Once one level is satisfied, another emerges and requires attention. Maslow believed five levels of needs exist. These needs, as shown in the following diagram, are physiological, safety, social, esteem and finally self-actualization. There may always be a proportion of a level that is left unsatisfied.

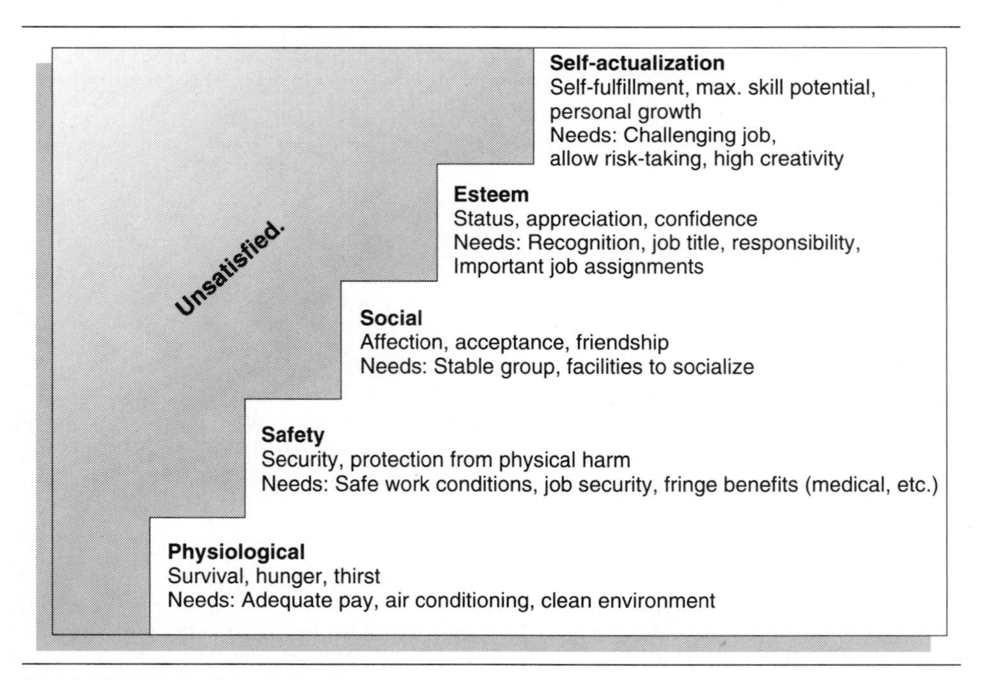

Maslow's Hierarchy of Needs *Figure 18.10*

Physiological needs relate to the basic requirements of pay and a good office environment. These will dominate when left unsatisfied. At the next level, safety needs include job security. We all know how demotivated people are when there is a possibility of retrenchment. The third level is social need which links to team-building and social functions in the organization. Esteem needs, the fourth level, comprise both the awareness of the person's importance to others (self-esteem) and the actual esteem of others. Satisfying this need leads to self-confidence and prestige. Recognizing good performers and permitting employees to work on their own to complete challenging and meaningful jobs will help to satisfy this need. The top level in the hierarchy is self-actualization. Maslow describes this as becoming more and more what one is - to become everything one is capable of becoming. Satisfying this need enables the individual to fully realize his potential. The organization climate required to satisfy this need includes the encouragement of creativity and allowing risk-taking in decision-making. While Maslow's Hierarchy of Needs is very acceptable to managers, there is little evidence to support it's accuracy.

Frederick Herzberg developed his motivation theory in the 1960's, based on a study of engineers and accountants. This is called the two-factor theory of motivation, or the Motivation-Hygiene Theory. Herzberg divides the factors which motivate employees into two main categories: hygiene factors and motivating factors. Hygiene factors (or maintenance factors) are so called because they relate to the job environment and not to the job itself. In the same way as public hygienic measures do not make people healthy but prevent them from being unhealthy, so these factors do not make people happy and satisfied at work, but they do prevent them from being unhappy and dissatisfied.

The main hygiene factors are:

- company policies and administration
- type of supervision
- working conditions
- interpersonal relations
- money status and security

The motivating factors, on the other hand, are found in the job itself and include:

- opportunities for achievement, i.e., the feeling of personal accomplishment at completing a job, solving a problem and seeing the results of one's efforts
- recognition for a job well done
- responsibility and authority, which is the degree to which a person has control over his own job and responsibility for the work of others
- interest in the job itself
- growth, which is the chance to learn new skills and knowledge to advance to a more challenging job

Both hygiene and motivational needs must be satisfied. One need cannot replace the other, but rather a balance is needed between the two. It is important to understand that hygiene factors refer to how you treat an employee and motivating factors refer to how you use an employee. Both factors have equal importance. The hygiene factors have a specific function of stopping dissatisfaction and subsequent reduction in output, while the motivating factors have the specific function of improving performance. Note also that hygiene factors have escalating endpoints, i.e., there is no one level of hygiene factor that will keep employees satisfied forever. People become accustomed to a level fairly quickly and subsequently want more. Consider pay as a motivator. There is no pay level that will keep employees satisfied indefinitely. A hygiene factor must be given for a specific purpose, i.e., where there is a need for it. Hygiene factors should be used to remove a specific dissatisfaction. If this is not done, we merely increase the source of potential dissatisfaction. Motivating behavior must be reinforced with a motivator.

Supplying the Motivators

Job enrichment is a practical method of applying motivators. Job enrichment attempts to restructure the job to give the employee greater responsibility, increased opportunity for achievement, recognition for achievement and more interesting work. Job enrichment should be separated from job enlargement, which involves giving an employee more tasks to do which require similar abilities. Herzberg feels that this activity merely enlarges the meaninglessness of the job. However, it can sometimes temporarily reduce boredom and broaden the employee's perspective, thereby preparing him for job enrichment. To increase job enrichment, introduce more difficult tasks not previously handled. Or give a complete unit of work, which is both meaningful and large, along with as much autonomy as possible.

Alternatively provide the opportunity to learn and increase knowledge and skills. Most of us want to achieve and grow. When an individual achieves, give more responsible tasks as recognition of those achievements. Good work should always be praised and promotion should be dependent upon personal effort and successful performance.

Another way to enrich a job is to increase responsibility and authority. This can be done with capable employees by removing certain controls, such as continually checking their work. Delegate as much as possible and make individuals accountable for their own work. This accountability can also involve the training and development of other staff. Job enrichment means providing more interesting work and a larger variety of tasks. The key to job enrichment is to give employees a complete job, involving planning, organizing, doing and controlling. This effectively means that individuals set their own objectives, determine resources, set priorities, implement the plan, followed by measuring, evaluating and correcting their own performance. Many jobs only involve the doing stage and incorporate very little planning and controlling. It is the control phase that tells the individual how effective the plan was and allows him to re-plan. The control phase provides direct feedback. Feedback gives work its meaning and absence of feedback is a common cause of complaint amongst I.S. professionals.

Motivating I.S. Personnel

Motivation of I.S. personnel has been researched by Dr. Daniel Couger, a distinguished professor at the University of Colorado. His national and international studies, carried out since 1980, are based on a model of motivation developed by psychologists in the early 1970's. He also uses Herzberg's theory. In the many studies done throughout the world, Couger identifies the most important motivating factor in the I.S. industry, irrespective of age, sex, position, job and culture, is The Job Itself. Based on this, he uses the Job Characteristics Model (figure 18.11) to identify motivation factors.

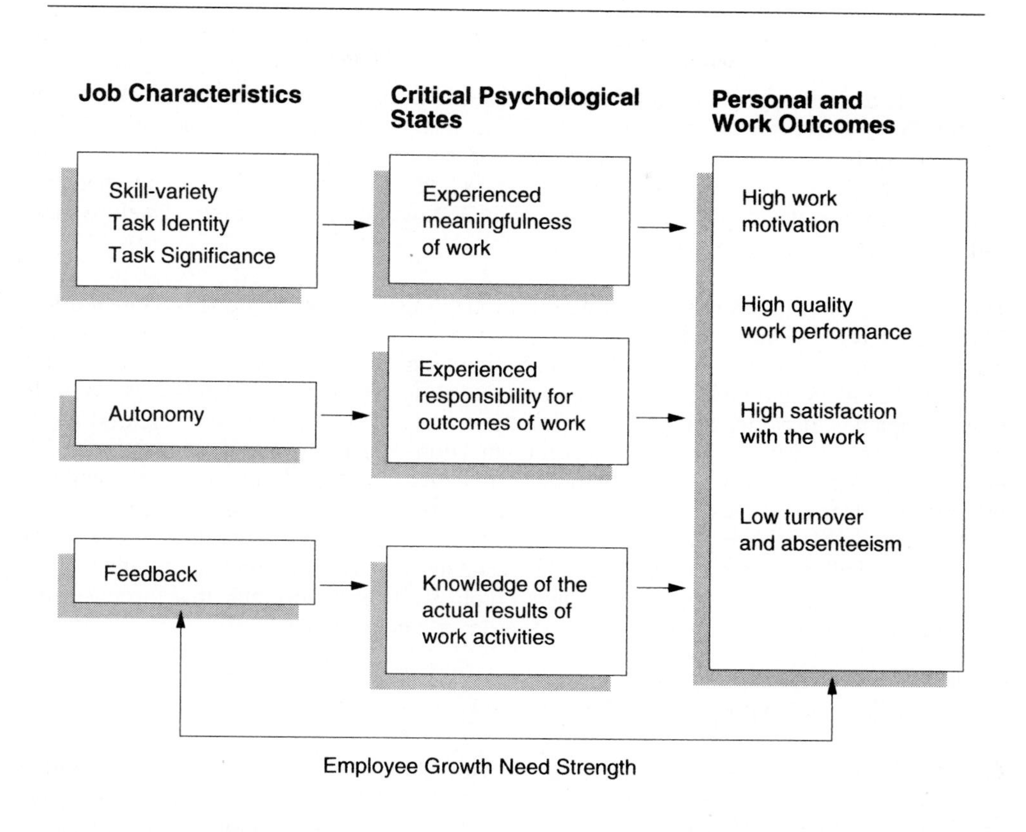

Motivating Potential of Job *Figure 18.11*

The core job dimensions specified in the model are those most sensitive to motivating staff. These comprise skill variety, task identity, task significance, autonomy and feedback. These are the aspects of the work itself found to be most important in motivating employees. The presence of these five key variables contribute to an employee's feeling of meaningfulness from the work, responsibility for its outcome and knowledge of the results of the work. By addressing these factors a job's motivating potential can be enhanced. The personal and work outcomes of these job dimensions leads to high motivation, high quality work and high performance of the individual, with the possible outcome of low turnover and absenteeism. All these outcomes would be highly beneficial to any organization. A questionnaire is used to measure these dimensions for a particular individual. The completed questionnaire identifies an employee's Growth Need Strength (GNS). This is a measure of the career growth and challenge required by the individual. Obviously some employees have a higher growth need than others and want "rich" jobs to achieve these needs, while others do not have the same requirements. The I.S. industry appears to attract people with a significantly higher GNS than any other professional group. This means that an individual's need for growth and challenge should be a project manager's focus when considering new tasks.

The research also identifies the Social Need Strength (SNS) of individuals. This is the employee's need for social interaction. Couger has shown that analysts and programmers have the lowest need for social interaction of all the professions, although they also have the highest need for growth. This conclusion is based on a study of 500 different occupations. The SNS is the individual's need to interact with others. Thus, given the choice of working on their PC or attending a meeting, it looks like the analyst would rather do the former! This is a problem if one considers the high interaction needed with users and the project team. However, the project manager can use the person's high GNS and send him on interpersonal training courses to develop these skills. The high GNS will ensure the person develops these skills even though they may feel a little uncomfortable.

The five core job dimensions determine the Motivating Potential Score (MPS) of the job. This is a measure of the potential of the current job to motivate the individual. The MPS of the job is then compared to the GNS of the individual. If there is a major discrepancy between the MPS and GNS, then there is a motivation problem that requires attention. In large companies, it is increasingly difficult to motivate staff because of the increased levels of management and the problems with communication. Where there is a mismatch, the job dimensions (skill variety, task identity, task significance, autonomy and feedback) should be investigated. Deficient areas are identified and jobs re-specified or formal training courses recommended. Managers must ensure that jobs are matched to individual's needs and abilities - a small investment compared to the resultant increase in individual productivity. In an industry notorious for staff shortages, high staff turnover and low productivity, improvements through highly motivated staff are worth considering.

Team Selection and Productivity

I.S. organizations use the word "team" very loosely. Any group of people who are brought together to perform a common task is called a team. What these groups of people have missing is a definition of group success or an identifiable group mission. When a group of people "jel" together to form a team, the whole is greater than the sum of the parts. The individuals in a jelled team seem to enjoy themselves and the likelihood of successful completion of the task goes up dramatically. Managing such a team is a pleasure as the team has its own momentum and tends to be self-motivating. How a manager can achieve this is obviously of great interest to us all! If one thinks of a top sports team, one can ask why this team is consistently so good. One aspect is quite clear - the team has a common goal and that goal is of ultimate importance to each member.

A team goal is not necessarily the same as a corporate goal. A normal condition of employment is to accept and follow company goals. However, this is slightly different to a team goal. While senior management may be very excited about increasing profits by a certain percentage, the professionals in the I.S. department probably consider these objectives as being very low on their priority list. However, when a team decides on a goal then things start to happen. The achievement of the goal leads to joint success and personal pleasure in achieving the goal as a team.

What the project manager has to do is to link organizational goals to the team goal. Compare this with the goals in sports which are totally arbitrary. Whether West Germany or Argentina get more balls between the posts is a totally arbitrary thing, yet a large number of people get very involved and emotional about the outcome! Their involvement is a function of the social units they belong to. Those on the periphery of these social units may show some interest in whether the teams succeed or fail, but their interests are minimal compared to those of the social units. The people involved in those social units are so psyched up that one would almost see it as an overreaction. What managers have to realize is that although individuals perform certain tasks and attain individual goals, the team is the ideal device to get people all pulling in the same direction.

A team that is working well together has a low turnover and a strong sense of identity. These teams stay together over lunch or at their favorite pub. They also have a sense of elitism and feel they are part of something unique. A further distinction is that the team has a joint ownership of the product that they are building. They want their names attached to the product and are eager to discuss their successes and failures. This leads to an enjoyment that each team member gets from working in the team, where the interactions are confident and healthy. Note the difference here between a team and a clique. The clique represents a threat to an organization, whereas the team may be somewhat irritating and exclusive, but it does assist the company in achieving its objectives.

So how is a strong and jelled team built? One very important measure is trust. Obviously, you might have more experience and better judgment than your staff. What is important is to allow staff to take responsibility and to sometimes make mistakes. Allowing staff to make mistakes means they can apply judgment to a situation and means they know you trust their ability to make decisions. People who feel they cannot be trusted have little interest in merging into a productive team. Another inhibitor is paperwork. The advent of structured methodologies has led to a considerable increase in paperwork needed to develop and install an information system. Paperwork generally stops people from doing productive work. Furthermore, making products under ridiculous cost pressures (which prevent quality) cannot create a sense of accomplishment in a team. When the eventual inferior product is produced, the team members will try desperately to go in different directions and distance themselves from the product and the team. The same applies to artificial project deadlines, where the date is often impossible to meet, so that successful completion is totally impossible.

Another major performance inhibitor is the work environment. Research has shown that people can improve their performance considerably if they have a quiet work area. This is difficult in open-plan areas which have high noise levels and lots of interruptions. You should look for ways to get the project team into a self-contained area to reduce the disruptions and increase productivity.

I.S. organizations that achieve high quality products are not just lucky. They follow good healthy management strategies. The first one is to get a handle on quality. However, turning out quality products costs money and strong teams tend to produce better products than the user is expecting. Quality products have to be protected from ridiculous time-scales and naive budgets. If the manager is really keen to make the team more productive and goal-oriented he will have to give up some of his control. This distinguishes the good manager from the poor one. All it takes is everyone moving in the same direction (i.e., following the same objective) and then firing them up to do it. The team that works well together is often made up of individuals with a lot in common. However, this does not mean the team is totally homogeneous. Often different "personality types" will improve the odds of the team working well together.

A fully functioning team is energized and takes great delight in achieving deadlines.

Personality Difference in Teams

When we discussed motivation, the growth need strength (GNS) and social need strength (SNS) of I.S. personnel were analyzed. The conclusion reached was that systems development staff have a very high growth need strength and a very low social need strength compared to other professionals. In terms of the Myers-Briggs Type Indicator this means that systems development staff are generally introverted, sensing, thinking and judgmental. These characteristics imply that development staff are more suited to working on self-contained projects. Project managers should recognize that they are more likely to be managing a group of individuals rather than a team. Because of the low SNS, people may be less inclined to communicate than the users they will have to deal with. On the other hand, a high GNS means that the individual will require a job that gives him a high motivating environment.

If systems development staff are essentially thinking types as opposed to feeling types, then this lack of feeling in the team can be an important contributor to the failure of the project. Research has shown that a mix of thinking and feeling personality types leads to a better outcome for the project and a more successful team in general. The feeling types in the successful projects should be distributed among the systems development staff and the end-user representatives in the project structure.

Team Size

The days of large monolithic project teams are over. Most organizations break large projects into a series of smaller self-contained ones. These should be small enough to be developed by a project team of five or six staff over six-month time periods. Note that this team does not work as a true team in practice. If properly planned, much of the work can be done in self-contained units by individuals. The purpose of grouping the work is to ensure that everyone is committed to achieving the overall objective of developing a successful system and the team leader must ensure that individual goals are aligned with those of the team.

Individuals brought together in a systems development team do not form a close-knit unit immediately. Teams go through their own stages of development as shown in figure 18.12.

The stages in the team development process are characterized by different behavior and team performance. The project manager must recognize and reduce the impact of the earlier phases of team development so that the team can progress as quickly as possible to the more productive phases. It has been found that teams made up of unlike individual personalities work better during the earlier project phases when the amount of routine work is small. Such teams are good for creative, problem-solving tasks and for tasks involving complex decision-making. This is because the team members stimulate each other and produce a high level of performance and quality. Teams made up of unlike individuals can however create a great deal of conflict.

Teams consisting of people of similar personalities work best on simple routine tasks. Thus teams made up of these people will be more appropriate during the later development phases when routine is greatest. The formation of a balanced team requires consideration of more than just the technical expertise. The personality of individual members and the need to

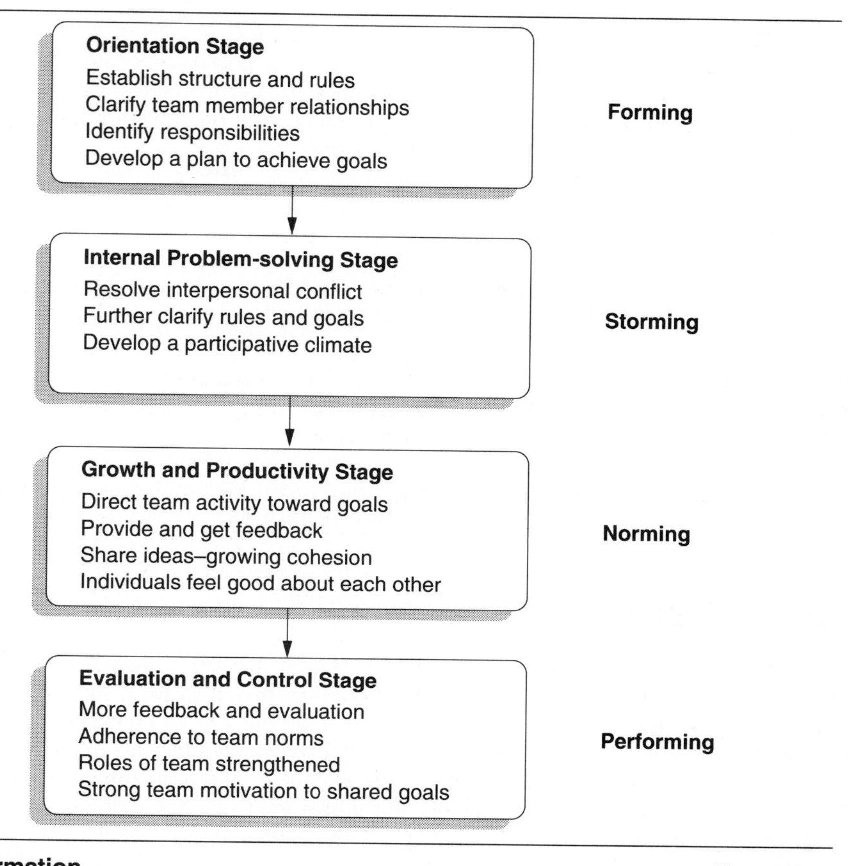

Team Formation *Figure 18.12*

change the team composition in the later stages of the development project are also important factors.

Career Development

Because typical I.S. professionals have high growth needs and low social needs, and often leave their jobs because of lack of career growth, career planning is vital.

The concept of career has many meanings. The popular one is the notion of upward mobility - making more money, having increased status, more power and more responsibility. A useful definition of career is

> ***the individually perceived sequence of attitudes and behaviors associated with work-related experiences and activities over the span of the person's life.***

The term "career" does not imply success or failure, except in the judgment of the individual. It is also interesting that a career consists of both attitudes and behaviors. A person's personal life also plays an important role in a career. For example, the attitudes of a 40-year-old mid-career project manager concerning a possible promotion to a systems development manager can be very different than a project manager nearing retirement. A batchelor's reaction to a promotion involving moving from one location to another is likely to be different than a father with school-age children.

Career Effectiveness

Career effectiveness is a concern of both the organization and the individual and can be measured by performance and by attitudes. Sometimes an organization does not fully recognize performance because it has a staff-appraisal system that cannot reward at the appropriate time. The organization may also feel disappointed with an employee's performance because the individual is satisfied with career performance while the organization considers the person as underachieving.

Effective careers in project management are more likely for individuals with high levels of performance, as well as positive career attitudes, adaptability and clear career goals.

Career Stages

Individuals go through distinct career stages. These are the prework stage (school and tertiary education); the initial work stage (moving from job to job); the stable work stage (maintaining one job) and the retirement stage (leaving active employment). Needs and expectations change through these stages. Employees tend to be very concerned about security in the early part of the stable work stage (called the establishment substage), but are more concerned with advancement in the latter part of the same stage (called the advancement substage). These two substages normally span the ages between 25 and 45.

Two career stages are critical - namely the recent hiree and the mid-career person. Early career difficulties relate to frustration with a low-demanding job and poor performance feedback. To overcome these problems, "excellent" companies are using "realistic job previews" where recruits are given opportunities to learn about the benefits and the drawbacks of the job and the organization before being employed. This approach has helped to reduce initial job frustration. Other approaches include giving new employees challenging, initial assignments and allocating them to demanding supervisors.

The mid-career employee is normally established at work and in society. Often these employees have reached a career plateau and feel stifled in their jobs. This can lead to depression and ill health and ultimately to resignation. These problems can be addressed by professional counseling or lateral transfers to user areas.

I.S. Careers

Traditionally, there have been four career paths in a typical I.S. organization, namely: operations, systems, applications and specialists. Most people entering the operations path seem to stay in the Facilities area. Those entering the applications programming path move into the systems area. This makes the applications path, and especially the analyst/programmer position, an important transition point in all I.S. careers. Most people entering the specialist path (e.g., systems programmer) seem to stay in that technically-oriented area. Programmers and analysts seem willing to move in and out of different career paths frequently. This implies two things. First, it is possible that they do not understand which career is the most suitable. Second, it is likely that organizations lack flexible career paths and individuals move in and out of different paths by joining different organizations.

A trainee joining an organization will see a career path over a period of time. This may not be suitable for some, as it is probably aimed at the average performer. An alternative is to use a career model developed in the company which removes this "straitjacket" approach. A career model is designed to guide the organization and provide a road map that shows the various routes and destinations that an individual can follow. The road map can be used by the manager and the individual to decide on specific development strategies in the short term. A typical process is where a trainee can move through stages and grow in terms of technical, managerial, communications and company skills and knowledge. Each stage can have several job levels which require more demanding technical work or alternatively managerial function.

The usual progression involves an initial trainee moving from programming through analysis, into project management and then into senior I.S. management positions. The first real management level encountered is that of project manager and this is an important step. Moving into project management is a critical career choice as this now brands the individual as a manager as opposed to a technical professional. This is because managing people becomes more important than managing technology. Note how different these jobs are and the difficulty individuals must have in following such a career path - especially if they were recruited solely for their technical ability.

During each of the career stages there are several performance checkpoints that must be discussed. These checkpoints can be standard performance appraisals where strengths and weaknesses can be highlighted, and action plans developed. Only once a particular career stage has been mastered, can the next career move be discussed. At this point the subordinate's interest is the controlling aspect in the career decision along with the manager's judgment of the subordinate's readiness and, of course, the organization's need. None of these facts can be overlooked, least of all the possibility that the individual may have plateaued for a time or for good. The individual may even be better suited to a completely different job. If the decision is to move up rather than out or not move, the career decision discussion then focuses on the preparation necessary for the selected upward move. This may require training or secondments to ensure readiness.

Although a career plan does not have a time frame, most expectations by companies and individuals reveal that the lower stages should take between one and three years. Thus an entry-level trainee could become a project manager in as little as five years on a fast track, although seven to ten years would be the normal progression. Not every employee is expected to move through the stages at this speed and reach the top. What is needed in every organization is a set of career paths to cover a technical orientation and a management orientation. These paths should be continually assessed and should have typical time-frames for lateral and horizontal movement.

Leadership

Being a good project manager requires more than management - it requires leadership.

Leadership is that quality of an individual that motivates others to willingly participate to achieve goals which they come to share with the leader

Whereas a manager uses his position to get results, the leader gets results through cooperation and free will by influencing staff. In project teams, a common goal is developed and plans are created to achieve the goal. The strong project leader must focus on three areas - the task needs, the people needs and the team needs. All three must be satisfied to ensure success. Blanchard, Zigarmi and Zigarmi present a model of leadership (figure 18.13) to help the manager with different staff situations. They claim a manager's approach should depend on the managed individual's maturity in the job. At the outset, when the subordinate lacks experience and ability to complete a task, a "directing" approach is used with strong controls. As the subordinate takes on more responsibility, so the leader gives him more authority to make decisions, adopting a "coaching" style. In the third quadrant, the individual is encouraged to become self-sufficient, with the support of the manager. Once into the fourth quadrant of the model, subordinates are allowed to manage themselves and resolve their own conflicts.

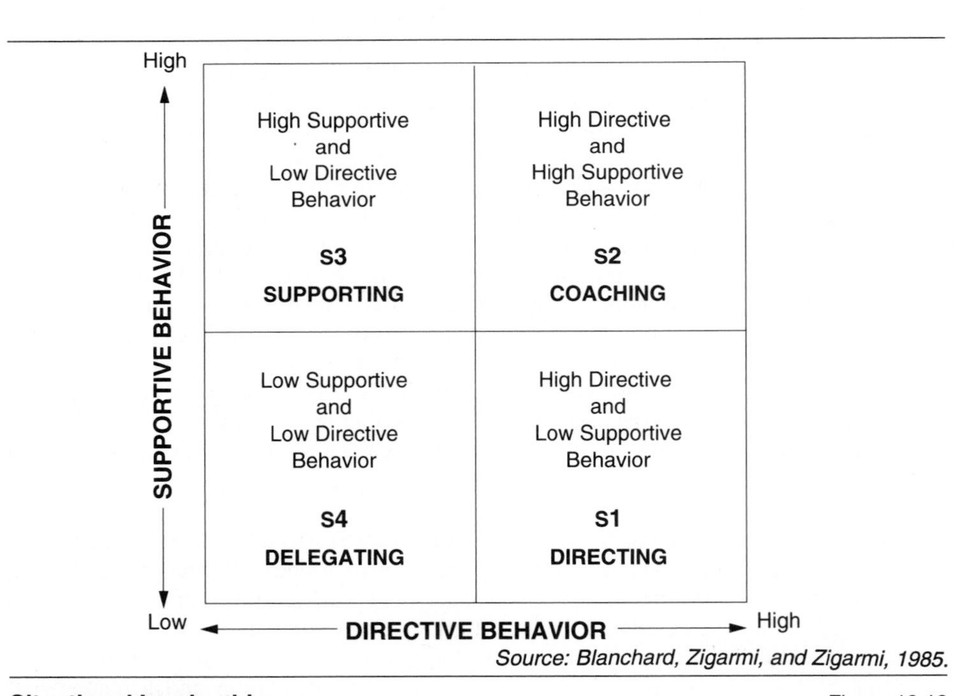

Source: Blanchard, Zigarmi, and Zigarmi, 1985.

Situational Leadership *Figure 18.13*

We should realize that an employee can be simultaneously in several quadrants with respect to different activities. For example, an Analyst/Programmer could be in quadrant 4 as regards designing a program, in quadrant 2 as regards writing user documentation, and in quadrant 1 in terms of preparing and giving a presentation to users.

The model assumes that managers can switch their leadership styles dynamically - not an easy thing to do! The model reinforces the idea that, as effective managers, we must delegate work to staff as soon as possible in their careers and provide them with enough support to achieve high-performance goals. This will make them competent to grow in their careers - and will make you a successful project manager.

Case Questions

MyWay Organizer

Q18.1

What type of personality profile might suit the role of an analyst programmer in the software house environment developing the Organizer product? Why? (10 mins)

Q18.2

Draft personal objectives for an analyst recruited to your team. Her name is Theresa. She has been with the company for two years, previously working in the Medical Practice Administration package team. She has a bachelor's degree in computer science, and five years' experience overall. She is people-oriented and outgoing, getting on well with users and colleagues alike. She can be forgetful and ignore (or dismiss) detail which later turns out to be important. She has previously worked only on unix-based systems and has no experience of PC systems, LANS or client server. Graphical User Interfaces are also new to her. Her responsibilities on the team will be the specification of the second release of the Organizer, including the incorporation of an Internet access module. For each objective, list how you would assess or measure achievement. (15 mins)

Q18.3

Divide yourselves into groups of four, with access to an IBM compatible personal computer for each group. Run the software provided by the instructor to determine the MBTI profile of each member in the group. Summarize the profiles of the four individuals on one graph. Identify strengths and weaknesses in the combination of profiles. (30 mins)

Gleam Stores

Q18.4

What type of individual would be suitable to lead the Gleam Stores implementation project as it spreads out to install the new system in all branches country-wide. What attributes would assist or hinder this person in achieving a successful implementation? Can you suggest suitable MBTI profiles? (20 mins)

Handover Trust

Q18.5

One of the client server projects has run into severe difficulties. The project team complains that the current project manager (Gordon) is autocratic and domineering, as well as a "male chauvinist pig". He is a senior and highly regarded person among the Handover middle management fraternity. The complaints have come from a business analyst (Trisha) and a user interface designer/prototyper (Cecil) on the team. Gordon previously ran the system software group which until recently also looked after the DBA function. He has 22 years' experience, dating back to early mainframes programmed in assembler.

You have spoken to him and he, in turn, has expressed his dissatisfaction with the team. "These youngsters don't know what goes on inside a machine - they write horribly inefficient code. They also haven't learned to really work yet."

What might the sources of conflict be in this situation? How can you resolve things so that personal relationships and feelings are not damaged and the project can get back on track?

(20 mins)

ThoughtWell Books

Q18.6

Use the definitions of Job Characteristics provided by the instructor. Consider the team which you chose to tackle the ThoughtWell Project in Q5.5. By considering the team members and the characteristics, determine what you as project leader can do to increase the motivating potential of each team member's job or role. (30 mins)

Q18.7

The productivity of team members working on-site at ThoughtWell is substantially lower than that which is normally delivered at your offices, even for the same individuals. This has led to overruns relative to original estimates. The team is also frustrated by the amount of pressure they are under at this late stage of the project. They complain that they cannot concentrate in the kiosk provided as an office at ThoughtWell, where there are several order-taking telephones and clerks in and out to retrieve call orders. Even though they are working ten hours a day, they claim that they only get about two uninterrupted hours. The people on-site are finding that Jane frequently changes her mind about system features. They find this very frustrating. How can you address this situation immediately and prevent any further problems on the subsequent phase of the project? (40 mins)

19 *Implementation*

Preparing for Implementation

The completed system has to be handed over and installed in the user's premises. This means preparing the work environment, doing the data conversion, interfacing the system into the user's work procedures and tuning the system. The planning for this was all done in the systems development phase when activities like user training, procedure development, implementation planning and data conversion planning were all completed.

The implementation phase is the most difficult one, because a technical product now has to be fitted into a human organization. The new system has already led to suspicion and fear while under development. Now it is going to change the way people work and think. This stage must not be played down but must be completed in a positive, enthusiastic manner.

Site preparation does not necessarily involve false floors, air conditioning and clean electrical power. However, even a computer terminal on a desk linked to the outside world needs careful consideration. Someone has to use this equipment effectively as a major part of the information system that has just been developed, so put some effort into locating it correctly! Systems have failed because of the poor work setting of equipment.

Conversion of the data involves setting up the files and databases needed by the system. This area is problematical because one-off programs have to be written, tested and implemented to capture manual and automated data from various sources into the system. Obviously the data take-on must be complete and validated. The controls and checking to do this exercise must not be underestimated.

Although a lot of documentation has already been produced (including user manuals), user operating procedures are needed to guide the user in the proper use of the system. These procedures include how to start up and close down the system, how to recover from problems and whom to contact when problems cannot be resolved. Despite the training and documentation the user has now received, there will still be a lot of hand-holding required in the early working sessions before user competence levels are reached. These tasks take time and patience - both of which are in short supply at this stage of the project!

The introduction of the system can be implemented using different strategies, depending on cost, risk and the users.

An immediate cutover (or "big bang") from the old to the new system, as shown in figure 19.1, requires a high degree of confidence in the new system by everyone.

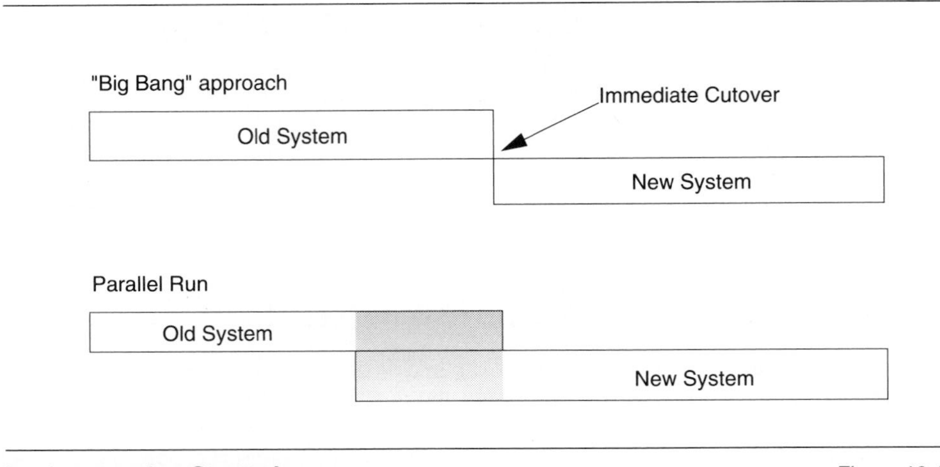

Implementation Strategies *Figure 19.1*

Although this is the least costly and fastest method, it usually creates a period of chaos in the organization which can lead to total rejection of the system. Because, once implemented, there is no going back, this method requires considerable planning and is often used when replacing mainframe computers or large computer components. However, because it is a very high-risk option, the project manager should avoid it unless there is no alternative.

Parallel running of the old and new system, shown in figure 19.1, is when both systems are run together over a period of some months.

This popular method is used to ensure that the results from the new system are reliable by reconciling output from one with the other. If things go wrong, the old system is still available, and both systems can continue to be run until the new system works correctly in the user's hands. The major disadvantage is that there is a cost attached to this method, but more importantly, there is a considerable increase in the amount of work required by the user and also considerable confusion trying to run two different systems together.

Phased implementation, as shown in figure 19.2, is when a system can be divided logically into subsystems and implemented as such.

Each subsystem can be introduced and assimilated into the organization before the next one is introduced. This reduces disruption and sometimes lowers the cost of implementation.

A further option is the pilot implementation approach. This is particularly useful where there are multiple sites. The new system is implemented at a site were the users are more receptive to change. Teething problems are resolved before implementing at the other sites. The users at the pilot site automatically become the salesmen for the new system.

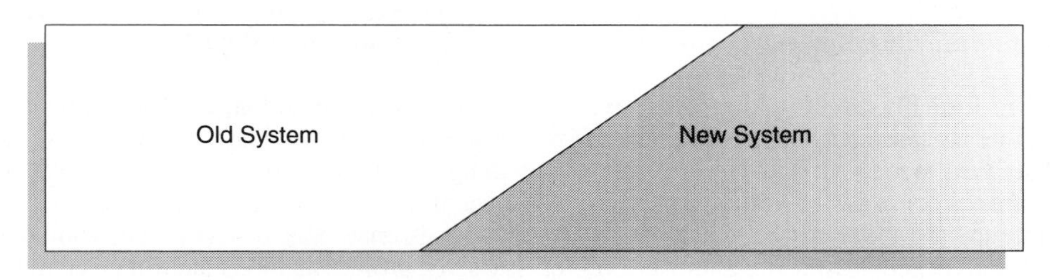

Could be functionality based, geographically based or number of users based

Phased Approach *Figure 19.2*

Having determined the implementation method, the final consideration is timing. Most accounting systems have to be implemented after the completion of a month-end, but quarter- and year-ends should be avoided if possible due to the extra workload and extra problems that could occur.

Implementation Problems

There are three main reasons why systems fail at this late stage of the project - a poor quality system, a lack of commitment to implement the system from user management, and resistance from the end-user.

Given that we can develop a tested system that matches user requirements, how can we raise management awareness of the problems of system implementation? Clearly management must provide the time and resources for adequate training and parallel running (if appropriate). Users need time to adjust to major changes in work patterns and management must not force the pace.

So, the most serious implementation problem is resistance to change. This resistance, which is quite natural, can lead to poor use of the new system or even total rejection. Resistance can be reduced by encouraging maximum project participation by as many users as possible and designing a user-friendly interface to the system.

To help minimize resistance to change, we use approaches from the field of psychology. Lewin-Schein propose a three-phase approach. Firstly to unfreeze people (get them involved and used to the idea that changes are coming so that they understand the implications), followed by making the change (implementing the new system) and finally refreezing (giving support and assistance with the new environment).

The tuning and debugging of the system that occurs after cutover is an ideal time to work on the refreezing process with the users.

Based on the career planning that is done on an ongoing basis, the project team will now be looking for another challenge to get their teeth into! Planning what is going to happen to team members should have already been communicated by this time and the project manager will be tying up the loose ends and releasing staff to other assignments.

Before formally handing the system over to the maintenance function, the team should get together to carry out a project review. This is a means of identifying what went right and what went wrong with a view to improving things next time around. As a project is a learning process for everyone, it is worthwhile trying to formalize the lessons learned and communicate them to the whole team. Areas like estimating and scheduling should be analyzed as should staffing and user problems. Use of project methods and tools should be evaluated for effectiveness - especially if new ones have been used. This information should be documented and circulated to all I.S. staff. After all, you have just completed a major endeavor successfully and you should tell everyone how you did it! In any event, another team will carry out a post-implementation audit in a short while so you had better get in first!

Case Questions

Gleam Stores

Q19.1

Would it be appropriate to involve users (line managers) in the Gleam Stores implementation project? How could this be achieved? What role should they assume? What difficulties might arise with systems staff? (20 minutes)

Handover Trust

Q19.2

We are nearing the end of integration testing of the New Business system. This has generally gone well, although not all interfaces can be tested in the development environment (some of the systems we need to talk to are in old technology which is only available in the production configuration). Management has asked you to look at the options for installation, including:

- Parallel running of the new system with the old applications processing system. This will necessitate a variety of one-time interfaces to synchronize data across systems
- A "big bang" approach, where things are tested thoroughly and then implemented as a cutover from the old to the new. This minimizes the interface complexity, but has a high risk if there are any glitches with the new system. There is just one weekend in the next three months when this can be done, because of closing financial and tax years.

The system's correct functioning is critical to the business. A failure for one day could cost several hundred thousand dollars.

Decide on an implementation strategy. Document the phase plan in Gantt format. Discuss how you will handle the negative aspects of the approach chosen. Can you think of other creative ways to address the problem?
(40 mins)

ThoughtWell Books

Q19.3

Given the staff profiles from Q5.5, whom do you think would be the best candidate to lead the ThoughtWell installation, given that you have been promoted and will be leaving the project? How would you prepare your successor to handle the installation? How would you handle this development with the client? What role(s) would you see for client staff in the process? (30 mins)

20 *Multiple Project Coordination*

Integrating Plans

Our project is seldom the only one running in the organization. This almost inevitably means that, at some point, we will be competing for resources. A typical example is a number of parallel development projects that need the services of a central Database Administration function at about the same time. If our needs for resources external to the project team are not carefully identified and planned for, chances are that we will encounter unforeseen delays when we need to call upon them. One way to try to alleviate these clashes, and to reduce the load on shared resources, is to try and integrate the plans from the various teams so that a macrolevel picture can be obtained. This would be very difficult if the plans were all prepared differently. Fortunately we have seen that we can use common frameworks for the plan regardless of the type of project, and nearly identical plans (in terms of structure) for similar types of project. Plans will obviously differ with respect to the real detail tasks at the lowest level, but can have a very high degree of similarity at summary levels. These concepts are shown in figure 20.1, implemented via a development support group.

Eliminating Bottlenecks

We can enhance the degree of commonality by starting from a common base. One approach that we have used with several clients is to prepare a set of sample "skeleton" project plans for the types of projects tackled by the organization: one for custom mainframe development, one for End User Computing projects, one for package implementation, etc. By using the configuration management framework, we can ensure that at the phase-level summary, the naming and structure of all these plans are consistent. By introducing standards for naming of common resources (e.g., DBA, Capacity Planning, Facilities) we can ensure that resources outside the team are easily identified. The standard plans should be maintained and distributed from a central group. In our case, this was a "development support" function.

Individual project managers then modify the plans and add detail, before submitting a copy to the central coordinating function. The plans returning from the teams are integrated together to identify any resource conflicts. This can be done manually, or using a project management package, where the team's plans are treated as subprojects within a higher level

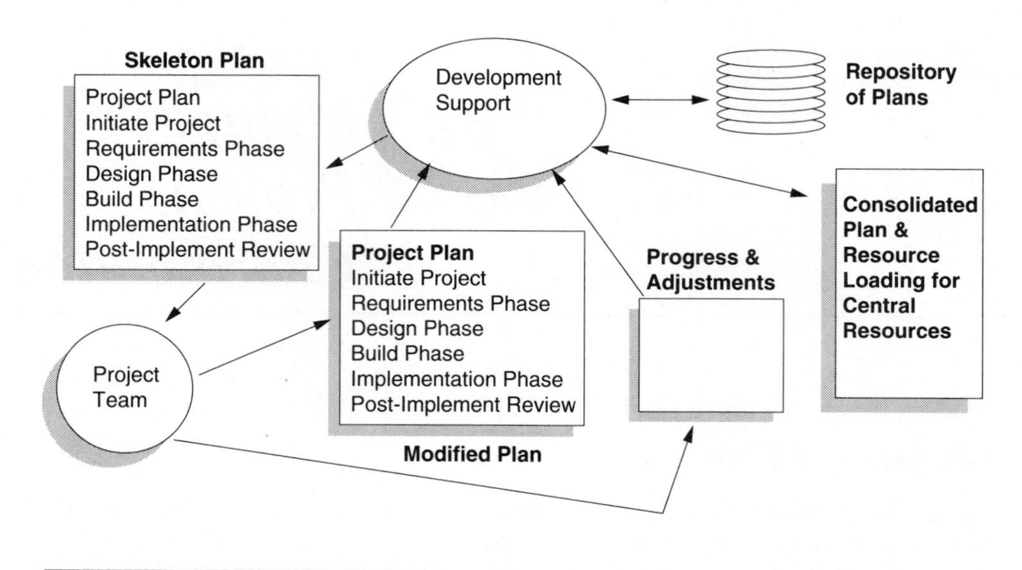

Multiple Project Coordination *Figure 20.1*

plan. The overall resource picture is then evaluated. We can pick up excessive workload and move activities around in consultation with the teams and central groups to alleviate problems. In other words, we are doing "resource smoothing" at a macrolevel, not just within a team.

If the teams also feed back their actual progress and resource consumption, as well as amendments to their plan for future estimates, the central group can act as an early warning center to other related projects. For example, we may see that the Order Processing Project is dependent upon the Mainframe Upgrade and Product Maintenance projects. If one of these provides revised estimates indicating a significant slippage, we can alert the Order Processing team so that they can take appropriate action. A further benefit of the central coordinating team is the issue of quality. It is remarkable how the quality of project plans improves when

- teams are given a good example to begin with
- project plans are visible and worked with by another group outside the team
- actuals are tracked and visible outside the team
- standard project plans are updated and revised to reflect the experience of teams in the organization

Boundary Management

A further worry across multiple projects is the issue of scope. We may initiate related projects whose results will need to dovetail. An example would be a project which will generate a system supporting Point of Sale (POS) gathering of sales data in a retail organization, and a related Sales Analysis Project. These could easily become confused in terms of defining requirements, liaising with users, and exactly who is responsible for what. The most dangerous thing here is to make assumptions. One view of this is that they "make an ASS out of U and ME". It is all too easy to expect that the other team is doing something, when in fact they are not. This can leave "holes" where requirements have gone through the cracks, so to speak. Equally problematic is the situation where the teams encroach on each other's territory. This can lead to redundant work, arguments, and very confused users.

To avoid these problems, we must scope our projects carefully and manage the boundaries from inception until after installation. If our projects are involved with system development, a good technique is to use the context diagram introduced in an earlier chapter. We can in fact produce a very high level architectural diagram (see figure 20.2) which shows the related systems, and then mark out the context of each on this diagram. Wherever possible, we should try to make a clear interface by means of data. If we can identify the groups and elements flowing across the boundaries, then we can control the scope of each project precisely. This is easier said than done when we are dealing with a shared database. In this case, we may need to explicitly define which system is responsible for which state changes or transactions. Having a good data dictionary or CASE product can greatly facilitate this. Some of the more modern development languages/environments, particularly the Object Oriented ones, will offer facilities to automatically create cross references as we develop from prototypes to production code. If you do not have this kind of environment, you should

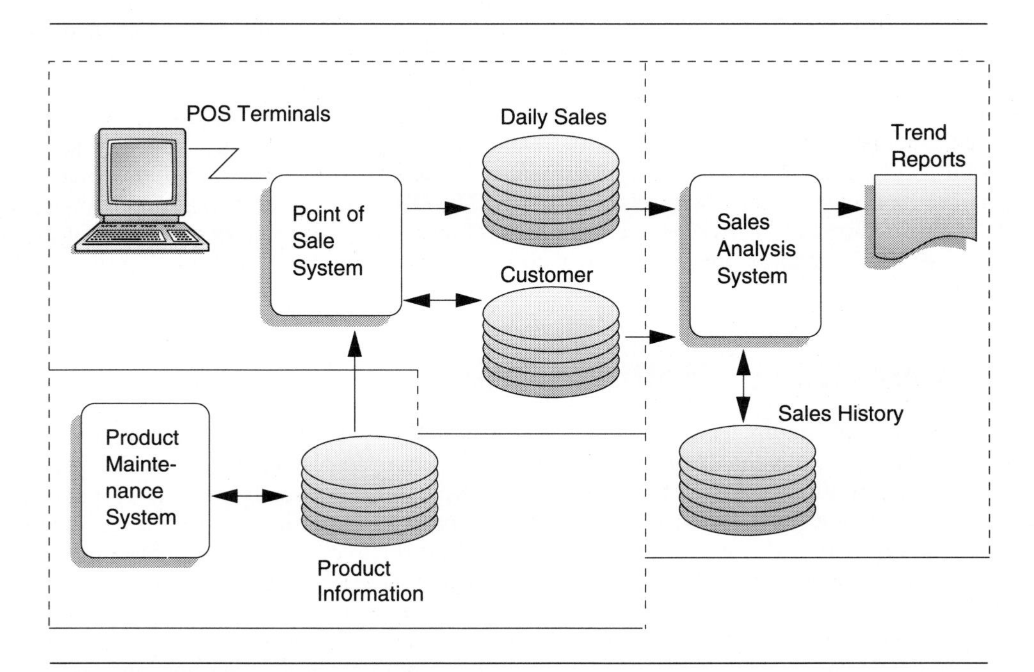

Managing Scope via Boundaries *Figure 20.2*

consider using a configuration management tool such as CCC (mainframe) or PVCS (personal computers and LANs).

The Role of Reviews

Reviews can prevent major problems by identifying scope creep in projects. If we use the configuration management approach, the check against original criteria should ensure that projects are not under-delivering on scope. If the same resources are used to perform reviews across the related projects, clashes and differing assumptions can be easily spotted, especially where a consistent methodology is used. A good place to position these resources is in the development support group.

Reviews are a safety net, however, and should not be relied upon as our primary mechanism for keeping things straight. We should use good analysis techniques which are consistently applied across teams. Teams should also bear a responsibility to liaise regularly and notify each other of any issues which might create problems for the related groups.

Sharing Resources

Where a single person is to "belong to" and function as a member of two or more teams, we must make allowances for the conflicts and stress that this can cause. Even if the personal relationships are all cordial and functioning well, the individual still undergoes a "context switch" each time they shift from one team to the other. We can attest from personal experience that this can be very exhausting. If the interpersonal relationships are not running smoothly, or if more than one team is under pressure and expecting all its members to put in extra effort, the "shared" individuals come under considerable pressure. Great stress can also be caused by split loyalties and conflicting demands from the different team leaders. Generally, this type of arrangement should be avoided, unless the individuals concerned are mature and senior enough to resolve these issues for themselves.

Other resources which are typically shared include those which cut across project and functional boundaries. Typical of these is the Database Administration team. One way to alleviate bottlenecks in this area is to train teams in the skills necessary and allow them to perform their own database analysis and design. The DBA group then reviews and amends the completed design. This gives the team higher ownership of their design, while ensuring that corporate standards are met and that the resulting design will have adequate performance.

Consistency of Methodology

Where there are multiple projects running in parallel, it will be much easier to manage successfully if the methodology that we use across them is consistent. This should apply to the structure of plans (naming of phases, reviews, etc.), to the deliverables that are produced (as far as possible) and to the manner of project progress reporting. If we can get these consistent, we will find it much easier to assess projects relative to one another since we will not be comparing apples with oranges. A further benefit is that project staff can move more easily from one project to another. We had one instance which illustrated this particularly well. We were involved with a project at a major insurer, where we had implemented a system development and project management methodology. We transferred an Analyst

Programmer onto the project from another organization using the same methods and technology, but in a totally unrelated industry. Because the project was under time pressure, the person was given five program specifications within the first day. At the end of the week, she had completed these, and approached us for more work to carry on with!

Mixed Messages

We should take particular care in our liaison with groups outside the project environment. They may well feel uneasy already and are far less used to change than we are. After all, we are usually the change agents. If they see different people from different teams and get conflicting reports or different messages, they will become confused, anxious and start to loose trust and faith. We should carefully coordinate our interaction with them and ensure that we have our ducks in a row before making presentations, or issuing documentation. A good principle here is that all user/management contact outside the team should go through the project manager or a designated analyst. These persons should regularly meet with their counterparts on other related teams to ensure coordination.

Stay Business Focused

With all the technical complexities of managing multiple efforts, it is easy to get bogged down in the interfaces and details. We should periodically step back, remind ourselves what the overall *business objectives* are, and then refocus our activities to deliver these. We will frequently find that the technical issues can be avoided, subcontracted or otherwise dealt with when we remember what it is we are trying to do in the first place.

21 *Subcontractors*

Using Subcontractors

Using subcontractors to provide vital resources in a project is a common strategy. It makes sense to employ contractors and consultants to add certain short-term skills to a project team. It is also important to use software and hardware suppliers to provide products and services where these can help solve the user's problem. This chapter looks at when to use subcontractors and how to choose them. It also discusses the tendering and negotiation process, how to draw up contracts, and the project manager's problems dealing with the subcontractors.

When to Use Subcontractors

The people resources needed in a project vary over the lifecycle in terms of skills and numbers. Sometimes the skills and/or the numbers are not available in-house. In these situations, the project manager may have to resort to outside help. These decisions are not easy as outsiders sometimes create short- and long-term problems. Thus extreme care is required when deciding on, first whether to use a consultant/contractor and second, which organization to approach. The normal approach is to specify your requirements, identify the supplier you want to deal with, and then select the specific individuals.

How to Choose Subcontractors

Subcontractors can provide several different services to a project. Service bureaus, software houses and software package vendors are categories of subcontractors used in project development. If a software house is developing a part or all of the system for you, you must manage that process as a project following the PLC in the same way as an inhouse project. Planning may be a combined responsibility with the software house, but you must be solely responsible for control. Consultants and contractors are often used to assist in the development process. Consultants can be hired from accounting firms or from consulting firms specializing in the I.T. industry. Contractors can also be recruited from software contract houses and personnel recruitment agencies. In all these cases, the organization you deal with should be one you would wish to do business with on an ongoing basis. The actual individual selection process should be the same as the selection process for permanent staff. Don't let the outside firm do the selection for you, as it is your ultimate responsibility to ensure you have competent, motivated staff. The selection criteria you adopt should include an individual's proven skills, the individual's personality, his charge rate, and the supplier's credibility and track record.

The Tendering and Negotiation Process

The tendering process begins when the project requires services which can only be provided from outside the organization. For example, software suppliers may be requested to offer their packaged solutions to satisfy certain business problems. The tendering process starts with the project manager drawing up a Request for Proposal (RFP) document. This document specifies the business problem in detail and requests a solution to that problem from the supplier. The process includes the following steps:

- Draw up the RFP
- Determine selection/evaluation criteria
- Select possible suppliers
- Send RFP to suppliers
- Evaluate RFPs and develop a supplier short list
- Carry out an in-depth evaluation
- Select the supplier
- Develop a mutually agreeable contract
- Take delivery of the product, ensure quality and then pay

Request for Proposal

Name of Company Contact Person
Company Background
Project Scope
System Functions and Outputs
System Performance
Possible System Growth
Operating Environment
Interfaces to other systems
Reliability and Availability
Maintenance and Support Requirements
Documentation and Training
List of Current Customers
Terms and Conditions

Request for Proposal *Figure 21.1*

We will look at the contract issues in a later section. An RFP can be a one page outline or a large comprehensive document. In principle, it should contain enough information to allow the supplier to understand the problem. It is recommended that the detail in figure 21.1 be considered.

This RFP should be discussed with senior I.S. management to ensure completeness and synergy with longer term I.S. plans. It should then be sent with a covering letter to established suppliers of the required services with the date when proposals should be returned. Allow plenty of time for this exercise as, in many cases, suppliers will need to discuss your requirements with you in more detail and prepare comprehensive documentation. The selection criteria include a list of essential and important features that you require. These criteria are "weighted" in importance. For example, if you were tendering for a project management software package, you may have developed the list of factors with their associated weightings as shown in figure 21.2.

When the tenders are received from the potential suppliers, they are evaluated objectively using the weighted criteria above. For each supplier, a raw score (out of say 10) is allocated to each criterion based on your evaluation of the product (figure 21.3).

Factor	Weighting
1. COST	**(40%)**
Price (80%)	.80*.40 =.32
Implementation (20%)	.20*.40 =.08
2. SUPPORT	**(20%)**
Maintenance (30%)	.30*.20 =.06
Training (30%)	.30*.20 =.06
Installation (40%)	.40*.20 =.08
3. FEATURES	**(30%)**
Resource-leveling (25%)	.25*.30 =.075
Resource Gantt (20%)	.20*.30 =.06
Goal-seeking (20%)	.20*.30 =.06
Windows support (35%)	.35*.30 =.105
4. CAPACITY	**(10%)**
Multiple project support (40%)	.40*.10 =.04
250 activities per project (30%)	.30*.10 =.03
Fast calculation (30%)	.30*.10 =.03

Factors and Weightings *Figure 21.2*

The raw score for each criterion is then multiplied by its weighting factor, and the adjusted score added together to give a grand total for each supplier as shown in figure 21.4.

These totals are guidelines to assist the decision process. Thus they can be used to eliminate suppliers before an in depth investigation of short-listed suppliers commences. Once a supplier is chosen, the real work of negotiation begins.

Factor	Weighting	Product A	Product B	Product C
1. COST				
Price	.32	7	6	9
Implementation	.08	6	9	5
2. SUPPORT				
Maintenance	.06	5	7	9
Training	.06	8	5	5
Installation	.08	8	7	5
3. FEATURES				
Res.-leveling	.075	9	9	6
Resource Gantt	.06	8	8	5
Goal-seeking	.06	2	6	5
Windows support	.105	3	3	8
4. CAPACITY				
Multiple project	.04	7	6	7
>250 activities	.03	8	4	6
Fast calculation	.03	4	9	6

Product or Service Rankings *Figure 21.3*

Factor	Weighting	Product A	Product B	Product C
1. COST				
Price	.32	2.24	1.92	2.88
Implementation	.08	0.48	0.72	0.40
2. SUPPORT				
Maintenance	.06	0.30	0.42	0.54
Training	.06	0.48	0.30	0.30
Installation	.08	0.64	0.56	0.40
3. FEATURES				
Resource-leveling	.075	0.675	0.675	0.45
Resource Gantt	.06	0.48	0.48	0.30
Goal-seeking	.06	0.12	0.36	0.30
Windows support	.105	0.315	0.315	0.84
4. CAPACITY				
Multiple project	.04	0.28	0.24	0.28
>250 activities	.03	0.24	0.12	0.18
Fast calculation	.03	0.12	0.27	0.18
GRAND TOTAL		6.37	6.38	7.05

Final Scores *Figure 21.4*

Negotiating

A project manager has to negotiate at different levels both inside and outside the company. Negotiating is an invaluable skill in project management. Successful negotiation requires that you know the facts. It is quite normal for your management to try and reduce your cost estimate for your project. If you have broken the project down into small components and estimated each one, you are in a position to discuss a cost-reduction strategy by requesting which component can be reduced realistically.

A well-prepared negotiator will know what is absolutely necessary and what can be given up or reduced. You should also anticipate how much negotiation there could be. In an internal project, the three negotiables are time (price), quality and functions. Take heed of the old project manager's saying, "You can have it cheap, fast or good: pick two." Negotiating for outside services presents the same problem. Thus if you accept the lowest bid for software services, that company may have underbid to get the job and you may have to pay more when they overrun their budget. (You may say you wouldn't because you would have a fixed price contract, but it doesn't help you when the software company can't pay its staff and the job cannot easily be transferred in midflight to another company). Basically, you get what you pay for in our industry. Another familiar scenario is where a senior user manager demands that a project be completed in a certain time at a certain price despite the fact that you know it will cost a lot more and take longer. When asked on what basis these amounts have been set, the typical reply is that they have already been promised to users or "agreed" by senior management! Because we all know that a 12-month project cannot be done in 6 months, you must educate your management that your approach to estimating and planning follows sound management methods and that they can trust you to provide a quality system if they pay a fair price and wait patiently. This applies to your service suppliers as well!

Contracts

After negotiations are complete, a contract is drawn up specifying price, delivery date and deliverables. In addition, warranties, user's responsibilities and escape clauses may also be included. Afixed-price contract is the most common type of contract. This is where the supplier quotes a total project cost at the beginning of the project. Because the project manager carries the risk of things going wrong, this approach should only be used where changes are unlikely and the hardware/software platform is familiar. The cost-plus (or time and materials) project is based on an hourly (or daily) rate plus direct costs. This method is used where the risks mentioned above are high.

Managing Subcontractors

Managing outside contractors requires careful attention. Apart from the process of selection, negotiation and contracts, ongoing management control is vital. Particular attention must be paid to how contractors interact with the rest of the project team, how product quality will be measured to conform with contract terms, and how system components will be handed over to the team for later integration into the final system. Given that you have a good project plan, you should be in a strong position to assign tasks to any new team member and to monitor progress using the project management approaches discussed in previous chapters.

Care should be taken to ensure that contractors understand the standards and quality requirements and produce all required deliverables, especially documentation. They will most likely not be there when the maintenance queries arise. Where possible, quality requirements of all output should be built into contracts to avoid ambiguity and confusion. These requirements will spell out the criteria for acceptance testing the software products. These will generally be linked to the method of payment and the percentage payment to be withheld until the system has been finally accepted. One problem is deciding what tasks should be given to subcontractors. On the one hand, demotivation of permanent staff will occur if "rich" tasks are given to "outsiders". On the other hand "outsiders" may be the only ones with the skills to do the job. To ensure continuity after the subcontractor has completed the task and left the project, a permanent staff member should always work closely with the subcontractor, to ensure continuity in the maintenance phase. This approach will also provide staff growth and development - an ongoing responsibility of the project manager. Giving "rich" tasks to "outsiders" requires careful justification by you - and valid reasons for your actions should be communicated to the project team.

Another common problem when using subcontractors is the different levels of productivity achieved. Often a subcontractor is more productive than a permanent staff equivalent. Higher productivity by the subcontractor can cause rivalry and jealously in a poorly managed project. A good manager will anticipate this situation and get permanent staff "fired" up to perform at similar performance levels. This can only occur if your permanent staff are assigned to the project full-time and do not carry other loads like maintaining other systems or providing ongoing user support. Permanent staff may also complain that their salaries are significantly less than the contractors. Avoid this negative and misleading discussion by justifying to project team members in the early stages that contractors will be getting a fixed income only without the company benefits of annual leave, sick leave, pensions or training. In fact, most contractors do it for a period of time, e.g., three to four years, and then feel the need to take up permanent employment again so that they can develop their careers further.

22 *Program Management*

The Very Large Project

We have seen many reasons throughout the text why we should keep projects small. These include:

- Manageable time frames which people can relate to
- Overhead of communication in large teams
- Increasing complexity and difficulty of testing large systems
- Reduced risk to the organization should the project fail

What do you do if you are assigned a real monster and management insists that it really is necessary and must be done? The answer is that we have to break it into smaller, more manageable chunks.

Tight Deadlines

To compound our woes further, these monsters often are crucial to the organization and come with tight deadlines. These can prevent us from using an approach where we break the functionality up and deliver in phases. Let us say we are commissioned to write a management system for all the events and results at the next Olympic games. The system is extremely complex, must be very reliable (20 billion viewers will be watching its output) and must be ready in 20 months. By the way, it also has to control and interface to a large number of special timing and telemetry devices. Some of these will not be physically available to us until 6 months before delivery. You will also notice that it is virtually impossible to test the system under "live" conditions. How can we succeed?

Fortunately for us, people in the military and aerospace industries have already faced many similar challenges. We may just have to make you a rocket scientist! Seriously though, there are a number of techniques from the discipline of Program Management that we can borrow.

The Role of Architecture

One of our chief weapons will be the use of architectures. An architecture is a high-level conceptual design or blueprint of how something will fit together and how the interrelated parts will work. It allows us to focus initially on the objectives and not on the details; to try various alternatives and approaches without committing to them. We are already familiar with architectural models in the building and construction industry. The architect might build several different models to try out client response to them, assess integration with the environment, or difficulty of construction. When an approach is decided upon, the model can guide the selection of materials and provide a shared vision for the various contributors. We need such a model, but what does it look like for a hardware/software solution?

You are probably already familiar with a variety of modeling techniques:

- Entity Models for depicting the data used by an organization or application
- Function or Process models which show the handling of events and transactions
- Prototypes which show how the system will behave in operation, and how the user community will interact with it

What we need to do is to adapt these to provide us with the tools necessary to architect our very large project. We particularly need to identify *responsibilities and interfaces.*

Responsibilities

To get anything done in the project, you need someone, or some thing to do it. To have anything work in our delivered product, some hardware or software component must deliver a certain function. Responsibility implies that someone or something can be relied upon to perform a given function. If we are to accomplish complex things reliably and quickly, we need this kind of assurance. To identify responsibilities, we might proceed as follows:

- Identify the overall *business* goal of the project
- Identify the overall *business* goal of the product
- Use functional decomposition techniques to break each of these goals down to subgoals and further subgoals until we can identify small enough components to allocate to specific teams, individuals, or components

We should come up with two lists of actions:

- Those derived from the product goal, which will be things which the various components of the delivered solution must perform
- Those derived from the project goal, which will be things which the teams, individuals and subcontractors to the overall project will take responsibility to deliver

We need to go further: for the product goals, we need to define in what time the component must be capable of performing its function. For the project goals, we need to define completion dates. We can represent the goal hierarchies as we did for Product Models and Work Breakdown Models earlier. We can add to the models the response time, or deadline criteria.

Next we can examine dependencies between goals. We should try to partition responsibilities to reduce these. This will reduce risk and interdependencies. Unfortunately, it also tends to push up the size of tasks or components, thus defeating our objective of breaking down the job into smaller components. A balance must be reached between level of dependencies and size of component or activity. This process is not unlike trying to optimize modularity in a structured program design. What we want is maximum cohesiveness within each component or task, and minimum interaction with other tasks, or components.

It can be very useful to use the human brain's capacity for organizing complexity and finding patterns. We can facilitate this by drawing charts which show the interactions of components visually. Components which share a high degree of interaction with each other, but few interactions with other parts of the system, may be merged (size permitting) or grouped as a subsystem with a common interface. This allows us to simplify the interaction of other parts of the system with these components. It also insulates users of the subsystem's functionality from changes in the implementation of the subsystem. This provides future flexibility and reduced maintenance effort. Aguideline which we can use to group functionality is to look for things which utilize the same data. This implies that we have constructed a data model describing the problem space. This should be at an appropriate level of detail. Components which interact with the same data are good candidates for combination into subsystems. This principle is now becoming widely recognized and implemented via object oriented techniques which store the procedures (methods) which manipulate data with the data itself, thus creating *objects*.

A similar process can be used to group related tasks and activities together. We can map out the whole set of tasks required, which may be a very complex set of interrelationships, and then simplify it by grouping tightly related and dependent tasks into subprojects which can be managed separately. A further guideline will be a matrix mapping tasks against resources. Those tasks which require the same resources can be more easily grouped into subprojects.

Interfaces

Once the components, subsystems and subprojects are identified, we can concentrate on the interfaces. These are the areas where responsibility passes from one unit to another. The interface between a software system and a hardware device is the object code definition. The interface between an application program and a DBMS is the set of Database Manipulation Language (DML) instructions which are documented as available to the programmer. The interface between a user application under Windows™ and the Windows™ system software is the published Application Program Interface (API). This guarantees that, if an application is written according to the API guidelines and standards, the environment will perform the requested services as intended by the programmer.

A key thing here is that the interface and the results returned to the user are guaranteed, not the implementation. It is common for hardware manufacturers to radically change their underlying machine architectures to achieve better price/performance ratios, while users and user applications programs are oblivious of the change in implementation. On one processor adding two floating point numbers together may translate to twelve instructions while on the next model it may translate to three quite different instructions. This does not concern the user who will still achieve the same result in the same way, albeit now more quickly. We need to achieve the same stable interface behaviour within our own architectures.

I was once involved in implementing a development support infrastructure in the Adabas Natural development environment. The idea was to save application programmers the problems associated with performing commonly required functions, such as handling large and complex programs within the severe source line number limits of Natural 1.2; providing context-sensitive help on screens, intercepting function keys, trapping and logging errors, handling menus and fast paths, ensuring user authorization and so on. This was complicated by the fact that the client was already proceeding with development of several systems to run in the new environment. Our first task was to agree on what services the infrastructure would provide to the application programmers. This was based purely on feasibility and no detailed design. Next, we defined the manner in which the user programs would interact with the facilities provided by the infrastructure. Using this definition, project teams were able to proceed with their own designs and implementation, trusting that we would deliver the services required. The development of the infrastructure proceeded in parallel. As components were ready, the "dummy" components were replaced with fully functional ones, transparently to the application developers. Their applications were automatically endowed with the necessary services. This clean architecture has allowed sites using it to enrich and enhance the behavior of their systems over an extended period without requiring application changes. It has also enabled them to take advantage of later versions of the environment in a global fashion.

GM

There are various types of interfaces that we might be interested in:

- *Data Interfaces* between software components. These may take the form of parameters passed, files exchanged, or shared access to a database (where it is useful to identify views). These interfaces are normally asynchronous, i.e., a component which alters a data value will not normally wait for or expect a specific response from the next component to use the data

- *Control Interfaces* where one component is requesting an action or operation from another and expects a certain response. These are normally synchronous: i.e., the sender or requester will normally wait for a response from the receiver (or server)

- *Resource Interfaces* where one component must wait for a resource to be released by another component before it can proceed

We can regard project task dependencies as special cases of the data interface (where a deliverable from one task is the input for a subsequent task) or the resource interface (share limited resources are needed for multiple tasks).

We can identify interfaces by examining the interaction between components. We need to stabilize interfaces by identifying and defining:

- Data groups and elements which are shared between components

- Agreed sets of requests (protocols) which client subsystems will expect server subsystems to handle correctly

- Shared resource conflicts and how they will be resolved. In software, this may be a locking mechanism or semaphore signal. In projects, we may choose to assign priorities to subprojects which will allow them to claim resources in proportion to their importance to the overall goals

Defining interfaces is a crucial job and should be performed by the highest skilled persons available. Above all, we want to keep interfaces stable, since any change at this level will have severe ripple implications throughout the macroproject.

Breaking Up the Work

There are two ways in which we can divide up the product. We can draw a high-level architectural picture (e.g., a data flow or event model) and partition subsystems which appear cohesive and have minimal interaction to other areas of the model. We illustrated this approach in chapter 20. We can think of this as *horizontal partitioning*. We are dividing up the total functionality required into separate interacting components at a single level. There is another way in which we can control complexity: this is to work top-down. We can architect the top level model at a very low level of detail, and then decompose components in successively higher degrees of detail. Let us use a portable radio to illustrate the concept, figure 22.1.

Several points should be noted in the figure:

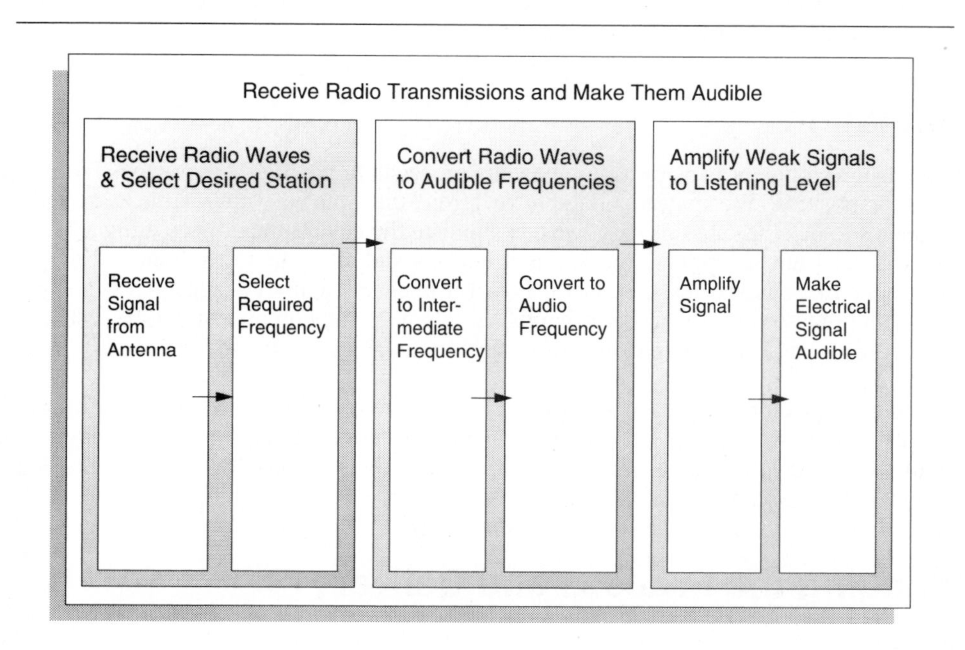

Levels of Abstraction in Architecture *Figure 22.1*

- Each component can be individually designed, constructed and tested without reference to the others, provided we know what inputs it can rely upon, and what it is expected to produce
- The implementation of any component can be changed completely without affecting any other component
- The architecture can be appreciated at a high level by anyone, even someone who has no technical knowledge
- There are successively higher levels of abstraction to the top of the model, and higher levels of detail to the bottom of the model

The Role of Abstraction

Abstraction is an extremely useful organizing principle. We can use it to achieve a *vertical partitioning* in addition to the horizontal partitioning discussed earlier. While it is common for the horizontally partitioned components to have serial dependencies, these are less common with vertical partitioning. This can translate to higher levels of parallelism across projects within our program. Before we can take advantage of this, however, we must define the interfaces clearly. To do this safely requires an expert knowledge of feasibility or, alternatively, technical prototypes to prove the concept. Architecting in this way has an added advantage of allowing us to achieve high levels of reuse of underlying service components, and to make relatively large changes in delivered user functionality by rearranging our use of fundamental components. To return to our engineering analogy, we might use the same timer chip in a watch or a microwave oven. What does this look like for a software system? Typically, we would end up with user-level transactions at the top level, data related or input/output oriented application services below that, and infrastructural or operating system services below that. See figure 22.2.

Simulation

While our development of various components proceeds in parallel, we may wish to begin testing components. We may not be able to afford the time to wait for the surrounding components to be ready. In this case, we can simulate the environment by creating a harness which will feed our component the inputs it expects and allow us to examine the outputs. This is much like an electrical engineer using instruments to apply voltages to test points on a circuit and read the results at other points. In this way, the component can be tested and certified functional, even though the surrounding components are not yet ready.

The whole architectural approach and idea of simulation as a way of managing the lifecycle is deeply rooted in the Object Oriented paradigm now gaining ground rapidly. We recommend you investigate this for further ideas. Good texts to consult are those from Grady Booch and Martin/Odell.

Managing the Process and Delivery

Managing the parallel projects in a program requires special attention. We need to be explicitly aware of the interdependencies and the impact that problems in one area can have on other areas. We can use our architectural model to help us control this. If we are advised

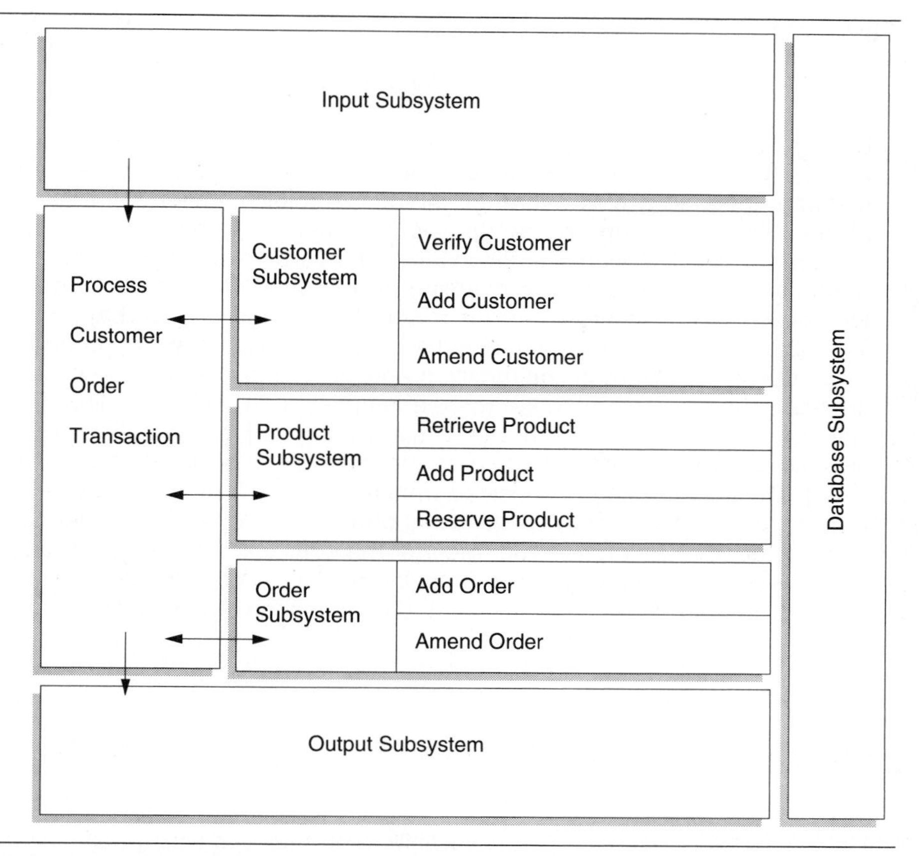

Architecture for an Order-Processing System *Figure 22.2*

by the leader of the team developing the Customer Services module that they will not be able to keep the interface stable, we can immediately see that this will affect our Order Processing module. This can alert us to get the two teams together to resolve the issue. We can easily see which interfaces are affected, and which other teams will be impacted.

As regards delivery, we mentioned earlier that we could record deadlines for the various activities on a Work Breakdown model. We can track actual delivery against this. If there are any slippages, we can immediately determine which tasks are dependent upon this date and advise affected parties accordingly.

Logistics

Because of the ripple or "knock on" effect of any problems, we need to pay careful attention to logistics. In the military sense, this involves everything to make sure the troops at the front stay fighting. In our project environment, it means that equipment and software tools arrive on time, that resources are added to teams when we promised them, that scheduled training takes place as planned, that central resources are there to assist when necessary, that the coffee machine stays working and a hundred other details. Nothing is too menial here.

The NCR 9300 Program

During my career at NCR, I was involved in the release preparation and activity for several new products. One stands out in my memory. The company needed a new midrange interactive minicomputer. An engineer at the California plant came up with an idea to shrink the previous mainframe 8400 design literally board for board to a set of chips and produce an equivalent power machine at a fraction of the cost. The idea was presented to management as a set of overhead foils and approved. The 9300 program had begun. Before it was finished, it would produce a new chip set, a new machine family, a new operating system, a new set of firmware, and a new database management system. The incredible thing was the timeframes - from the initial conceptual presentation to the simultaneous world-wide release of the new machine in forty countries was just fifteen months. I received training at the 12-month mark, and it was incredible to see how all the components developed in parallel were coming together as architected. Most of the software we used for training was running under simulation on an older machine. When the first systems shipped, they worked exactly as we had experienced they should.

GM

Remember that as a manager, your job is to achieve results through other people. Anything you can do to make your teams more successful adds to your success.

Summary

Managing multiple interrelated projects delivering a complex product within tight deadlines and to stringent quality levels is arguably one of the greatest challenges you will face in an I.T. career. It is not for the faint-hearted . Done right it can give you a fantastic sense of achievement. Go to it, and let us know how you do!

Introduction to Case Studies

Background

During our years in industry, we have encountered a great many interesting, daunting and illuminating project situations. There is no doubt in our minds that, in project management, there is no substitute for experience. This is very difficult to obtain from an academic or self-study program. We have tried to imbue the book with practical examples and anecdotes to help you gain an insight into the techniques, concepts and philosophies presented. To further enhance the opportunity for you to gain practice with the techniques and an appreciation for the subtleties of their application (which may not be apparent from the descriptions in the text) we are including several case studies. These will be introduced in this section. These may give you an appreciation of the type of situations in which you might apply the ideas in the body of the book. As we progress through the chapters in the main text, we will return to these cases and set relevant questions. A background knowledge of the cases will thus be useful before you proceed.

The cases range from a small, self-contained system with a well-defined set of requirements (the MyWay Organizer) to quite extensive ones involving major transition and several projects in a large corporate environment (e.g., Handover Trust). Questions will be set at a number of levels, ranging from something you might tackle in class in ten to fifteen minutes, to those with considerable depth and subtlety requiring between forty minutes and an hour to complete. The more trivial questions will deal with techniques and the mechanics and representation issues, while the longer ones will focus more on testing understanding and application as well as more management-oriented issues. Finally, you will probably detect that the questions evolve from the early chapters dealing with techniques and "harder" issues, to the middle chapters dealing with management topics such as risk and quality, and finally to people and organizational issues toward the end.

In writing the cases, we have assumed a computer industry and systems background. If you have difficulty with any of the terms or acronyms, please consult the Glossary at the end of the book.

We hope you enjoy the cases!

MyWay Organizer

Background

The MyWay Organizer is a planned personal computer product to run under Microsoft Windows™ or OS/2™. The goal is to provide a very user-friendly piece of software which will support an executive or marketing individual in planning and organizing his time, as well as staying in touch with colleagues, associates, clients and others.

Requirements

Functions which the product will provide include:

- A diary, with the ability to view by day, week or month
- An appointment scheduler which integrates with the diary
- An address and telephone book function, with a telephone dialer
- A free-form database where the user can specify the items of information to store for each entry
- A planner which provides a year to view and allows the scheduling of blocks of time. For example, you could record leave, training and specific projects or trips
- A "to do" list, which will carry forward unfinished items and allow sorting by priority, location, etc.
- Ability to mark areas or entries as "private". This will allow the executive to give the organizer to a secretary or colleague without fear of disclosing sensitive information

Technology

Most users will use the product on a pen computer or a notebook. Import and export features are planned to allow interoperation with leading desktop products, including spreadsheets, databases and word processors.

The organizer should typically be loaded as a "startup and stay resident" program which will pop up in response to a user-defined key combination. It should save the screen area over which it appears and restore this when it is closed.

The product will be sold as a "shrink wrap" application through retail dealerships and direct through mail order. A market of some 10 to 20 000 copies is envisaged. It should be efficient to be relatively undemanding of hardware resources. To this end it will be written in C++, making use of existing database, communications and Graphical User Interface (GUI) libraries.

The package should include everything a PC-literate user needs to install the software and to teach himself to use the package without recourse to our support organization or expensive third-party training.

Your Role

You are a project manager in Doublon Software, the creators of the MyWay Organizer. You have been assigned the project for the specification, development and testing (up to the initial public release) of the product. You have access to technical specialists, analysts, programmers and a marketing and legal department.

Gleam Stores

The Company

Gleam Stores is a medium-scale retailer with some 200 small furniture and appliance outlets across the country. Established in 1950, the chain has done well, offering a blend of attractive quality merchandise coupled with easy credit terms. Most sales are on credit, with the company running its own installment sale scheme. Clients pay on a monthly basis, either via check or through debit order. Client loyalty is encouraged through special discounts to repeat clients, focused marketing, and a "Value Club" which sends clients a monthly magazine/catalogue with details of the latest merchandise and special offers.

Current Systems

The group has operated a Unisys™ mainframe at the head office for the past 15 years. The machine runs the MP operating system, although it is also capable of running Unix. This system has coped well with high-volume batch processing needed to handle the generation of monthly statements, and the mailing of club catalogs. All central applications are written in COBOL and run against a Codasyl network database, with the exception of general ledger and salaries systems, which are application packages purchased from Unisys. These still use indexed files. There are currently about 200 000 client and past client records on the system with associated credit, purchase and payment history. Applications include: Client Details Maintenance, Credit Authorization, Contract Processing, Stock Control, Statements, Receipt Processing, Debit Order Processing, Club System, the packages mentioned, and Sales Consolidation.

Seven years ago, a system was implemented to consolidate sales information from the stores on a daily basis. This system also calculates commissions for sales staff and feeds this information through to the salaries system. Each store has one or more programmable terminals. These are NCR machines specifically designed for credit authorization and receipting. They read their programs from magnetic tape cassettes, and can store receipting information on another similar cassette. They are based on the Intel 8080 chip and are programmed via a dedicated proprietary language (TRANCON) compiled on an NCR minicomputer at head office. Object code is downloaded to the branches via dial-up communications. Where a store has multiple terminals, one terminal will act as the master and will store the software for the other machines in the store, which are connected via a local area network. The master unit in each store has communications capabilities to request credit authorization online from Head Office. Since the volume of authorizations required on a daily basis is low, most stores make use of dial-up facilities. When an authorization is requested, the terminal will automatically dial Head Office. Some of the bigger branches have dedicated lines. At the end of the day, the central mainframe runs a job which dials up each store in turn and requests all receipt and sales details which have been stored on the cassette units. Once all the collection is complete, a batch process consolidates all sales and receipts, updates client accounts, and calculates commissions.

New Developments

The chain is expanding rapidly, and while the current systems work well, the dedicated terminals have become expensive relative to standard personal computers (which also provide much more functionality). It is not cost effective to install new dedicated terminals in new branches. There is an increasing demand for personal productivity and other applications from the stores. Store managers are looking for budgeting and quota management systems which they can administer at the store level. They also want more access to and control over client records, to allow them to perform more intelligent marketing and follow-up. Payment history and account information is also required to facilitate collections. A pilot has been installed where PC's have been used in a new store to perform similar functions to the dedicated machines in other stores. These machines have also been used for budgeting and quota management using spreadsheet and word-processing packages.

Your consulting organization has been retained by Gleam to recommend how to approach the next step in moving to standard "commodity" technology at the branches with the following aims:

- Reduce cost of hardware in new stores
- Improve serviceability of branch equipment via inhouse expertise
- Decentralize client information in support of local marketing and collections
- Integrate with personal productivity applications, including spreadsheets, word processing and presentation graphics software

Credit and contract approval is to be retained centrally for the forseeable future.

A draft proposal has been made with the following components:

- Full documentation of the software system that has been piloted, including suitable user manuals for field deployment
- Development of a training package, including classroom materials, student manuals, and worked examples for hands-on practice. Two days of training is envisaged for store-level operators, and four days for in-store supervisors
- Stress testing of the developed software to ensure that it can cope with production volumes for all stores and peak trading periods, as well as to ensure that it recovers correctly from equipment and other failures. Any necessary modifications or tuning of the application to achieve better performance are to be made
- An initial implementation phase involving developers and users from H.O. which will roll out the system to the other stores in the same region
- A user-led implementation of the system to all other stores

Handover Trust

Background

Handover Trust is a major life assurance company. Established in 1880, they now service some 4 million policyholders in seven countries. Head office is based in London and includes divisions dealing with individual policyholders, group schemes for corporate clients (medical aid, pension, etc.) and investments (e.g., unit trusts). There is also a services division providing infrastructural support to the other divisions including accounting, auditing, personnel management, etc. Head Office employs some 1200 staff. There are approximately 26 branch offices, comprising some 2000 staff divided into 12 regions.

Applications

The organization has over the years built a great many computer applications to support the business, including: New Business (setting up of contracts accepted from applications), Policy Administration, Claims Processing, Quotations and Commissions systems. They also operate packaged software for Salaries, Assets Register and General Ledger. All of the above systems run on an IBM Mainframe, installed some seven years ago. Most systems are written in COBOL. About half make use of the TOTAL network database management system. All online systems use the CICS transaction processing monitor. In all, there are some 8000 modules. The applications have become increasingly unwieldy, unreliable and difficult to maintain. The company is also experiencing difficulty finding TOTAL skills in the marketplace.

New Developments

In the last five years, personal computers have been purchased by departments on an ad-hoc basis. These were initially used for word processing and spreadsheet analysis, as well as producing business presentations. Increasingly, they are being used to circumvent the clumsy mainframe systems. Systems have sprung up in Lotus 1-2-3™, Microsoft Access™, and dBase IV™. This is not coordinated, and there have been numerous instances of lost data and information corruption. There was recently a fraud scare as well.

The new General Manager, Mr. Renfrew, has initiated a strategy to replace all existing systems with flexible applications developed on new technology over the next five years. The plan calls for the use of Client Server technology to achieve rapid development, scalability and cost effectiveness. You have been assigned to manage the overall project. You have been briefed by Mr. Renfrew and his management team. They are looking for rapid application development, responsive, easily maintained systems, a "seamless as possible" migration from the legacy technology, and extreme reliability in applications which are the lifeblood of the business.

Options

Options being considered include the use of relational database products (Sybase, Oracle, SQL Server) and graphical development tools (Powerbuilder, SQL Windows). Local area networks will probably use Novell servers, while wide area networks will use the industry standard TCP/IP protocol and Unix servers.

Staff

From a staffing perspective, Handover has a centralized development team of 180 staff, from application area managers down to junior programmers. There is a related support group under the Data Base Administrator which provides development support, database support and network support services. This group numbers some 28 people. The development staff have good traditional mainframe skills, network database skills, some online transaction processing experience and exposure to rigorous structured methodologies (although these are not followed strictly, or on all projects). There are currently very few skills in graphical user interfaces, event-driven programming, client server and relational database.

ThoughtWell Books

Background

You are employed as a senior project manager by **MacroSoft**, a software house based in Stockholm, Sweden, specializing in leading edge software development projects. MacroSoft, founded in 1980, has a proud reputation for delivering systems meeting requirements to tight deadlines. Much of the work currently undertaken involves networked microcomputers and client server technology, which is also your background.

You have met with Jane Ostin, the marketing manager of a new client, **ThoughtWell Books**, a major supplier of technical and specialist books to academics, students, industry and the public. She has briefed you on a project which your management is keen to undertake on their behalf.

The Problem and Opportunity

Jane has told you that obtaining technical books currently not in stock in the country has always been a problematic process. The problem arises because books are ordered in bulk from clearinghouses in the United States and United Kingdom. They are wholesalers and will only ship in bulk. This means a consignment of at least 100 books, with a value not less than 2 000 US dollars. They also ship via surface mail which can take between 6 and 10 weeks to arrive in Sweden. Coupled with the delay to assemble an order of the required size, the delay to a customer wanting an unusual book can often extend to between 3 and 4 months.

Jane recently found a clearinghouse in the United States which is offering a new service. They will accept orders for small volumes of books (minimum order 100 US dollars), and ship per air ex-stock as soon as funds are cleared. Orders can be placed via e-mail on the Compunerve network, once an account has been set up. Payment is normally made through an approved international credit card (Vista, MonsterCard, American Excess). Jane sees a major market opportunity to offer a new, rapid order service. Total turnaround time for an order placed electronically and shipped by air could be as short as four to five days. This will attract customers who are more interested in speedy delivery than the price of the books concerned. The sale price will need to be some 30 to 40 percent higher than standard to accommodate the air delivery, but clients in high-technology fields have indicated that this would be acceptable.

Existing Computerization

ThoughtWell Books currently operates several branches: 4 in Sweden, 3 in Norway, 2 in Denmark, and 1 each in Greenland and Finland. There is a distribution center in Göteborg, Sweden. Each branch is equipped with a small number of Point of Sale (POS) terminals which serve as cash registers while capturing details of payment, stock codes sold, discounts allowed, and tax collected. POS terminals are connected to a 386 PC acting as a server in each branch. At the Head Office in Stockholm, there is a 486 server which is linked to the branches via dial-up lines. This server establishes connection with each branch in turn in the morning, and again in the evening. In the morning, sales figures from the previous day are collected for consolidation. In the evening, any new pricing details are transferred from Head Office to the branches. There is a remote printer at the distribution center, but no other computer equipment. All servers run Novell local area networking software. Wide area communications use the TCP/IP protocol.

Software at the various locations includes:

Each Branch:

Stock System
This contains details of all books at the branch, including cost price and selling price. It is updated by the central system transmitting details of books transferred to the branch from the distribution center. If any book prices subsequently change, the price change is received from H.O. and the records updated accordingly. A report notifies the branch staff of what stock to expect from the distribution center, and pricing changes.

Sales System
Runs in the POS terminals and retrieves prices from the stock database. Updates stock on hand. Collects sales statistics.

Head Office:

Publication Ordering System
Orders in bulk from conventional suppliers. Payments are authorized through the system when consignments are received and reconciled to orders.

Payment System
Driven by the Publication Ordering System and the Salaries system. Issues checks and generates bank transfer transactions on a magnetic tape.

Sales Consolidation System
Collects sales statistics from the branches and produces management reports.

Stock Distribution System
Determines, based on historical data regarding sales categories and volumes, what proportion of newly arrived stock to distribute to which branches. Generates distribution instructions to the distribution center, and notifies branches to expect arriving consignments. Establishes selling prices based on cost prices and other parameters provided by marketing manager.

Price Change System
Allows H.O. to change the selling price of a stock item. Transmits the change to the branches to which stock of this kind has been sent.

Communications Service System
Provides reliable communications between H.O. and branches. Acts as a service system to business applications.

All of the above systems have been custom written using the dBase IV relational database and application development product.

The salaries system is a package, previously provided by MacroSoft. This is written in Clipper with compatible database formats to dBase. There are also a variety of PC packages used by H.O. staff, including the Word 6 word processor, the Lotus 1-2-3™ spreadsheet and the Pegasus e-mail package.

Requirements

Jane is keen to implement a system which will:

- Allow online entry of orders for international books not in stock, including client details if these are not yet on file. Client credit card details are to be checked, or entered. This facility should be available on at least one POS terminal per branch, as well as at Head Office.
- Route these orders from the branch to H.O.
- Check that the requested items are not in stock within another branch, or on order already.
- If the item ordered is not already in stock or on order, pass the order on to a clerk at H.O. who will enter the cost price of the book from an international catalogue on microfiche, and a selling price based on profit margins and projected shipping price.
- Perform an online inquiry to the bank system to verify availability of funds on the client's credit card.
- Collect orders from the branches until they reach a value exceeding one hundred US dollars. When this occurs, automatically generate an e-mail order to the US clearing house, and transmit this via modem. The overseas clearing house will debit a credit card account held by ThoughtWell Books.
- Notify the branch that the book has been ordered and what the selling price should be.
- Upon receipt of the order, debit the client credit card via online connection to the bank system. Notify the branch that the book has arrived and is being sent on via airmail.

Jane is keen to see a system in operation as soon as possible. A business associate had told her your organization would be the best partner in achieving this. She is particularly concerned that no competing bookstore should bring out a similar service before ThoughtWell.

Glossary of Terms

Adversary teams
A technique where a separate team is set up to try and break the code of a system at the acceptance testing stage.

BANG
An estimating technique, developed by DeMarco, which assigns a technology-independent functional weight to a system.

Baseline
The baseline is a snapshot of the status of all the deliverables and documents in place at the termination of a particular phase in the development process.

CASE
(see Computer Assisted Software Engineering)

Change control
A management framework to ensure that adverse effects of changes requested and made to systems are minimized.

COCOMO
A software estimating technique developed by Barry Boehm to estimate project cost, effort, schedule and staffing. Among other things, the technique requires the number of source lines of code (SLOCs).

Code inspections
A technique involving meetings where prepared participants review programs with a view to increasing the quality of the program.

Computer Assisted Software/Systems Engineering (CASE)
Use of an automated software package (CASE tool) to assist in the analysis and design of an information system using modeling techniques and a repository.

CASE tool
An automated tool which supports the work of system engineers/developers. Normally provides a dictionary/repository and diagram editors. Integrated CASE tools support several phases of the lifecycle and code generation from specifications.

Concession
Part of a requirement that could not be delivered and which the user concedes can be left out.

Configuration management
A technique used to help the project manager coordinate the work products of the team. Documents the interrelationships of components, and their composition. Useful for change management, estimation and quality assurance.

Context diagram
A high-level data flow diagram highlighting the scope and boundaries of a system.

Contingency
A budget amount in addition to the official project budget which is used to compensate for unexpected problems. It is normally managed separately.

CPM
(see Critical Path Method)

Critical path
A path flowing through the activities with the longest estimated duration (relative to other activities occurring in parallel) in a project network plan. The sum of the estimated durations of activities on the critical path represents the minimum time it will take to complete the project. Should any activity on the critical path take longer than estimated, the entire project will slip (be delivered late). These are therefore the critical activities in the project.

Critical Path Method (CPM)
A network technique to define the tasks and dependencies in a project and to determine which ones are on the critical path.

Data dictionary
A software package to capture, index and cross-reference meta data (data about data). Would normally define the data elements, relations, relationships, aliases and role names of a data model. Some are extensive and support CASE tools. These may contain the data for various types of models, including process models, data models, and prototypes.

Data model
A model of data groups and their relationships. High-level models are normally represented as a collection of entities and relationships. Detailed logical models are normally represented as relational data models. Physical data models describe the physical structure of the database as it would be defined to a database management system.

Deliverable
A tangible output which is produced by performing a task. This may be a document, a model, a program, a control script or other piece of work.

DeMarco
An author and one of the fathers of structured analysis techniques.

Deviations
Where the specification or design differs from the requirements and is unacceptable to the user.

Earned value
The value of the products that are complete and have been quality approved in the project.

Entity
An entity is a thing about which we want to record data. It may be a physical thing, such as a Product, a conceptual thing such as a Sales Category, or a record of a transaction, such as an Order or Payment. Entities can also represent abstractions or aggregations of things: e.g., Asset.

Entity Relationship Model
A model of the entities in a system and the relationships between them.

EQF
(see Estimating Quality Factor)

ERM
(see Entity Relationship Model)

Estimating group
A separate team of experienced analysts who assist the project team to estimate the project size and duration of tasks.

Estimating Quality Factor (EQF)
The EQF, developed by DeMarco, is a measure of how effective the estimating process is.

Faculty Training Institute
A private training establishment offering Information System courses.

Feasibility study
An early stage in the SDLC to determine the technical, economic and business feasibility of developing a proposed system, or mounting a given project.

FPA
(see Function Point Analysis)

Function Point Analysis (FPA)
A technique, developed at IBM by Albrecht, to measure the value of a system to its users by quantifying its functionality.

Function points
A technology-independent score to assist in sizing a system.

Function(al) model
A model which describes the functions which a system will (or does) perform. May be shown as a hierarchy (functional decomposition chart) or a set of data flow diagrams using top-down decomposition.

Functional Decomposition Chart (FDC)
A top-down model of the functions needed in a system with detailed processes at the lowest level. Often used for packaging the functionality of a system, i.e., choosing functions to implement in particular programs. A more detailed variant can be used for program design.

Functional specification
A document produced at the end of the analysis phase of the SDLC showing the functions the user requires developed.

Gane & Sarson
The originators of a widely used notation for data flow diagrams.

Growth Need Strength (GNS)
A measure of the perceived need of an individual to be challenged by, and to achieve fulfillment in, the job.

IFPUG (International Function Point Users Group)
An influential group which evolves and publishes standards on how to calculate function points.

Information Engineering (IE)
Developed by I.T. guru James Martin together with Clive Finkelstein, this is an holistic set of tools, techniques and methods to support enterprise wide planning, data modeling and systems development.

Inspired
The name of the I.T. consultancy of one of the authors.

Iterative lifecycle
A lifecycle where subsets of the system are built and tested as early as possible and progressively expanded until the requirements are satisfied.

JAD
(see Joint Application Development)

Joint Application Development
Originally called Joint Application Design. A technique where intensive meetings, run by a trained facilitator and with the participation of all stakeholders, are used to accelerate and improve the quality of the analysis and design phases of the SDLC where prototypes, process models and data models are built in real-time. CASE tools can be employed to assist the process.

Ladder
A term used to describe the presence of a lead and lag in a network diagram.

Lag
A forced wait before an event can occur.

Lead
A forced wait before a task can commence.

Manageable Unit of Work
The longest planned duration allowed in a project plan, typically one week. An MUW is often the lowest level task in a WBS. The duration will be increased where the risk is low, the project is noncritical and the staff are competent. If the risk is high, the project is critical or the staff are inexperienced, it should be reduced.

Matrix organization
An organization structure where staff from various functional departments work on a project under the control of a project manager while their functional managers still retain control over them.

Merise
A system development methodology favored by the French government.

Method/1
SDLC and PLC methodologies developed by Andersen Consulting which make extensive use of automated tools.

Metrics
The techniques used to measure productivity, quality and effectiveness in a project.

Milestone chart
A high-level Gantt chart summarizing the key dates in a project.

Milestones
Key deliveries planned for achievement on specific dates. Normally associated with completion of key deliverables.

Motivating Potential Score (MPS)
A measurement of the richness of a job as perceived by the incumbent.

MUW
(see Manageable Unit of Work)

Object Oriented Technologies
The technology supporting a new approach to systems development where systems are viewed as a set of reusable objects, containing both data and behaviors, akin to building blocks.

OOT
(see Object Oriented Technologies)

Optimistic time
Used in PERT to determine a realistic, yet optimistic, estimate to complete a task.

Optional dependencies
Used in a network diagram to depict when a task can commence based on the completion of one of the preceding tasks. This is a special case: a network diagram normally implies that all preceding tasks must complete before the dependent task can proceed.

Pareto's Law
The 80/20 rule. For example, 80 percent of the systems developed in the world are completed by 20 percent of the project team members. The other 80 percent of the project team members only produce 20 percent of the output (but probably also a lot of the disasters!).

PBM
(see Product Breakdown model)

PERT
(see Program Evaluation & Review Technique)

Pessimistic time
Part of the estimating process in PERT. An estimate of the longest time it will take to complete a task.

Phased delivery
Also called evolutionary development. Small but useful components of the system are built and delivered to the user, giving them the impression that they are getting usable subsystems every few months.

PLC
(see Project Lifecycle)

Product Breakdown Model (PBM)
A hierarchical chart showing the products the project will deliver.

Product Structure Model (PSM)
A hierarchical breakdown of all the deliverables to be produced for the system. This can be a subset of the PBM.

Program Evaluation & Review Technique (PERT)
A technique used to model the activities in a project with their durations and dependencies. Specifically, PERT provides techniques to predict the likelihood of meeting particular estimates. It also allows prediction of optimistic and pessimistic times for delivery.

Project feasibility
The perceived chances of success for a project to meet its goals within constraints of time, cost and resources.

Project Lifecycle (PLC)
The set of activities that are required, along with the SDLC activities, to manage a project.

Project partitioning
The breakdown of a large project into smaller, manageable chunks.

Project risk
The chances of encountering problems which would prevent achievement of the project goals.

Project scope
The defined boundaries of the system requiring development. This can be specified in terms of functionality and interfaces.

Prototyping
A technique where successively more accurate models are used to assist in designing the user interface, data model and functionality. It can help refine ambiguous user requirements prior to delivering the final system.

PSM
(see Product Structure Model)

QA
(see Quality Assurance)

QC
(see Quality Control)

Quality Assurance (QA)
The use of tools, techniques, standards and methods to deliver a quality product or service.

Quality Control (QC)
The final checking, prior to handover, to ensure a product conforms to requirements.

RAD
(see Rapid Application Development)

Rapid Application Development
The use of advanced lifecycles and techniques to deliver applications more quickly than can be achieved using conventional techniques. Often used in Information Engineering context to describe a lifecycle using prototyping, JAD and CASE.

Repository
A sophisticated data dictionary containing both data and knowledge about the data. Often underlying CASE tools, sophisticated project management packages and configuration management packages.

Request for Proposal
A document specifying the business problem to a potential supplier of I.T. services.

Resource leveling
An approach where workload peaks are identified and work is reallocated or rescheduled to ensure that project team members have a realistic daily workload.

RFP
(see Request for Proposal)

SDLC
(see Systems Development Lifecycle)

Slack time
A project network term used to measure the delay which could occur for a particular activity without delaying the overall project.

Slip chart
A chart used to visually depict project slippage by plotting planned delivery time against actual delivery time.

SLOC
(see Source Lines of Code)

Social Need Strength (SNS)
A measure of an individual's need for social interaction.

Software Metrics
(see Metrics)

Source Lines of Code (SLOC)
The number of statements in a source program. Variants: KSLOC = thousand source lines of code, ESLOC = effective SLOC (i.e., minus comments).

SPC
(see Statistical Process Control)

SSADM
Structured Systems Analysis and Design Method. A systems development methodology favored by the British government.

Statistical Process Control
A process incorporating a methodology, measurements, and change control to ensure that the development process is repeatable and has predictable outputs. It allows us to make gradual, sustained improvements in the process, leading to ever higher quality levels. Also allows managing the impact of changes in methods and technology.

Steering committee
Managers who are responsible for the success of the project and meet on a regular basis to discuss progress and problems.

Systems Development Lifecycle (SDLC)
The overall model of activities, deliverables and controls needed to produce a system. Normally describes the modeling and technical tasks.

Systems Engineering Lifecycle
A rigorous SDLC used to develop extremely reliable systems using configuration management techniques.

Task-in-node
Network notation where task descriptions are placed in nodes, and lines are only used to indicate dependencies. Also called a precedence diagram.

Task-on-line
Project network notation used when an activity is recorded on a line and the event at a node.

Total Quality Management (TQM)
Encompasses QA and QC along with the company's overall quality environment.

TQM (see Total Quality Management)

User requirements
The original requirements whose satisfaction required the mounting of a project. Normally specified by users assisted by business and/or systems analysts.

Value of Work Complete (VWC)
The value of the work delivered to date according to the original plan. Only work which has been quality controlled is counted.

Version control
The retention of several historical versions of the project plan to facilitate analysis of the current project estimates and problems and to identify improvements for future projects.

VWC
(see Value of Work Complete)

WBS
(see Work Breakdown Structure)

Wide Band Delphi
A problem-solving technique used to gain consensus, using a group of experts. Useful where empirical methods are not available.

Work Breakdown Structure (WBS)
A hierarchical model which represents the entire project at the top level and subdivides the project into tasks at the lower levels. Used to identify detailed tasks which can then be estimated prior to producing a project budget. Can also be used to record progress against plan and to calculate Value of Work Complete.

Z
A formal specification language not mentioned anywhere in the book :-)

Bibliography

Albrecht, A.J. : **Measuring Application Development Productivity,** 1979, Proceedings IBM Share/GUIDE symposium, GUIDE International Corp., Chicago.

The seminal paper introducing the concept of measuring size of systems using functionality delivered to a user (Function Points) rather than internal measures such as code size.

Bennis, Warren & Nanus, Burt : **Leaders - The Strategies for Taking Charge**, 1985, Harper & Row, New York.

Definitive book on leadership. Practical with extensive examples. Accessible.

Bergen, S. A. : **Project Management: An Introduction to Issues in Industrial Research and Development**, 1986, Basil Blackwell, Oxford, U.K.

Useful coverage of research and development type projects. Emphasis on product management. Coverage of contractual and motivational issues.

Blanchard, Kenneth; Zigarmi, Patricia & Zigarmi, Drea : **Leadership and the One Minute Manager**, 1985, Fontana Collins, U.K.

A very accessible and easy reading text covering leadership and situational leadership in a management context.

Boehm, B.W. : **Software Engineering Economics**, 1981, Prentice-Hall, Englewood Cliffs, New Jersey.

A definitive text on the economics of software production. Includes many statistics and cases. Relates work done by the U.S. Dept. of Defense and defence contractors such as TRW. Empirical treatment of software estimating, including COCOMO model.

Booch, Grady : **Object Oriented Design with Applications**, 1991, Benjamin Cummings, Redwood, CA.

A seminal work on object oriented analysis and design as applied in embedded systems and systems engineering work.

Brooks, Fred : **The Mythical Man-month**, 1975, Addison-Wesley.

An account of the problematic OS360 project at IBM. One of the largest software projects ever tackled. Many "home truth" lessons which seem obvious after the fact. A classic.

Card, D.N.; Clark T.L.; Berg R.A. : **Improving Software Quality and Productivity**, 1987, Systems, vol. 29 no. 5 June.

An excellent article discussing the application of statistical process control to the software process, with empirical results from Computer Sciences Corp.

Chen, P. : **The Entity Relationship Model: Towards a Unified View of Data**, 1976, ACM Transactions on Data Base Systems, Vol. 1 No. 1.

The seminal paper introducing the concept of entity relationship modeling.

Crosby, P. : **Quality Without Tears, The art of hassle free management**, 1984, McGraw-Hill, New York.

The most accessible and easy-reading text covering the total quality management philosophy. Crosby is the originator of many of the techniques now widely practiced. Covers the prevention philosophy in detail. Provides extensive guidance on introducing Quality Management programs.

Couger, J.D. and Zawacki R.A. : **Motivating and Managing Computer Personnel**, 1980, Wiley-Interscience, New York.

The theories in this book have been tested successfully over the years in many countries and published research has proven that I.S. personnel worldwide are different from other professional groups and can be managed in ways to enhance performance and motivation.

DeMarco, Tom : **Controlling Software Projects: Management, Measurement and Estimation**, 1982, Yourdon Press, New York.

A book that every project manager will enjoy reading. Although over a decade old, the key issues are still relevant. Arguing for better planning and control, DeMarco offers tools and techniques which can be used immediately. The authors view this book as essential reading.

DeMarco, Tom and Lister, Timothy : **Peopleware: Productive Projects and Teams**, 1987, Dorset House, New York.

This book is fast becoming a classic text on managing the people aspects of project management. Written in a light-hearted way, the highly respected authors develop sound advice for project managers to get staff more motivated and more productive. Much of this advice can be implemented at the project level. The book is relatively short and easy to read.

Downs, E., Clare, P. & Coe, I. : **Structured Systems Analysis and Design Method (SSADM) - Application and Context**, 1981, Prentice Hall, U.K.

Clear and concise exposition of the U.K. government mandated SSADM development methodology.

Ewusi-Mensah, K. & Przasnyski, Z.H. : **On Information Systems Project Abandonment: An Exploratory Study,** 1991, MIS Quarterly March, pp 66-85

Empirical investigation of why projects fail.

Fenton, N.E. : **Software Metrics: A Rigorous Approach**, 1991, Chapman & Hall, London.

This academic book takes the SDLC and applies measurement to all the important areas. After explaining measurement theory, the author builds a framework to guide the reader through the initial establishment of a measurement programme and, once established, the measurement tools and techniques required for control. Many of the techniques are drawn from research in the Esprit projects in Europe.

Flood, R.L. : **Beyond TQM**, 1993, John Wiley, Chichester, U.K.

A systems-oriented view of Total Quality Management. Fairly easy reading.

Franch, J.V. : **METHOD/1**, 1987, Auerbach Publishers, pp 37-1-10.

Quick summary of the huge Andersen Consulting METHOD/1 methodology.

Gane, Chris, and Sarson, Trish : **Structured Systems Analysis**, 1979, Prentice Hall, Englewood Cliffs, NJ.

An excellent work which introduced the most widely used data flow diagramming standards and techniques.

Gilb, Tom : **Principles of Software Engineering Management**, 1988, Addison-Wesley, Wokingham.

Full of practical ideas, the author, an eminent consultant, identifies practical approaches to managing and controlling software engineering projects. The book is strong on ideas and practical techniques which are all still relatively current.

Handy, Charles : **The Age of Unreason**, 1995, Arrow Books, London.

A systemic view of people and organizations in the future and how to handle some of the complex problems leaders will be facing.

Humphrey, Watts S. : **Managing the Software Process**, 1989, Addison-Wesley, Reading, Mass.

An up to date treatment of the software development process and the associated economic issues. Introduces a valuable model for determining the degree of maturity of an organization with respect to maturity in the software production process. Covers innovations such as CASE, I-CASE and the concept of Statistical Process Control.

IEEE., **Draft Standard for Software Project Management Plans - IEEE** p1058, 1987, Institute of Electrical and Electronic Engineers Inc.

The standard which applies the rigorous engineering discipline of configuration management to the software production process.

IFPUG (Robert Ragland, Chairman), **Function Point Counting Practices Manual**, Release 3.4, 1992, International Function Point Users Group. Westerville, Ohio.

The definitive guideline on the counting of function points. Covers the original Albrecht approach, expands on this in more detail, and also introduces a more modern approach based upon data modeling. Has guidelines for development and enhancement projects.

Ivancevitch, J.M., Donnelly, J.H. and Gibson, J.L. : **Management: Principles and Functions**, 1989, Fourth Edition, BPI/Irwin, Boston.

This is a comprehensive introduction to management. Designed for any management position, it develops a broad perspective on the management role by emphasizing the planning, organizing, leading and controlling functions in great detail. A 700-page book that covers the theories and techniques underpinning the need for all of us to become effective managers.

Kerzner, H. : **Project Management: A systems approach to planning, scheduling and controlling,** 1992, fourth edition, van Nostrand Reinhold, New York.

The classic text covering all aspects (1000 pages) of project management as applied in engineering disciplines. This edition has expanded coverage of management and people aspects, including conflict management and team building. Extensive engineering examples and case studies. Of relevance is the section on concurrent engineering which complements our discussion of program management.

Lustman, F. : **Managing Computer Projects**, 1984, Reston Publishing, Reston, Virginia.

This is one of the few texts specifically directed at computer systems development projects. It has good coverage of the project design aspect.

Martin, James & Finkelstein, Clive : **Information Engineering, Volume 1**, 1981, Savant Research Studies, Carnforth, Lancashire, U.K.

The landmark work in which Martin and Finkelstein introduced the Information Engineering methodology which took data modeling as its central tenet in contrast to the process centered methods predominant at the time. It also introduced the notion of enterprise-wide, rather than application or project specific modeling and planning.

Martin, James : **Recommended Diagramming Standards for Analysts and Programmers**, 1987, Prentice-Hall, Englewood Cliffs, NJ

This book, developed with Dr Carma McClure, details a comprehensive but easily usable and consistent set of diagramming standards for all common structured analysis/I.E. analysis and design deliverables, including: Entity Relationship diagrams, Decomposition charts, process models, structure charts, action diagrams. These are the standards followed by most I.E. CASE tools.

Martin, James : **Principles of Object Oriented Analysis and Design** ,1993, Prentice Hall, Englewood Cliffs, NJ.

Extends Information Engineering to bring it right up to date with the Object Oriented revolution. Developed with methodologist James Odell, provides a bridge from structured techniques and I.E. knowledge to the OO world.

McFarlan, F.W. : **Portfolio Approach to Information Systems**, 1981, Harvard Business Review, Sept-Oct pp 142-150.

Useful background to project feasibility study. Introduces the idea of a balanced portfolio of projects, with some high-yield high-risk projects, but keeping the overall portfolio risk acceptable.

McFarlan & McKenney : **Corporate Information Systems Management: The Issues Facing Senior Executives**, 1983, Richard D Irwin, Homewood, IL.

Good managerial text covering critical issues in I.S. planning and strategy.

Metzger, P.W. : **Managing a Programming Project**, 1973, Prentice-Hall.

Very readable approach to project management in I.S. Somewhat dated now.

Moder, Joseph; Philips, Cecil & Davis, Edward : **Project Management with CPM, PERT and Precedence Diagramming**, third edition, 1983, Van Nostrand Reinhold, NY.

Very good coverage of the mechanics and calculations underlying the CPM and PERT techniques.

Norden, see: Putnam

Olle, T.W.; Sol, H.G.; Verrrijn-Stuart, A.A. (Editors) : **Information Systems Design Methodologies: A Comparative Review,** 1982, North Holland, Amsterdam.

A comparison of popular system development methodologies. A Eurocentric perspective. Refreshing after the predominantly American views we normally get.

Opperman, Piet : **Gold, the Comcon Project Management Methodology**, 1987, Comcon/QData, Johannesburg.

An excellent proprietary methodology incorporating many of the concepts used in this work. Links to the Tetrarch strategic planning and development methodologies. Good integration of management by objectives and thorough coverage of quality.

Parr, F.N. : **An Alternative to the Rayleigh Curve Model for Software Development Effort**, 1980, IEEE Transactions on Software Eng., vol. SE-6 no. 3, May, pp 291-296

Describes Parr's alternative to the Putnam-Norden model. This is better suited to small, commercial projects and projects where some resources are already in place at the start.

Putnam, L.H. : **Example of an Early Sizing, Cost and Schedule Estimate for an Application Software System**, 1978, Proceedings COMPSAC '78 , IEEE, NY

Putnam introduces the Norden/Putnam model which describes the behaviour of software projects with respect to resources and delivery date. This is the model underlying the SLIM software estimation method and the PADS and Butler Cox (now CSC Index) databases.

Rakos, J.J. : **Software Project Management for Small to Medium Sized Projects**, 1990, Prentice-Hall, New Jersey.

An excellent book produced by an ex-DEC employee who describes in detail how the organization developed excellent computer software very successfully over many years. Covering all areas of project management, the book uses lots of examples to help the newcomer understand the problems and techniques. Somewhat lacking in the softer management problems, the author nevertheless gives an excellent overview of the process from a pragmatic viewpoint.

Schach, S.R. : **Software Engineering**, 1990, Irwin, Boston.

A good software engineering text. Good coverage of options for system development and discussion of various lifecycles. Also extensive coverage of quality management in software development. Coverage of some quantitative estimating techniques.

Sommerville, I. : **Software Engineering,** 1992, Addison-Wesley Publishing Co, Wokingham

An excellent and comprehensive text covering the full field of software engineering. Good coverage of concepts related to configuration management.

Stewart, Rosemary : **Managing Today & Tomorrow**, 1991, Macmillan, Hampshire, U.K.

A text covering contemporary management theory. Well researched, current and well written.

Tajima D. & Matsubara T. : **The Computer Software Industry in Japan**, 1981, Computer, pp 89-96.

Fascinating account of how Hitachi software achieved very high quality levels in software production.

Tetrarch International : **The Tetrarch/1 Strategic Planning Methodology**, 1987, Tetrarch International, Holland.

An excellent I.E.-style strategic planning methodology now marketed by P.A. Consulting.

Tetrarch International : **The Tetrarch/2 System Development Methodology**, 1985, Comcon, Johannesburg.

An information engineering approach to structured systems development incorporating a unique process design and performance prediction component. Integrates with the GOLD project management methodology. Both products now distributed by Q-Data Consulting.

Yeates, D. : **Project Management for Information Systems**, 1991, Pitman Publishing, London.

A useful overview of project management with some emphasis on the project management methodology PRINCE - the U.K. govt. "standard". Useful chapter on managing the implementation of projects.

Yourdon, E. : **Managing the System Life Cycle: A Software Development Methodology Overview**, 1982, Yourdon Press.

Useful review of the structured systems life cycle.

Index

A

B

C

D

E

F

G

H

I

J

K

L

M

N

O

R

S

T

U

V

W